ANALYTICAL TECHNIQUES in DNA SEQUENCING

ANALYTICAL TECHNIQUES in DNA SEQUENCING

Edited by
Brian K. Nunnally

Taylor & Francis
Taylor & Francis Group

Boca Raton   London   New York   Singapore

A CRC title, part of the Taylor & Francis imprint, a member of the
Taylor & Francis Group, the academic division of T&F Informa plc.

FIRST INDIAN REPRINT 2009

Published in 2005 by
Taylor & Francis Group
6000 Broken Sound Parkway NW, Suite 300
Boca Raton, FL 33487-2742

© 2005 by Taylor & Francis Group, LLC

Printed in India by Saurabh Printers Pvt. Ltd.

International Standard Book Number-10: 0-8247-5342-9 (Hardcover)
International Standard Book Number-13: 978-0-8247-5342-9 (Hardcover)
Library of Congress Card Number 2004066429

Library of Congress Cataloging-in-Publication Data

Analytical techniques in DNA sequencing / edited by Brian K. Nunnally.
    p. cm.
  Includes bibliographical references and index.
  ISBN 0-8247-5342-9 (alk. paper)
  1. Nucleotide sequence. I. Nunnally, Brian K.

QP625.N89A53 2005
611'.01816--dc22   2004066429

**Visit the Taylor & Francis Web site at
http://www.taylorandfrancis.com**

FOR SALE IN SOUTH ASIA ONLY

Preface

The modern era of DNA sequencing began in late 1977 with the introduction of the most common DNA sequencing method, the Sanger method. The Sanger method involves the use of radioactive dideoxynucleotides, a deoxynucleotide with the 3′ hydroxyl group from the deoxyribose sugar removed.[1] The Sanger method relies on statistics to create fragments that are terminated at every position of the DNA. The presence of a band indicates the base position and identity.

Fluorescence-based sequencing was introduced in 1986 by L.M. Smith, L.E. Hood, and coworkers.[2] Four different fluorescent dyes were attached to the dideoxynucleotides allowing for spectral discrimination of the fragments. Fluorescence shows comparable performance and is an attractive alternative due to its relative safety, real-time capability, and ease of automation, as well as the ability to multiplex. In addition, the fluorescent dyes do not significantly affect the fidelity of the enzymes.[3]

The first multiplex fluorescence-based sequencing systems used a four-channel approach, similar to the radioactive-based sequencing.[4] Smith, Hood, and coworkers[2] used a set of four dyes with different emission maxima. The signal was selected using different interference filters based on the different dye emission maxima. The first system of dyes included the following: fluorescein isothiocyanate (λ_{em} = 516 nm), NBD-aminohexanoic acid (λ_{em} = 540 nm), tetramethylrhodamine isothiocyanate (λ_{em} = 582 nm), and Texas Red (λ_{em} = 612 nm). This system was later commercialized by Applied Biosystems (ABI) using a revised set of dyes: FAM (λ_{em} = 521 nm), JOE (λ_{em} = 555 nm), TAMRA (λ_{em} = 580 nm), and ROX (λ_{em} = 605 nm).[5] This commercial system allowed analytical sequencing to become a popular and routine technique for many laboratories. Mobility corrections are needed for this dye system. Ju et al.[6] developed a novel energy transfer system that addressed some of the failings of this system, which allowed for more efficient excitation with equal mobilities.

The original DNA sequencing systems were based on the standard slab polyacrylamide gel electrophoresis equipment, which allowed numerous samples to be analyzed on the same gel. Not long after the introduction of the slab-gel sequencing systems, a capillary electrophoresis (CE)-based sequencing system was developed. The CE system permitted increased speed, ease of use, and increased accuracy, although the CE system had a much lower throughput than the slab-gel system until the development of multicapillary systems. These systems are now commercially available and use from 8 to 96 capillaries in large arrays. Other techniques such as MALDI MS have been tried with modest success, but have no significant application. The future of DNA sequencing may lie in the use of microfabricated sequencing systems. These chip-based techniques will allow DNA sequencing to expand into a variety of new environments.

From the early days of radioisotope sequencing, a wide variety of new techniques have emerged to meet the needs of biotechnology. Techniques aimed at reducing the amount of sample needed, improving the accuracy, and reducing the amount of time needed to generate a sequence have been employed. The focus of this book is discussion of the different analytical DNA sequencing techniques, as well as some of the exciting applications of DNA sequencing. Numerous applications are discussed in this book, including microbiological identification, forensic DNA sequencing, and ancient DNA sequencing.

REFERENCES

1. F Sanger, S Nicklen, AR Coulson. Proc Natl Acad Sci USA 74, 5463–5467, 1977.
2. LM Smith, JZ Sander, RJ Kaiser, P Hughes, C Dodd, CR Connell, C Heiner, SBH Kent, LE Hood. Nature 321, 674–679, 1986.
3. LM Smith, S Fung, MW Hunkapiller, TJ Hunkapiller, LE Hood. Nucleic Acids Res 13, 2399–2412, 1985.
4. H Swerdlow, JZ Zhang, DY Chen, HR Harke, R Grey, S Wu, NJ Dovichi, C Fuller. Anal Chem 63, 2385–2841, 1991.
5. C Connell, S Fung, C Heiner, J Bridgham, V Chakerian, E Heron, B Jones, S Menchen, W Mordan, M Raff, M Recknor, L Smith, J Springer, S Woo, M Hunkapiller. BioTechniques 5, 342–348, 1987.
6. J Ju, AN Glazer, RA Mathies. Nat Med 2, 246–249, 1996.

Editor

Brian K. Nunnally, Sr., Ph.D., received his Ph.D. in chemistry and a certificate in molecular biophysics from Duke University, Durham, NC, in 1998. He is assistant director at Wyeth Laboratories in Sanford, NC, where he works in Vaccine Analytical Development. Dr. Nunnally's research interests lie in bioanalytical chemistry, including research on protein and polysaccharide therapeutics and BSE/TSE issues (BSE, bovine spongiform encephalopathy; TSE, transmissible spongiform encephalopathy). Prior to joining Wyeth, he worked for Eli Lilly and Company as a research scientist in quality control.

Dr. Nunnally has published numerous articles and lectured on a variety of analytical and pharmaceutical disciplines. His graduate work focused on the development of multiplex detection for bioanalytical separations using fluorescence lifetime. This included DNA sequencing fragment separations. In 2004, the book Dr. Nunnally coedited with Professor Ira Krull, titled *Prions and Mad Cow Disease* (Marcel Dekker, New York), was published.

Dr. Nunnally currently serves as assistant editor for *Analytical Letters* (CRC Press, Boca Raton, FL). He served as the cochair of the CE in the Biotechnology and Pharmaceutical Industries Conference in 2003 and 2004. Dr. Nunnally is a member of the Analytical Chemistry Division of the American Chemical Society and of the Society for Applied Spectroscopy. He was graduated with two honors degrees from University of South Carolina, Columbia. Dr. Nunnally is an Eagle Scout and father of two children, Brian Jr. and Annabelle.

Contributors

Robert G. Blazej
Department of Chemistry
  and UCSF/UCB Joint Graduate Group
  in Bioengineering
University of California
Berkeley, California

William Goodwin
Department of Forensic Medicine
  and Science
University of Glasgow
Glasgow, Scotland

Samuel A. Heath
Computer Science Department
Brown University
Providence, Rhode Island

Franz Hillenkamp
Institute for Medical Physics
  and Biophysics
University of Münster
Münster, Germany

Dorrie Main
Clemson University Genomics Institute
Clemson, South Carolina

Richard A. Mathies
Department of Chemistry
University of California
Berkeley, California

Terry Melton
Mitotyping Technologies, LLC
State College, Pennsylvania

Brian K. Nunnally
Wyeth Laboratories
Sanford, North Carolina

Brian M. Paegel
Department of Chemistry
University of California
Berkeley, California

Franco P. Preparata
Computer Science Department
Brown University
Providence, Rhode Island

Markus Sauer
Applied Laserphysics and
  Laserspectroscopy
University of Bielefeld
Bielefeld, Germany

Jeffrey P. Tomkins
Clemson University Genomics Institute
Clemson, South Carolina

Eli Upfal
Computer Science Department
Brown University
Providence, Rhode Island

Dirk van den Boom
Sequenom, Inc.
San Diego, California

Victor W. Weedn
Carnegie Mellon University
Mellon Institute
Pittsburgh, Pennsylvania

Kenneth D. Weston
Department of Chemistry
  and Biochemistry
Florida State University
Tallahassee, Florida

Todd C. Wood
Bryan College
Dayton, Tennessee

Edward S. Yeung
Ames Laboratory-USDOE
  and Department of Chemistry
Iowa State University
Ames, Iowa

Yonghua Zhang
Ames Laboratory-USDOE
  and Department of Chemistry
Iowa State University
Ames, Iowa

Contents

1 Introduction to DNA Sequencing: Sanger and Beyond

Brian K. Nunnally

CONTENTS

INTRODUCTION

The analytical chemistry of DNA sequencing is fascinating; the technology is impressive. When the Human Genome Project was commissioned in 1990, the goal was to complete the project in 15 years for less than U.S. \$3 billion. This was considered a difficult set of goals by the originators of the project. Not only was the project completed in 10 years, but it was also completed under budget, a rarity for any government endeavor. The lasting impact of the Human Genome Project will be not only the 3 billion DNA bases, but also the analytical technology that allowed the project to be completed faster than expected. From the early days of radioisotope sequencing, a wide variety of new techniques have emerged to meet the needs of biotechnology.

Techniques aimed at reducing the amount of sample needed, improving the accuracy, and reducing the amount of time needed to generate a sequence have been employed.

DNA sequencing involves a reaction, a separation, and detection and data analysis. The sequencing reactions can involve base-specific reactions or enzymatic extensions utilizing DNA polymerases. Separation methodology is commonly polyacrylamide gel electrophoresis (PAGE) or capillary electrophoresis (CE). The most common detection methodologies include fluorescence, although radioactivity has been used previously. Each of these steps is discussed in more detail.

REACTIONS

MAXAM–GILBERT METHOD

The Maxam–Gilbert sequencing method was actually published prior to the Sanger method (both were published in 1977). The Maxam–Gilbert method uses base-specific chemical degradation reactions to determine the sequence of an end-labeled DNA fragment. It is applicable to both single- and double-stranded DNA and requires no DNA polymerases. Four samples of radioactively end-labeled fragments are base-specifically chemically cleaved and separated electrophoretically in four separate lanes based on the specific reactions employed. A representative gel electropherogram is shown in Figure 1.1.

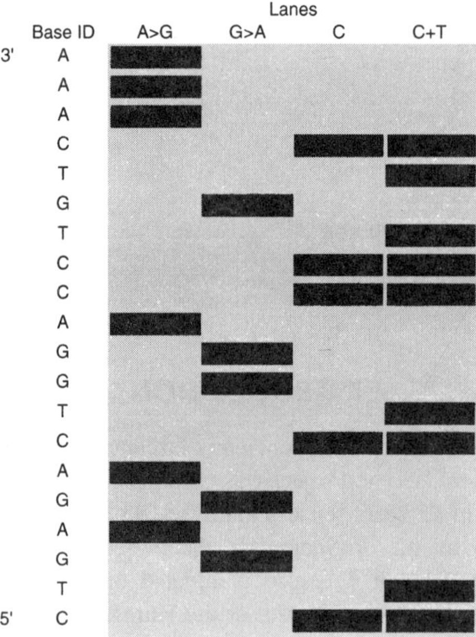

FIGURE 1.1 A diagram of the electropherogram resulting from the electrophoresis of the four Maxam–Gilbert sequencing reactions. Because cytosine is cleaved in two separate reactions, bands that appear in the cytosine (C) and cytosine/thymine (C + T) lanes indicate cytosine. The guanine (G > A) and adenine (A > G) lanes do not show any light bands due to the other base, although they can be present.[1] The sequence of this DNA is 5′-CTGAGACTGGACCTGTCAAA-3′.

The chemistry of the Maxam–Gilbert reactions is summarized in their 1977 publication.[1] Briefly, the adenines and guanines are methylated. The glycosidic bond is broken at neutral pH and then 1.0 M NaOH cleaves the sugar from the phosphate groups. The guanines are methylated fivefold faster than the adenines,[2] so the result is a dark guanine band and a weak adenine band (G > A). To enhance the adenine cleavage, the glycosidic bond is broken using dilute acid (0.5 M HCl). The result of cleavage with base is a dark adenine band and a weak guanine band (A > G). Cytosines and thymines are cleaved with hydrazine. The DNA is cleaved further with piperidine to give both cytosine and thymine bands. The addition of 2 M NaCl instead of water preferentially cleaves cytosine. Thus, the piperidine cleavages result in only cytosine bands. A diagram of the resulting electropherogram is shown in Figure 1.1.

SANGER METHOD

The modern era of DNA sequencing began in late 1977 with the introduction of the most popular DNA sequencing method, referred to as the Sanger method,[3] in honor of Fred Sanger, who was awarded a Nobel prize in 1980 for this work. This method starts with the denature of many copies of double-stranded DNA into single strands. The single-stranded DNA is mixed with a DNA polymerase, a DNA primer (a short single strand of DNA that will bind to the template DNA), deoxyribonucleotide triphosphates of all four bases (dNTPs, where N is the base), and a buffer (to minimize pH excursions). This mixture is aliquoted into four reaction tubes and a small amount of one of four dideoxynucleotide triphosphates (ddNTPs, where N is the base; the 3′ hydroxyl group from the deoxyribose sugar is removed); one ddNTP type (Figure 1.2) to each tube. The Sanger method involves the use of radioactive ddNTPs. These labeled dideoxynucleotides are added in addition to deoxynucleotides in a 1:100 ratio of each of the four reaction tubes. When each of the ddNTPs is incorporated into a sequence fragment, the fragment cannot be extended further. The Sanger method relies on statistics to create fragments that are terminated at every position of the DNA. The sequence is determined by comparing bands in each lane; only each band in one of the lanes of a DNA sequencing (i.e., PAGE) gel should appear at each position. The DNA sequencing gel is exposed to x-ray film, which is developed and the sequence is read from the bottom of the gel (5′ end) to the top of the gel (3′ end). The presence of a band indicates the base position and identity (Figure 1.3). Even in the first publication of this method, 300 bases from the primer site were determined.[3]

COMPARISON OF THE MAXAM–GILBERT METHOD
WITH THE SANGER METHOD

The Sanger method is widely used for large-scale sequencing projects. On the other hand, the Maxam–Gilbert method has not been utilized on a large scale for several reasons: base-specific fragments must be labeled and generated in two steps, labeling of DNA (and the whole technique, for that matter) is laborious and difficult, sequencing of ssDNA is thorny, and a primer walking strategy cannot be applied.[4] Chemical

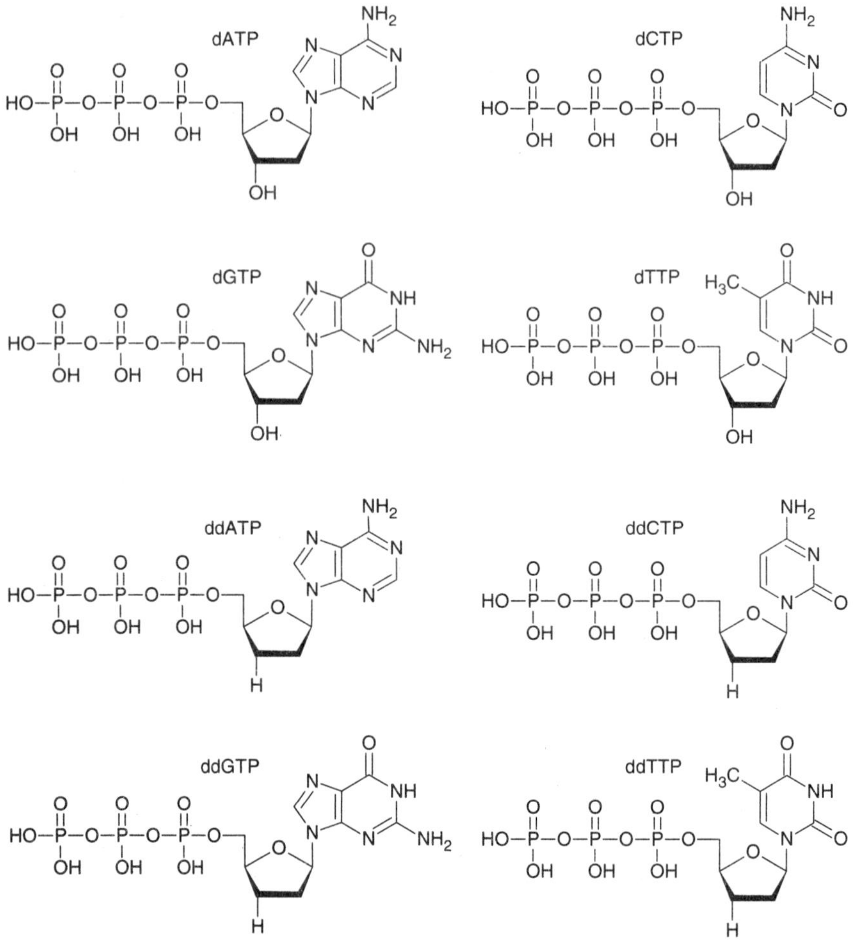

FIGURE 1.2 The deoxyribonucleotides, including deoxyadenosine 5′-triphosphate (dATP), deoxyguanosine 5′-triphosphate (dCTP), deoxycytidine 5′-triphosphate (dCTP), and deoxythymidine 5′-triphosphate (dTTP), as well as the dideoxyribonucleotides, including dideoxyadenosine 5′-triphosphate (ddATP), dideoxyguanosine 5′-triphosphate (ddCTP), dideoxycytidine 5′-triphosphate (ddCTP), and dideoxythymidine 5′-triphosphate (ddTTP). The preparation of ddTTP[26,27] was described previously in Sanger's 1977 publication.[3] The preparation of ddA was described previously[28] and was combined with other established methods to prepare ddATP.[29,30] The preparations of ddGTP and ddCTP were first described in the original Sanger method publication[3] based on the same techniques used to prepare ddATP.

degradation methodologies have proved to be useful for eukaryotic genomes, especially those with rich GC contents, and provide more uniform band intensities.[4] The Sanger method, because of its simplicity, has proved to be the "technique of choice" for DNA sequencing projects.

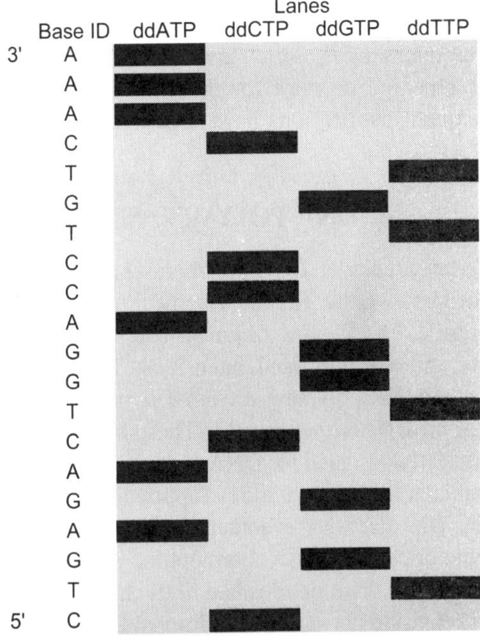

FIGURE 1.3 A diagram of the gel electropherogram resulting from the electrophoresis of the four Sanger sequencing reactions. The bases are determined by reading the sequence information vertically. The gel electropherogram is called a sequencing ladder for this reason. The sequence of this DNA is 5'-CTGAGACTGGACCTGTCAAA-3'.

OTHER SEQUENCING METHODS

Other sequencing methods have been developed since the initial development of the Sanger and Maxam–Gilbert methods. None has achieved the same level of usage as the Sanger method. In 1985, a single chemical cleavage method was developed.[5] Extensive treatment of DNA with aqueous piperidine at 90°C cleaves the DNA at each nucleotide. The relative intensity of the band determined the nucleotide. The order of reactivity is adenine > guanine > cytosine > thymine. In addition to the base cleavage differences, band spacing is used as an additional criterion for base determination. The relative difference in band separations are as follows, 1.3 guanine:1.2 thymine:1.1 adenine:1.0 cytosine. As noted in the article,[5] there are several advantages of this method compared with the conventional Maxam–Gilbert method. First, the procedure is greatly simplified, i.e., a single sample and a simplified reaction procedure. Second, dimethyl sulfate and hydrazine, two very toxic substances, are eliminated. Third, the amount of DNA needed is reduced, which is less of an issue with the introduction of polymerase chain reaction (PCR) and cycle sequencing. Finally, band ordering problems seen because of curvature due to intra-gel differences are eliminated. The method does not attain the level of accuracy obtained by either the Maxam–Gilbert or the Sanger methods.

In 1988, a new sequencing method based on the detection of inorganic phosphate (PP_i) was published.[6] A DNA polymerase catalyzes the reaction of a dNTP with a template/primer and releases PP_i, which is measured by light generation in the final step. The method appeared to work for the poly(dA·dT) sequence in the paper. However, no subsequent publications have appeared in the literature.

DNA POLYMERASES

Several excellent review articles on DNA polymerases are available.[7,8] The first enzyme used for DNA sequencing was the Klenow fragment,[3] a large fragment of *Escherichia coli* DNA polymerase I. The Klenow fragment had several problems including lack of thermal stability and variable band intensities. The isolation, purification, and characterization of the DNA polymerase from *Thermus aquaticus*, an extreme thermophile, revolutionized DNA sequencing.[9,10] These hyperthermophiles have evolved enzymes that are perfectly designed for DNA sequencing and PCR reactions. Because the enzyme was stable at high temperatures (optimum temperature of 80°C),[9] longer reads were possible. This discovery eventually initiated a quest to discover and characterize DNA polymerases from other thermophilic species. As of 2001, more than 50 DNA polymerases from thermophiles had been characterized,[8] including bacterial species from ocean vents and hot springs. Additional DNA polymerases are still being discovered and characterized.[11,12] Could these be the future ideal DNA polymerase for DNA sequencing?

The ideal DNA polymerase should have a rapid rate of dNTP incorporation, lack exonuclease activities, and not discriminate toward nucleotide analogues.[13] High fidelity is also important. The native enzymes of these bacteria were often non-ideal for the intended biotechnological use. Companies such as Amersham Biosciences, Stratagene, New England Biolabs, and Perkin-Elmer have used genetic engineering techniques to produce mutants of the enzyme that eliminated other deficiencies, such as reduced incorporation of ddNTPs, variable band intensities, and high error rates. The pace of research appears to have slowed, but work is still being done. Through random and site-directed mutagenesis, polymerases with higher fidelity and increased incorporation of labeled ddNTPs have been created. Azrezi and colleagues[14] reported using these techniques to find mutants that show 250-fold incorporation improvement of labeled ddNTPs. Patel and coworkers[15] used similar techniques to understand the source of fidelity in *T. aquaticus* DNA polymerase. Isoleucine at position 614, part of the hydrophobic pocket that binds the base, was found to be critical to the fidelity.[15]

SEPARATION

SLAB-GEL SEQUENCING

The original DNA sequencing systems utilized standard slab PAGE equipment for separation of the products of the sequencing reactions. The PAGE experiment is not a true "electrophoresis" experiment. Electrophoresis is based on the separation of molecules with different charges, and DNA molecules have, essentially, the same charge and extremely similar charge-to-mass ratios in the sequencing reaction separation. The PAGE experiment is actually a size-based separation. The polyacrylamide creates a plethora of different sized "pores." DNA molecules become entangled in the pores. The larger the DNA fragment, the more entangled it becomes and the slower it moves through the gel (i.e., it will stay toward the top of the gel). Conversely, smaller fragments will move more rapidly through the gel matrix and will be found toward the bottom of the gel. The gels are typically 6% acrylamide in 1 × TBE (tris-borate-EDTA) buffer.[16]

One of the early advantages of slab-gel electrophoresis over CE-based separations was the large throughput of the slab gels. In fact, 64-well combs were typically used for the sequencing gels allowing for as many as 16 different samples to be sequenced on one gel. Combs with more wells were also used, increasing the throughput even further. The small size of the bands in the gels with larger combs made them difficult to read. These throughput advantages have been largely overcome by the newer large-scale CE sequencing systems, which are described later. An excellent review and application article on slab-gel sequencing was published in the Methods in Enzymology series.[16] As detailed in the article, the most important parameters to control in slab-gel preparation are elimination of dust particles, careful handling to reduce mechanical stresses, and careful pouring to reduce bubble formation.

CE SEQUENCING

Not long after the introduction of the slab-gel sequencing systems, a CE-based sequencing system was developed. The CE system allowed for increased speed, ease of use, and improved accuracy, although the CE system had a much lower throughout than the slab-gel system, until the development of multicapillary systems. There are many excellent reviews of the subject. An older review of CE-based DNA sequencing is still an excellent consideration of the subject.[19] Dovichi co-wrote a later review as well.[17] Another more recent review focuses on the theoretical principles of the technique.[18]

CE separations offer several advantages over slab-gel-based sequencing systems.[19] First, capillary systems are able to use dynamic coatings, allowing for replacement of the sieving matrix between separations. Slab gels must be poured and polymerized between the glass plates. The gels are difficult to pour (especially without creating bubbles) and time-consuming to prepare. Second, the flexible capillaries are easily coupled to a microtiter plate. As discussed later, this includes 384-well microtiter plates. Finally, multicapillary systems can be created that greatly increase the throughput of a sequencing system, which is discussed in more detail in the next paragraph.

Multicapillary systems are now commercially available and use from 8 to 384 capillaries in large arrays. These systems are overcoming the advantages of the slab-gel-based sequencing systems and are rapidly replacing them as the primary systems for large-scale DNA sequencing. Many companies offer 96-capillary systems; SpectruMedix (State College, PA) and Nyxor Biotech (London, U.K.) have developed 192-capillary systems; and Amersham Biosciences introduced the MegaBACE™ 4000 in the fall of 2001. This system was the first commercially available 384-capillary system, which is currently the largest number of capillaries commercially available on a single instrument. The MegaBACE 4000 system utilizes linear polyacrylamide-filled, 75 μm (i.d.) × 40 cm capillaries allowing read lengths of up to 1000 bp in 3 h.[20] This translates to a theoretical yield of more than 3 million base pairs of DNA sequenced per day! With current microtiter plates available in 1536-well format, one could envision a 1536-capillary system (more than 12 million bases sequenced per day per instrument). The problem with these extremely large sequencing systems, a problem that will ultimately limit the size of the instrumentation, is generating the samples to run on them. One 96-array system being sold by CombiSep (Ames, IA) uses technology developed in Ed Yeung's laboratory at Iowa State University. The MCE 2000™ utilizes ultraviolet (UV) detection and a separation voltage of 250 V/cm.[21] All other multicapillary DNA sequencing systems utilize fluorescence detection; however, Zhong and Yeung published a recent paper describing the use of UV absorption for the detection of DNA sequencing fragments.[22] The advantages of using UV detection are mostly related to expense, in that fluorescence detection systems are expensive, as are the reagents.[23] In addition, the mobility of the fragments is shifted because of the addition of the dye to the sequencing fragment.[22] UV measurements are not without their drawbacks: UV is less sensitive, with detection limits that are inferior to fluorescence-based systems; in addition, four separate capillaries must be employed because there is no discrimination between DNA terminated with different ddNTPs. Because many of the common sieving matrices cannot be used when employing UV detection, a new dynamic sieving mechanism based on surfactant self-assembly was utilized by Zhong and Yeung.[22] An internal standard was used because of differences between migration times in the multiple capillaries. The read length in this paper was short (approximately 100 bp), but shows much promise for the future. A 96-capillary array system microfabricated from a microchannel plate was created by Paegel and coworkers at the University of California, Berkeley.[24] This system, which significantly reduces the amount of reagents and sample volume needed for the sequencing reactions, was able to obtain a read length of 430 bp. Further, the rate of sequencing information generated was approximately five times greater than current commercial capillary array systems.[24]

Consumables for sequencing reactions run approximately $6.93/reaction (this amount includes labeling reagents, linear polyacrylamide, CE buffer, capillary arrays, and disposable plasticware).[23] Recently, efforts have been made to minimize the amount of sequencing reagents used in an attempt to reduce the overall cost per reaction.[23] Azadan and coworkers[23] were able to reduce the amount of dye terminator used to 1:4 with no apparent statistical difference in read length or accuracy. More importantly, they were able to reduce the overall volume of the sequencing reaction to 5 μl without any reduction in read length or accuracy (although some additional variability in read length is noted at the lower volume). This information is summarized in Table 1.1.

TABLE 1.1
Average Read Length (with standard deviation) and Base Calling Accuracy (with standard deviation) for Various Sequencing Reactions, Including Dilutions of the Dye Terminator Premix and Reduced-Scale Sequencing Reactions

Description	Dye Terminator Dilution	Reaction Volume (μl)	Average Read Length (bp)	Standard Deviation	Base Calling Accuracy (%)	Standard Deviation
SOP reaction ("control")	1:1	20	761	26	98.3	0.6
Diluted terminator	1:2	20	676	45	98.4	0.2
Diluted terminator	1:4	20	645	63	98.2	0.3
Scaled-down reaction	1:1	10	732	27	98.1	0.8
Scaled-down reaction	1:1	5	741	41	98.7	0.4

Source: Adapted from Azadan et al.[23]

DETECTION

RADIOACTIVE

Originally, detection was accomplished by radioactive labels such as ^{32}P or ^{35}S. Radioactive labels were extremely effective for detection of DNA sequencing reaction products. The labeled reagents are no different in size or shape than the unlabeled reagents, so the DNA polymerases exhibit no preference or fidelity reductions. However, radioactive gels must be exposed to the x-ray film, taking upward of 24 to 36 h to develop to collect 500 bases of sequencing data.

FLUORESCENCE

The invention of fluorescence-based techniques has eliminated radioactive labels from almost all DNA sequencing. This is due to the increased safety, significantly decreased upkeep and disposal costs, ability to multiplex, and real-time data acquisition. Radioactivity requires special licensing and active management, elements that are unnecessary for fluorescence dyes. The ability to multiplex is a crucial parameter. Instead of using four different lanes of a PAGE gel, a single well with four labels is utilized. Finally, real-time data acquisition is possible, eliminating the need for off-line data collection (i.e., x-ray film exposure and development).

The first fluorescence data system was developed in Hood's laboratory in the mid-1980s.[25] The original dye system utilized four dyes — fluorescein isothiocyanate, NBD-aminohexanoic acid, tetramethylrhodamine isothiocyanate, and Texas

Red — each with a different emission maximum. NBD-aminohexanoic acid was later replaced by a dye with a higher quantum yield and increased signal. The use of fluorescent dyes for DNA sequencing fragment detection was first commercialized by Applied Biosystems utilizing the dye set of FAM, JOE, TAMRA, and ROX. This detection methodology was responsible for most of the sequence generated from the Human Genome Sequencing Project. Chapter 2 discusses recent advances in fluorescence detection of DNA sequencing fragments.

CONCLUSION: IMPACT OF SEQUENCING

DNA sequencing is not finished revolutionizing science. As of the end of 2003, more than 32,000 publications with the term "DNA sequencing" had been published (Figure 1.4). The new frontier is the ability to sequence DNA rapidly for use in medical diagnosis. Many challenges remain, although the technology is sufficiently advanced today to allow for increased use of this powerful diagnostic tool. The human genome, while sequenced, has not been completely deciphered, in that the functions of only a few genes are known. As medical science and molecular biology increase their knowledge, the use of DNA sequencing in medical diagnosis will increase. The future should allow for every individual's genome to be sequenced (although the ethical hurdles are not insignificant).

Techniques aimed at reducing the amount of sample needed, improving the accuracy, reducing the amount of time needed to generate a sequence have been employed. The focus of this book is to discuss some of the different analytical DNA sequencing techniques as well as some of the exciting applications of DNA sequencing. Advances in DNA separations through advanced CE and microchip sequencing are presented. Chapters in the book highlight improvements to the

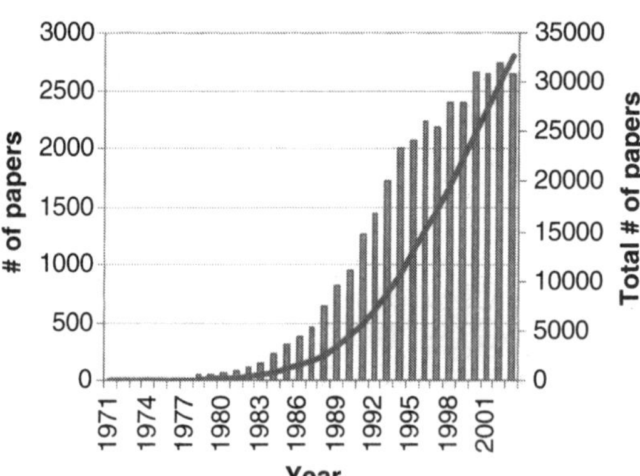

FIGURE 1.4 A graph of the total number of DNA sequencing-related publications according to MEDLINE by year. Numbers were compiled using PubMed (http://7.ncbi.nlm.nih.gov/entrez) based on a search for "DNA sequencing" limited by year. The first paper was published in 1972. A total of 32,697 papers have been published (as of 12/31/03). The trend is shown on the secondary y-axis. Data were compiled and were accurate as of 03/27/04.

detection of DNA sequencing fragments, including other types of fluorescence detection, the use of mass spectrometry, and single molecule detection. Chapters on the uses of DNA sequencing for forensic, ancient DNA analysis, and genome sequencing are included.

ACKNOWLEDGMENTS

I acknowledge Deedra Nunnally for her assistance with literature searching and thank my colleague Kun Yao for his review and comments on this chapter. Figures containing chemical structures were prepared using ACD/ChemSketch available from Advanced Chemistry Development, ACD/Labs. A freeware version can be downloaded from www.acdlabs.com/downloads. The author of this chapter has no financial ties to any of the companies mentioned in the chapter.

REFERENCES

1. AM Maxam, W Gilbert. A new method for sequencing DNA. Proc Natl Acad Sci USA 74:560–564, 1977.
2. PD Lawley, P Brookes. Further studies on the alkylation of nucleic acids and their constituent nucleic acids. Biochem J 89:127–138, 1963.
3. F Sanger, S Nicklen, AR Coulson. DNA sequencing with chain-terminating inhibitors. Proc Natl Acad Sci USA 74:5463–5467, 1977.
4. A Rosenthal, B Sproat, H Voss, J Stegemann, C Schwager, H Erfle, J Zimmerman, C Coutelle, W Ansorge. Automated sequencing of fluorescently labeled DNA by chemical degradation. DNA Sequence 1:63–71, 1990.
5. BJB Ambrose, RC Pless. Analysis of DNA sequences using a single cleavage procedure. Biochemistry 24:6194–6200, 1985.
6. ED Hyman. A new method of sequencing DNA. Anal Biochem 174:423–436, 1988.
7. SC Hamilton, JW Farchaus, MC Davis. DNA polymerases as engines for biotechnology. BioTechniques 31:370–383, 2001.
8. H Hogrefe, J Cline, AE Lovejoy, KB Nielson. DNA polymerases from hyperthermophiles. Method Enzymol 334:91–116, 2001.
9. A Chien, DB Edgar, JM Trela. Deoxyribonucleic acid polymerase from the extreme thermophile *Thermus aquaticus*. J Bacteriol 127:1550–1557, 1976.
10. MA Innis, KB Myambo, DH Gefland, MAD Brow. DNA sequencing with *Thermus aquaticus* DNA polymerase and direct sequencing of polymerase chain reaction-amplified DNA. Proc Natl Acad Sci USA 85:9436–9440, 1988.
11. Y Ishino, S Ishino. DNA polymerases from Euryarchaeota. Method Enzymol 334:249–260, 2001.
12. I Bruck, A Yuzhakov, O Yurieva, D Jeruzalmi, M Skangalis, J Kuriyan, M O'Donnell. Analysis of a multicomponent thermostable DNA polymerase 1 replicase from an extreme thermophile. J Biol Chem 277:17334–17348, 2002.
13. S Tabor, CC Richardson. DNA sequence analysis with a modified bacteriophage T7 DNA polymerase. Proc Natl Acad Sci USA 84:4767–4771, 1987.
14. B Arezi, CJ Hansen, HH Hogrefe. Efficient and high fidelity incorporation of dye-terminators by a novel Archaeal DNA polymerase mutant. J Mol Biol 322:719–729, 2002.

15. PH Patel, H Kawate, E Adman, M Ashbach, LA Loeb. A single highly mutable catalytic site amino acid is critical for DNA polymerase fidelity. J Biol Chem 276:5044–5051, 2001.

16. LM Smith, RL Brumley, Jr, EC Buxton, M Giddings, M Marchbanks, X Tong. High-speed automated DNA sequencing in ultrathin slab gels. Method Enzymol. 271:219–237, 1996.

17. NJ Dovichi, J Zhang. DNA sequencing by capillary array electrophoresis. Method Mol Biol 167:225–239, 2001.

18. C Heller. Principles of DNA separation with capillary electrophoresis. Electrophoresis 22:629–643, 2001.

19. NJ Dovichi. DNA sequencing by capillary electrophoresis. Electrophoresis 18:2393–2399, 1997.

20. http://www.amershambiosciences.com/aptrix/upp01077.nsf/content/6795588956BDB745 C1256BEC003023A3?OpenDocument&querytitle=&hometitle=search.

21. http://www.combisep.com/specs.html.

22. W Zhong, ES Yeung. Multiplexed capillary electrophoresis for DNA sequencing with ultra violet absorption detection. J Chromatogr A 960:229–239, 2002.

23. RJ Azadan, JC Fogleman, PB Danielson. Capillary electrophoresis sequencing: maximum read length at minimal cost. BioTechniques 32:24–28, 2002.

24. BM Paegel, CA Emrich, GJ Wedemayer, JR Scherer, RA Mathies. High throughput DNA sequencing with a microfabricated 96-lane capillary array electrophoresis bioprocessor. Proc Natl Acad Sci USA 99:574–579, 2002.

25. LM Smith, JZ Sander, RJ Kaiser, P Hughes, C Dodd, CR Connell, C Heiner, SBH Kent, LE Hood. Fluorescence detection in automated DNA sequence analysis. Nature 321:674–679, 1986.

26. AF Russell, JG Moffatt. Synthesis of some nucleotides derived from 3′-deoxythymidine. Biochemistry 8:4889–4896, 1969.

27. K Geider. DNA synthesis in nucleotide-permeable *Escherichia coli* cells. The effects of nucleotide analogues on DNA synthesis. Eur J Biochem 27:554–563, 1972.

28. JR McCarthy, MJ Robins, LB Townsend, RK Robins. Purine nucleosides. XIV. Unsaturated furanosyl adenine nucleosides prepared via base-catalyzed elimination reactions of 2′-deoxyadenosine derivatives. J Am Chem Soc 88:1549–1553, 1966.

29. GM Tener. 2-Cyanoethyl phosphate and its use in the synthesis of phosphate esters. J Am Chem Soc 83:159–168, 1961.

30. DE Hoard, DG Ott. Conversion of mono- and oligodeoxyribonucelotides to 5′-triphosphates. J Am Chem Soc 87:1785–1788, 1965.

2 Developments in the Detection of DNA Sequencing Fragments Using Fluorescence: Energy Transfer and Fluorescence Lifetimes

Brian K. Nunnally

CONTENTS

INTRODUCTION

Fluorescence has nearly eliminated radioactive-based sequencing as a detection technique for DNA sequencing fragments. The advantages of fluorescence are staggering. Fluorescence is cheaper (both to purchase and to dispose of), provides real-time data, allows for multiplex analysis (i.e., data on all four bases in the same lane), and is safer. There are several disadvantages of fluorescence-based detection that are not characteristic of radioactivity: dyes with fluorescent properties must be synthesized, the synthesized dyes must be conjugated to the biomolecule (either primer or dideoxynucleotide), and expensive detection systems must be employed.

These disadvantages have been steadily overcome through research, but improvements to the established methodology will be limited by these disadvantages.

Fluorescent dyes that are to be utilized for DNA sequence fragment detection must have several properties. First, the set of dyes must have spectral discrimination. In this chapter, dye sets with emission-based discrimination and fluorescence lifetime discrimination are discussed. Ideally, the dyes will have good quantum yields (0.8 or higher). Quantum yield can be thought of as the efficiency of converting excitation radiation into emission signal (the higher the efficiency, the higher the signal). Increased signal allows for lower detection limits, which allows for more sequencing data per run. Quantum yield limitations can be overcome, in some ways, by increasing the power of the excitation radiation. This is not a panacea, as photobleaching and other deleterious effects will result if the laser power is too high. Finally, the dye must have a high molar absorptivity. Light that is not absorbed cannot be converted to signal. Regardless of the spectral discrimination employed, quantum yields and molar absorptivity are critical parameters for fluorescence detection of DNA sequencing fragments.

ENERGY TRANSFER–BASED FLUORESCENCE SEQUENCING

It is extremely difficult to create a dye set with a single optimal excitation maximum, four well-resolvable emission spectra, and similar sizes to prevent mobility shifts. The invention of energy transfer primers solved many of these issues. The principle of energy transfer has been known for a long time, but was only applied to DNA sequencing in the mid-1990s. A "donor" dye is excited by a laser. The emission of the donor dye is used to excite a second dye (the "acceptor"). The efficiency of the energy transfer depends on the overlap of the donor emission spectrum and the acceptor excitation spectrum and the molecular distance between the donor and the acceptor. Two distinct approaches to energy transfer–based fluorescence sequencing have been undertaken. The first approach is to synthesize primers with the donor and acceptor dye separated by several DNA base pairs or by sugars. The second approach is to attach the donor and acceptor dyes together separated by a linker. The resulting dye dimer could be attached to either a primer or a dideoxynucleotide. Both approaches are discussed in more detail.

ENERGY TRANSFER PRIMERS

The first reported energy transfer (ET) primers utilized FAM as the donor dye and the FAM, JOE, TAMRA, and ROX dye set as the acceptors.[1] FAM is ideally excited by the 488-nm line of the argon ion laser and the four dyes have well-resolved emission maxima (525, 555, 580, and 605 nm, respectively). The contribution of FAM to the overall fluorescence signal in each channel was small and could be eliminated as a potential source of inaccuracy using filters. Data for the optimum primer set are shown in Table 2.1. The ET primers showed improved sensitivity relative to the standard single dye primer set. Normally, the TAMRA and ROX primers require threefold more template and twofold more

TABLE 2.1
Summary of Data for the Optimized Primer Set for the First Reported Fluorescence Energy Transfer DNA Sequencing Primers

Donor	Distance (bp)	Acceptor	Emission Maximum (nm)	Improvement	Efficiency (%)	Sensitivity (%)
FAM	10	FAM	525	1.8	NA	160
FAM	10	JOE	555	2.5	65	360
FAM	3	TAMRA	580	5.3	97	400
FAM	3	ROX	605	6.2	96	470

Note: The distance was the distance between the donor and acceptor in base pairs (bp). The improvement was determined relative to the single dye excited at 488 nm (for example, the FAM-10-JOE energy transfer primer has a 2.5 times higher fluorescence signal, normalized for DNA content, than a JOE-labeled primer excited at 488 nm). The efficiency was determined by comparison to the residual emission of FAM in the ET primers with a FAM-labeled primer of the same sequence and length. The sensitivity improvement was determined by plotting the band intensity vs. the quantity of template and comparing the slopes of the ET primers vs. the corresponding single-label primers.

Source: Adapted from data reported in Ju et al.[1]

primer to obtain comparable signals for the TAMRA and ROX lanes; however, the ET primers do not require the additional template or primer.[1] Another aspect of the optimization was the minimization of the mobility shifts needed for the standard single dye primer set. The mobility shift for the ET primers was reduced compared to the single dye labeled primers. The TAMRA and ROX labeled primers migrated nearly one nucleotide slower (compared to the FAM and JOE labeled primers) while the ET primers FAM-TAMRA and FAM-ROX migrated only one quarter of a nucleotide slower (compared to the ET primers for FAM-FAM and FAM-JOE)[1] In all, 500 bases of DNA were sequenced with 99.8% accuracy using these primers.

A set of 20 different primers was synthesized and characterized to determine the optimum primer set.[2] These primers varied in the position of the acceptor dyes. The tested distances between donor and acceptor were 1, 2, 3, 4, and 10 bp. Increasing the distance between the donor and acceptor dyes to 10 bp was determined to be better than the variable length described previously.[1] The fluorescence signal of the FAM-TAMRA and FAM-ROX primers with 10 bp differences were threefold higher than the FAM-TAMRA and FAM-ROX primers with 3 bp differences. The increased distance between the donor and acceptor did not adversely affect the mobility (approximately 0.2 bp between the four primers). One consequence of the increased distance between donor and acceptor is increased FAM emission. The emission maxima are sufficiently well resolved to prevent this from affecting the base calling accuracy.[2] The 500 bases of DNA were sequenced with 99.4% accuracy using these primers with a blind read. The improved signal strength allows for reduced template amount (when needed) or longer read lengths, if desired. This same dye set was

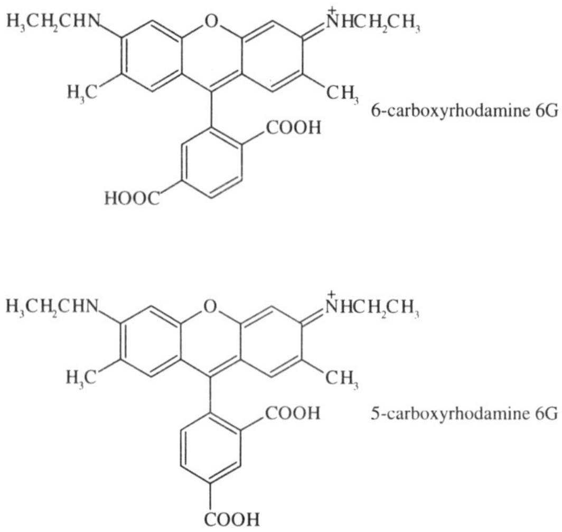

6-carboxyrhodamine 6G

5-carboxyrhodamine 6G

FIGURE 2.1 The structure of 5-carboxyrhodamine 6G and 6-carboxyrhodamine 6G.[4] All dyes are shown with carboxylic acid groups. The 5- and 6-carboxylic acids are changed to other functionalities, usually NHS esters, for conjugation to DNA primers or dideoxynucleotides.

used to sequence 600 bases with 100% accuracy and 850 bases with 98% accuracy, with the potential for ever longer reads.[3]

The emission of JOE overlaps with the emission of FAM. The use of a new dye, 5- or 6-carboxyrhodamine-6G (Figure 2.1) with a narrower emission spectrum (and thus less overlap) represents an improvement.[4] The emission maxima for both 5-carboxyrhodamine-6G (558 nm) and 6-carboxyrhodamine-6G (555 nm) were similar to the emission maximum for JOE (555 nm) with similar fluorescence intensities. The mobility of the new ET primers (donor dye = FAM, acceptor dye = 5- or 6-carboxyrhodamine-6G, and distance = 10 bp) were improved relative to a similar JOE primer (donor dye = FAM, acceptor dye = JOE, and distance = 10 bp). The FAM-5-carboxyrhodamine-6G and FAM-6-carboxyrhodamine-6G showed a mobility shift of less than 0.1 bp relative to the FAM-ROX ET primer while FAM-JOE showed a 0.2 bp mobility shift. Hung and coworkers[4] were able to sequence 620 bases with 99% accuracy without the need for a mobility correction.

In the effort to decrease the amount of signal obtained from the donor dye, a high-molar-absorptivity, low-fluorescence-quantum-yield dye was used as the donor dye.[5] A cyanine dye (CYA; see Figure 2.2 for structure) with a molar absorptivity at 488 nm of 142,000 $M^{-1}cm^{-1}$ was used as the donor dye for FAM, 6-carboxyrhodamine-6G (R6G), TAMRA, and ROX acceptor dyes. The four primers (CYA-FAM, CYA-R6G, CYA-TAMRA, and CYA-ROX) utilized the same donor-acceptor difference (10) as optimized previously.[3] The fluorescence intensity of the CYA-FAM primer was 80% of the fluorescence intensity of the FAM-FAM primer; the other primers showed increased fluorescence intensity when compared

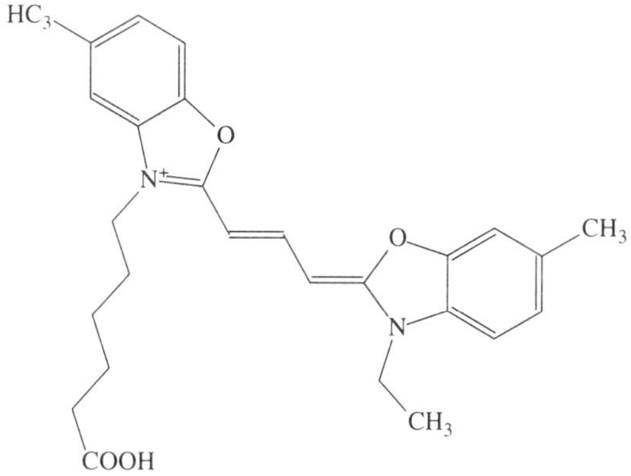

FIGURE 2.2 The structure of CYA (3-(ϵ-carboxypentyl)-3′-ethyl-5,5′-dimethyloxacarbo-cyanine).[5] The dye is shown with a carboxylic acid group. The carboxylic acid group is changed to other functionalities, usually a NHS ester, for conjugation to DNA primers or dideoxynucleotides.

with their corresponding FAM donor primers (CYA-R6G was 10% higher while CYA-TAMRA and CYA-ROX were each 70% higher). The new ET primers showed reduced crosstalk, as well as reduced mobility shifting. The 500 bases of DNA were sequenced with 100% accuracy using the new ET primer set.

A comprehensive study of 56 different ET primers, differing in the spacing of the donor and acceptor, the type of spacer, the primer sequence, and the donor and acceptor dyes identified an improved ET primer set.[6] A series of FAM-ROX primers with donor–acceptor differences of 1, 2, 3, 4, 6, 8, 10, and 12 bp were synthesized. The FAM-ROX primer with an eight pair difference gave the highest intensity. Primers with 10, 6, and 12 were fairly similar to each other and much higher than the primers with 4, 3, 2, or 1 bp differences. These results were similar to those generated previously.[2] A series of FAM-ROX primers utilizing six, seven, eight, nine, or ten sugars (1′,2′-dideoxyribose phosphate) as the spacers were synthesized. Utilizing sugars as the spacer for the ET primers allows any primer to be utilized rather than specially synthesized primers. The eight, seven, nine, and ten sugar spacer primers had similar fluorescence intensities (much higher than the six sugar spacer primer). Several common primers (SP6, T3, T7, M13 forward, and M13 reverse) were synthesized as ET primers. Curiously, the distance between the FAM (donor) and ROX (acceptor) was not consistent between the primers, preventing a full comparison. A new dye, 5- and 6-carboxyrhodamine-110 (R110; see Figure 2.3 for dye structures), was tested as a replacement for FAM as the acceptor dye. A set of primers utilizing CYA (Figure 2.2) as the donor dye and FAM, R6G, TAMRA, ROX, and R110 with donor–acceptor differences of 2, 4, 6, 8, 10, 12, 14, and 16 bp. The ideal (i.e., distance that gave the highest fluorescence intensity) donor–acceptor distance for all primers was 10.[6] The emission maximum for the CYA-R110 primer is 530 nm. The CYA-R110 primer had only 70% of the emission intensity of a CYA-FAM primer, but the fluorescence

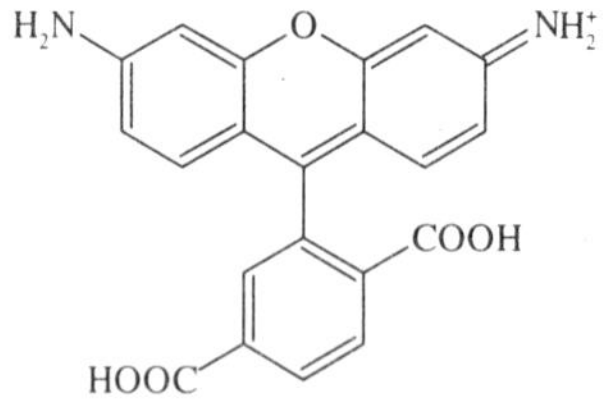

6-carboxyrhodamine 110

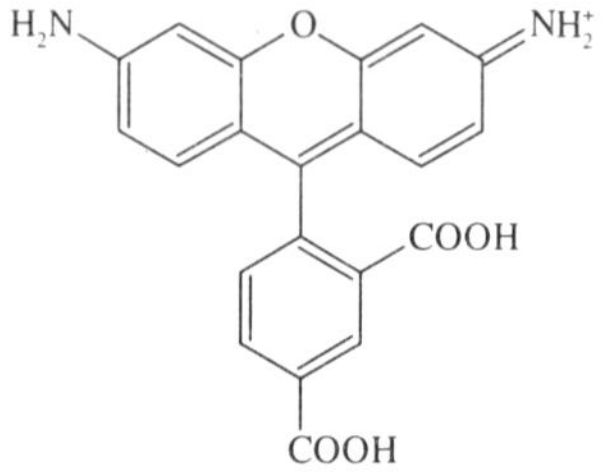

5-carboxyrhodamine 110

FIGURE 2.3 The structure of 5-carboxyrhodamine-110 and 6-carboxyrhodamine-110. All dyes are shown with carboxylic acid groups. The 5- and 6-carboxylic acids are changed to other functionalities, usually NHS esters, for conjugation to DNA primers or dideoxynucleotides. (Adapted from http://www.probes.com/servlets/structure?item=6479.)

intensity was still stronger (approximately 30%) than CYA-ROX, rendering it suitable for use in DNA sequencing.[6] The primer set of CYA-R110, CYA-R6G, CYA-TAMRA, and CYA-ROX (an entire rhodamine acceptor dye set) was utilized to sequence 600 bases with 100% accuracy and 850 bases with 98% accuracy. The mobility was also improved and well matched with the other primers.

BODIPY dyes have been used as donor–acceptor dyes for ET primers.[8] The narrow excitation and emission spectra combined with the high molar absorptivities of BODIPY dyes make them excellent dyes for use in DNA sequencing. BODIPY 503/512 (Figure 2.4) was utilized as the donor for all of the ET primers. The acceptor dyes were BODIPY 523/547, BODIPY 564/570, and BODIPY 581/591 (Figure 2.4). Amino linker arms (either a propyl or hexyl group) were used to separate the donor and acceptor dye linker arm. The length of the linker arms was fairly innocuous as less than a 5% fluorescence intensity difference between primers with the two different linker arms was noted. The overall intensity of the BODIPY-labeled primer was much lower than the fluorescence intensity of other ET primers. The FAM-ROX ET primer gave 2.7-fold more signal when compared to the BODIPY 503/512-BODIPY 581/591 primer when excited at 488 nm (both primers utilized a 3 bp separation). With excitation at 514 nm, the two ET primers gave similar fluorescence signals. A number of FAM-ROX primers could be created (by increasing the base pair separation) that exhibited severalfold fluorescence signal improvements vs. the optimal BODIPY 503/512-BODIPY 581/591 primer (BODIPY 503/512-BODIPY 581/591 with a 3 bp distance between donor and acceptor, a propyl amino linker, and excitation at 514 nm).

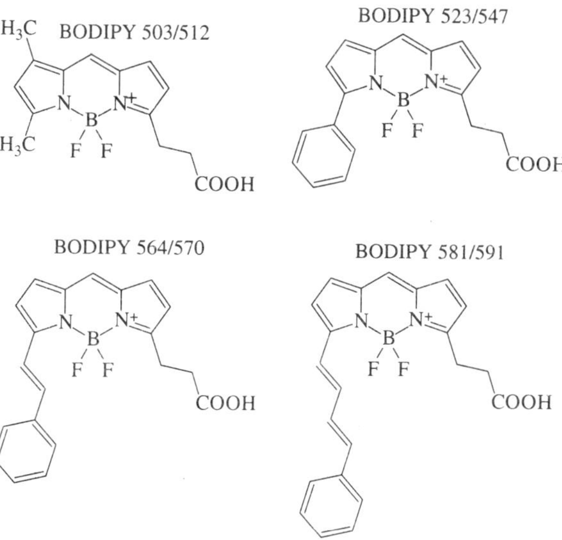

FIGURE 2.4 The structure of BODIPY dyes used for energy transfer primers.[8] All dyes are shown with carboxylic acid groups. These carboxylic acid groups are changed to other functionalities, usually NHS esters, for conjugation to DNA primers or dideoxynucleotides. (Adapted from Metzker et al.[15])

BigDye™ CHEMISTRY

In the 1980s, Molecular Probes synthesized an energy transfer dye by linking 5-carboxytetramethylrhodamine and 5-carboxyfluorescein.[9] The resulting dye had an absorption spectrum that was the combination of the two individual spectra, but an emission spectrum of only 5-tetramethyrhodamine. The approach of using energy transfer dyes consisting of a donor and acceptor dye tethered together, rather than separated by several nucleotides has proved effective. The first set of dyes, trademarked as BigDyes, utilized fluorescein and dichlorinated rhodamine dyes (Figure 2.5).[9] Dichlorinated rhodamine dyes have narrower emission profiles and slightly longer wavelength for their emission maxima relative to the same nonchlorinated rhodamine dyes.[9] The ET dyes synthesized here were only ~60% of maximal brightness, with both lower extinction coefficients and quantum yields than the rhodamine dyes alone. Despite these results, the ET dyes are brighter than the rhodamine dyes alone. The BigDyes were conjugated to a −21 M13 primer and utilized in a DNA sequencing run resulting in 700 bases of sequencing data before the first unambiguous call. The combination of dyes shown in Figure 2.5 required a mobility correction no larger than 0.4 bp.

The BigDyes dye set (Figure 2.5) was conjugated to dideoxynucelotides for use in DNA sequencing.[10] The average errors (to 720 bp), average read length at 98.0% accuracy, and signal strength for the BigDye dye terminators and two non-energy transfer dye sets are shown in Table 2.2. The average errors for the BigDye dye

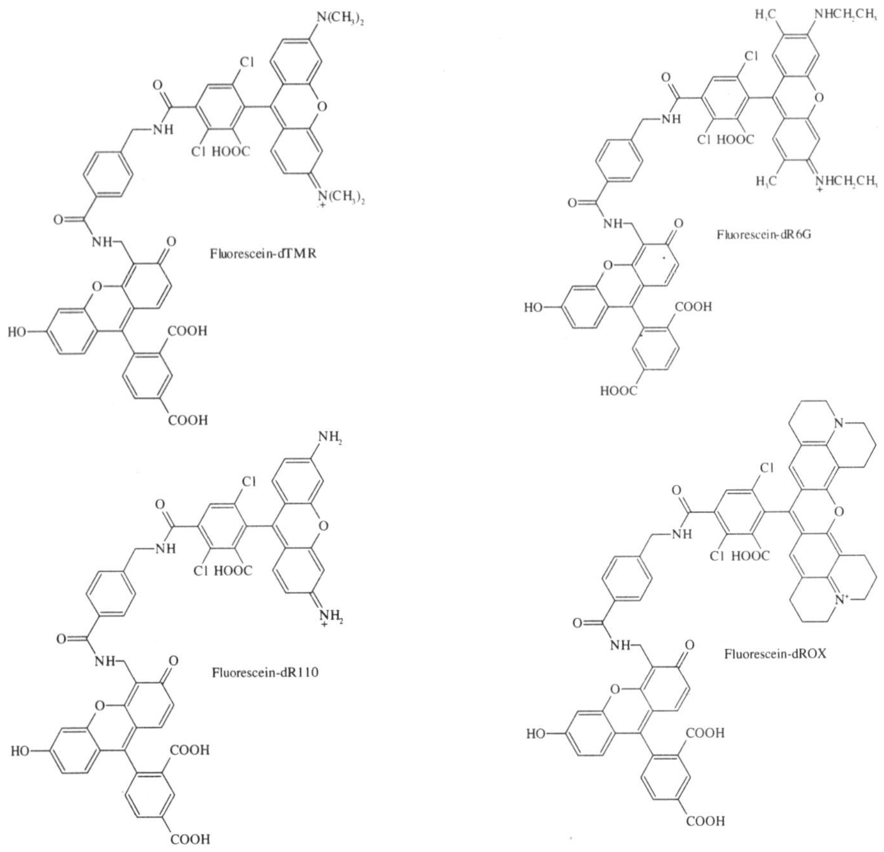

FIGURE 2.5 Structure of the optimal BigDye dye set. (Adapted from Lee et al.[9])

terminators and the dicholorhodamine dye terminators are significantly lower (at the 95% confidence level) than the average errors for the rhodamine dye terminators. The average read length for the BigDye dye terminators and the dicholorhodamine dye terminators is significantly higher (at the 95% confidence level) than the average errors for the rhodamine dye terminators. For both the average errors and average read length, there is no statistical difference between the dicholorhodamine dye terminators and the BigDye dye terminators. The BigDye dye terminators give significantly more signal strength than either the dicholorhodamine dye terminators or the rhodamine dye terminators.

DISCRIMINATION USING FLUORESCENCE LIFETIME

The dyes used for the detection of DNA sequencing fragments by fluorescence lifetime require some different characteristics than emission dyes. Similar to emission-based detection systems, fluorescence lifetime dyes require good quantum yields and high molar absorptivity. The spectral discrimination employed is

TABLE 2.2
Statistics for the BigDye Dye Terminators[a] Compared to Non-Energy Transfer Dye Sets

	Average Errors to 720 Bases	Error Standard Deviation	Average Read Length at 98.0% Accuracy	Read Length Standard Deviation	Signal Strength	Signal Strength Standard Deviation
Rhodamine dye set	20.39	14.99	662.9	133.7	2117.1	1005.6
Dichlororhodamine dye set	9.17	5.02	748.4	43.6	858.3	413.4
BigDye™ dye set	8.61	9.02	759.2	64.6	3081.7	1448.7

[a] See Figure 2.5 for the dye structures.

Note: The rhodamine dye set includes Rhodamine 6G, ROX, Tetramethylrhodamine, and Rhodamine 110. The dichlororhodamine dyes are the same dyes with chlorines substituted for hydrogens at the 4 and 7 positions.

Source: Adapted from data reported in Rosenblum et al.[10]

distinct fluorescence lifetimes. The emission maxima of the dyes can be exactly the same (in fact, it is helpful if they are). The theoretical fluorescence lifetime difference required is 20%.[14] The fluorescence lifetime measurement is independent of signal and is constant once the limit of detection has been overcome. The fluorescence lifetime should, ideally, be monoexponential. Multiexponential decay will make overlapping peaks difficult to resolve, a critical success factor for long reads. A large emission window is able to be used. Finding dyes with these characteristics has been a challenge.

There are two ways of measuring fluorescence lifetime, time domain and phase domain. An excellent review article on the use of fluorescence lifetime to detect DNA sequencing fragments has been published.[11] The use of fluorescence lifetime, both time domain and phase domain, is discussed in more detail below.

TIME DOMAIN

Time-domain measurements utilize a short (the shorter the better) excitation pulse to excite the dye molecules. The signal is measured as a function of time. The equation that governs this process is shown below.[22]

$$F(t) = N_0/\gamma e^{-t/\tau} \tag{2.1}$$

where $F(t)$ is the fluorescence intensity, N_0 is the initial population of fluorophores, γ is the emissive rate, τ is the lifetime, and t is the time.

The signal decays exponentially over time. The fluorescence lifetime is either determined from the slope of the plot of log $F(t)$ vs. time or by determining the time required for the signal to decay to $1/e$ of the original signal.[22]

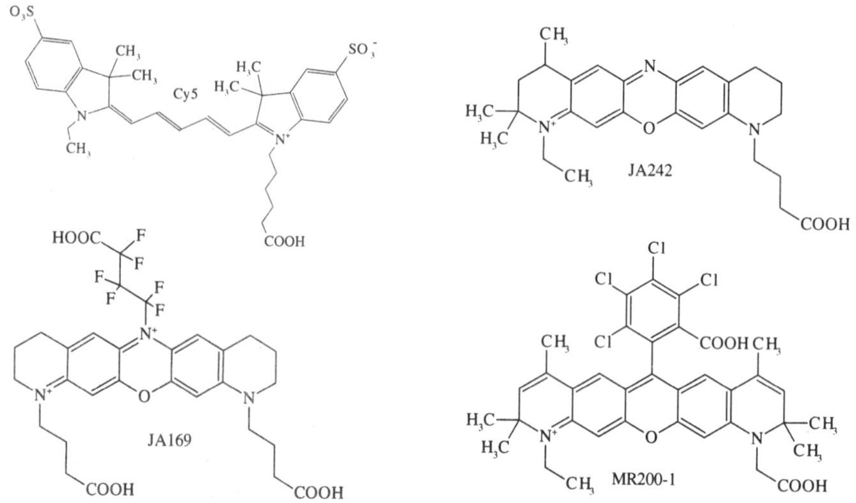

FIGURE 2.6 The dye set utilized for the first example of utilizing fluorescence lifetime for the detection of DNA sequencing fragments. (Adapted from Lieberwirth et al.[12])

The first demonstration of lifetime detection for DNA sequencing occurred in 1998.[12] Three new dyes were synthesized and combined with a commercially available cyanine dye (Figure 2.6) to create a set of dyes with distinct fluorescence lifetimes. The dyes had varied absorbance maxima (from 624 to 6669 nm). A semiconductor laser emitting at 630 nm was used for excitation. The average laser power was only 0.6 mW. High laser powers are not needed in the red region of the electromagnetic spectrum because there is low background signal in this region. The dyes were conjugated to a sequencing primer (5′-TGT17ACGACGGCCAGT-3′). The conjugated Cy5 exhibited a lifetime of 1.6 ns, conjugated JA242 exhibited a lifetime of 2.4 ns, conjugated JA169 exhibited a fluorescence lifetime of 2.9 ns, and conjugated MR200-1 exhibited a fluorescence lifetime of 3.7 ns. Utilizing this dye set, 660 bp were sequenced with 90% accuracy. The accuracy was adversely affected by two dyes, JA242 and JA169, exhibiting multiexponential decay. The mobility shift was small, less than 0.2 bp for the Cy5-labeled fragments. This low amount of mobility shift was achieved by introducing linkers between the dye and oligonucleotide and selecting a coupling position that minimizes the mobility shift.

Lassiter and coworkers[13] utilized fluorescence lifetime detection for detecting DNA sequencing fragments separated by a slab gel. Two commercially available fluorescent dyes in a two-lane scheme were employed. The average lifetime of IRD700 was 718 ps while Cy5.5 exhibited an average lifetime of 983 ps. With this approach, 670 bases of sequencing data were generated with 99.7% accuracy. An overlapping peak with as little resolution as 0.36 was able to be correctly identified.

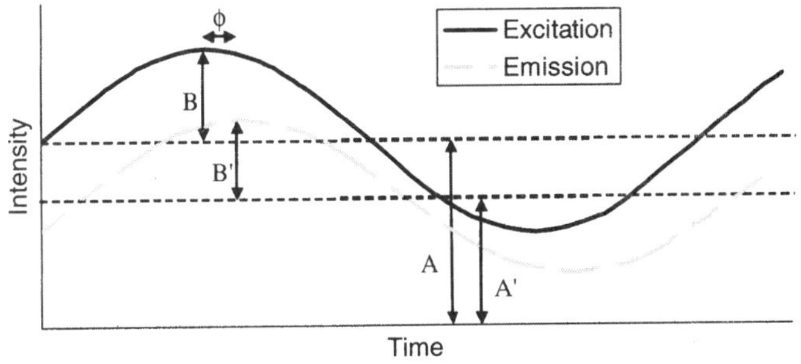

FIGURE 2.7 A theoretical excitation and emission signal used to determine the fluorescence lifetime at a single frequency is presented. The fluorescence lifetime is calculated utilizing Equations 2.2 and 2.3 based on the phase shift of the emission (relative to the excitation phase) and the demodulation (relative to the excitation amplitude). (Adapted from Nunnally.[17])

PHASE DOMAIN

The first fluorescence lifetime instruments utilized phase-domain.[16] In phase-domain measurements, the excitation light is sinusoidally modulated. When the molecules fluoresce, the emission is phase shifted and demodulated relative to the excitation (Figure 2.7). The equations used to calculate the lifetime (Equations 2.2 and 2.3) are shown below.[17]

$$\tau_p = \omega^{-1} \tan \phi \tag{2.2}$$

$$\tau_m = \omega^{-1}[(1/m^2) - 1]^{1/2} \tag{2.3}$$

where τ_p is the lifetime measured by the phase shift, ω is the angular modulation frequency applied to the excitation light, ϕ is the phase shift of the light, τ_m is the lifetime measured by the demodulation of the emission, m is the demodulation factor, calculated from the ratio of the change in amplitude for the excitation and emission signal, i.e., $m = (B/A)/(B'/A')$. Figure 2.7 provides a visual for the terms in these equations.

For a dye with monoexponential decay, the two lifetimes are equal. Modern instrumentation is capable of measuring multiple frequencies simultaneously. Data analysis software is able to deconvolute the data and provide a more accurate lifetime measurement than was capable with only a single frequency.

There has only been one reported use of phase domain measurements for the detection of DNA sequencing fragments.[18] The use of phase-domain fluorescence lifetime detection for capillary electrophoresis had been demonstrated previously.[19] After demonstrating that the technique was suitable for the detection of dye-labeled DNA primers,[20,21] He and McGown[18] sequenced nearly 200 bases of DNA with 96% accuracy using a dye system excited by a 488-nm argon ion laser. This detection scheme challenged the detection limit capability of the system and required some

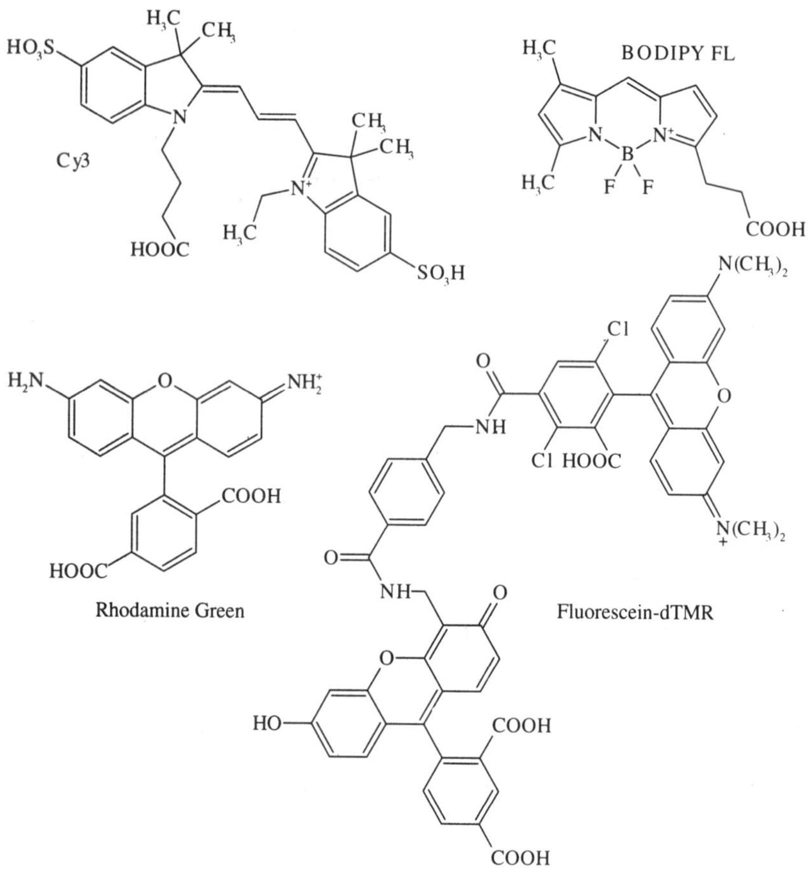

FIGURE 2.8 The 488-nm dye set used by He and McGown.[18] All dyes are shown with carboxylic acid groups. These carboxylic acid groups (the carboxylic acid group in the 6 position for Rhodamine Green and the 5 position of the fluorescein dye on the fluorescein-dTMR) are changed to other functionalities, usually NHS esters, for conjugation to DNA primers or dideoxynucleotides. (Figure adapted from Nunnally;[17] structure of fluorescein-dTMR adapted from Lee et al.[9])

optic modification to increase the signal collected from the capillary system. The 488-nm dye set used is shown in Figure 2.8. One of the dyes, fluorescein-dTMR, showed significant mobility shifts relative to the other dyes and required sequential injections to compensate for this issue. The lifetime of each of the 488-nm dyes was 1.7 ns (Cy3), 2.5 ns (fluorescein-dTMR), 2.9 ns (Rhodamine Green), and 3.9 ns (BODIPY-FL). The lifetimes of both Rhodamine Green and BODIPY-FL were much shorter than previously noted,[17] which may indicate the dye is being quenched by the gel matrix. He and McGown[18] attempted to use another dye set excited by the 514-nm line of an argon ion laser. This dye set is shown in Figure 2.9. The 514-nm dye set had good lifetime resolution, but one of the dyes (BODIPY–FL Br$_2$) had a

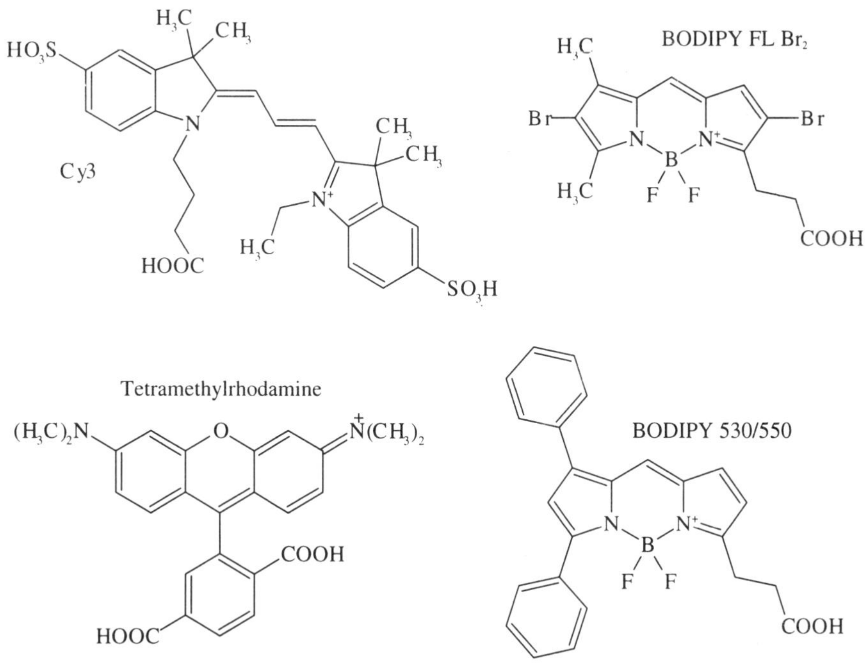

FIGURE 2.9 The 514-nm dye set used by He and McGown.[18] All dyes are shown with carboxylic acid groups. These carboxylic acid groups are changed to other functionalities, usually NHS esters, for conjugation to DNA primers or dideoxynucleotides. (Adapted from Nunnally.[17])

weaker signal and no replacement could be found. By using only the three dye combination (i.e., only three bases could be identified), the accuracy was improved to 98.5%. Li and McGown[23] found that the composition of the gel, including degree of cross-linking, type of cross-linker, and organic modifiers could affect the fluorescence lifetime offering the possibility of tuning or optimizing the fluorescence lifetimes for better discrimination in sequencing.

A new set of dyes, based on an acridone structure, was developed for use in phase-domain lifetime sequencing.[24] The structure of the acridone dyes are shown in Figure 2.10. These dyes have a great deal of homology, thus minimizing mobility shifts. The key property of these dyes is the difference in their fluorescence lifetimes. The four dyes have broad absorption spectra centered around 405 nm[24] and can be excited by a violet laser diode. The four dyes have distinctly different fluorescent lifetimes. The lifetimes of the dye-labeled primers detected on capillary were 4 ns (Dye I), 6 ns (Dye II), 11 ns (Dye 22), and 14 ns (Dye IV). These dyes are well suited to DNA sequencing utilizing fluorescence lifetime detection because of the magnitude of their lifetimes, lifetime differences between the dyes, and similar absorption maxima. No sequencing data have been reported using this dye system, but the early results are promising.

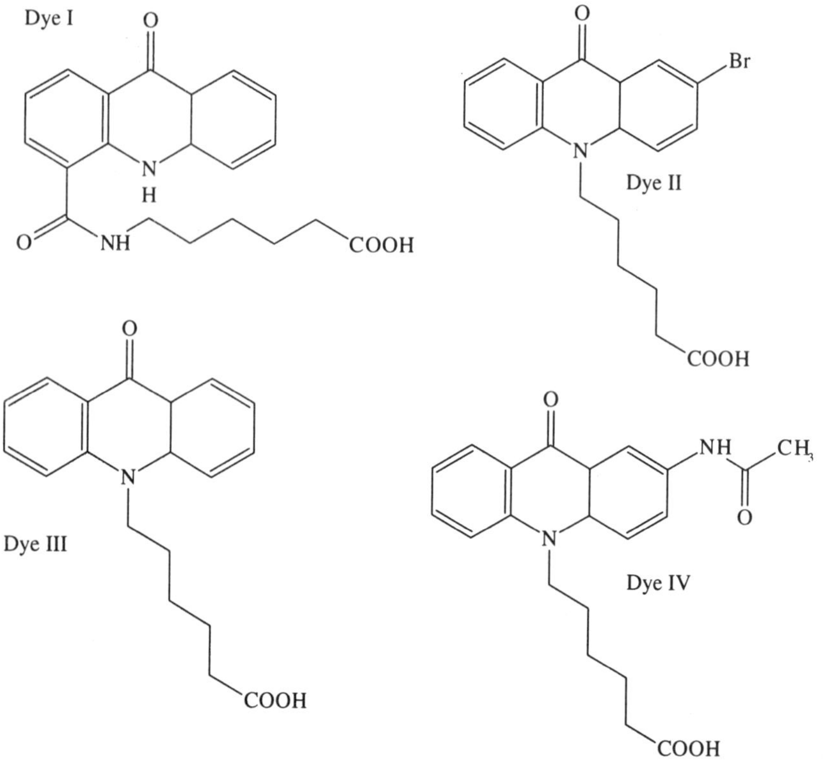

FIGURE 2.10 The structures of the acridone dye set investigated by Mihindukulasuriya et al.[24] All dyes are shown with carboxylic acid groups. These are changed to other functionalities, usually NHS esters, for conjugation to DNA primers or dideoxynucleotides.

CONCLUSION

Alternatives to the standard four-color detection scheme for DNA sequencing are varied. Two fluorescence approaches, energy transfer dye sets and fluorescence lifetime, maintain some of the intrinsic advantages of fluorescence while addressing some of the weaknesses of the standard fluorescence detection schemes. Energy transfer dye sets are widely in use, whereas fluorescence lifetime based sequencing has yet to reach its potential. Time will be the final judge of the utility of the fluorescence lifetime approach.

ACKNOWLEDGMENTS

I thank my colleague Kun Yao for his review and comments on this chapter. Figures containing chemical structures were prepared using ACD/ChemSketch available from Advanced Chemistry Development, ACD/Labs. A freeware version can be

REFERENCES

1. J Ju, C Ruan, CW Fuller, AN Glazer, RA Mathies. Fluorescence energy transfer dye-labeled primers for DNA sequencing and analysis. Proc Natl Acad Sci USA 92:4347–4351, 1995.
2. J Ju, I Kheterpal, JR Scherer, C Ruan, CW Fuller, AN Glazer, RA Mathies. Design and synthesis of fluorescence energy transfer dye labeled primers and their application for DNA sequencing and analysis. Anal Biochem 231:131–140, 1995.
3. J Ju, AN Glazer, RA Mathies. Energy transfer primers: a new fluorescence labeling paradigm for DNA sequencing and analysis. Nat Med 2:246–249, 1996.
4. SC Hung, J Ju, RA Mathies, AN Glazer. Energy transfer primers with 5- or 6-carboxyrhodamine 6G as acceptor chromophores. Anal Biochem 238:165–170, 1996.
5. SC Hung, J Ju, RA Mathies, AN Glazer. Cyanine dyes with high absorption cross section as donor chromophores in energy transfer primers. Anal Biochem 243:15–27, 1996.
6. SC Hung, RA Mathies, AN Glazer. Optimization of spectroscopic and electrophoretic properties of energy transfer primers. Anal Biochem 252:78–88, 1997.
7. http://www.probes.com/servlets/structure?item=6479.
8. SC Hung, RA Mathies, AN Glazer. Comparison of fluorescence energy transfer primers with different donor-acceptor dye combinations. Anal Biochem 255:32–38, 1998.
9. LG Lee, SL Spurgeon, CR Heiner, SC Benson, BB Rosenblum, SM Menchen, RJ Graham, A Constantinescu, KG Upadhya, JM Cassel. New energy transfer dyes for DNA sequencing. Nucleic Acids Res 25:2816–2822, 1997.
10. BB Rosenblum, LG Lee, SL Spurgeon, SH Khan, SM Menchen, CR Heiner, SM Chen. New dye-labeled terminators for improved DNA sequencing patterns. Nucleic Acids Res 25:4500–4504, 1997.
11. SJ Lassiter, WJ Stryjewski, Y Wang, SA Soper. Shedding light on DNA analysis. Spectroscopy 17:14–23, 2002.
12. U. Lieberwirth, J Arden-Jacob, KH Drexhage, DP Herten, R Müller, M Neumann, A Schulz, S Siebert, G Sagner, S Klingel, M Sauer, J Wolfrum. Anal Chem 70:4771–4779, 1998.
13. SJ Lassiter, WJ Stryjewski, BL Legendre, Jr., R Erdmann, M Wahl, J Wurm, R Peterson, L Midendorf, SA Soper. Time-resolved fluorescence imaging of slab gels for lifetime base-calling in DNA sequencing. Anal Chem 72:5373–5382, 2000.
14. MB Smalley, LB McGown. Limits of detection and resolution for on-the-fly fluorescence lifetime detection in HPLC. Anal Chem 67:1371–1376, 1995.
15. ML Metzker, J Lu, RA Gibbs. Electrophoretically uniform fluorescent dyes for automated DNA sequencing. Science 271:1420–1422, 1996.
16. Z Gaviola. Ein Fluorometer. Apparat zur Messung von Fluoreszenzabklingungszeiten. Z Phys 42:853–861, 1926.
17. BK Nunnally. Multiplex detection in capillary electrophoresis using fluorescence lifetime. PhD dissertation, Duke University, Durham, NC, 1998.
18. H He, LB McGown. DNA sequencing by capillary electrophoresis with four-decay fluorescence detection. Anal Chem 72:5865–5873, 2000.

19. LC Li, LB McGown. On-the-fly frequency-domain fluorescence lifetime detection in capillary electrophoresis. Anal Chem 68:2737–2743, 1996.
20. H He, BK Nunnally, LC Li, LB McGown. On-the-fly fluorescence lifetime detection of dye-labeled DNA primers for multiplex analysis. Anal Chem 70:3413–3418, 1998.
21. LC Li, H He, BK Nunnally, LB McGown. On-the-fly fluorescence lifetime detection of labeled DNA primers. J Chromatogr B 695:85–92, 1997.
22. JR Lackowicz. Principles of Fluorescence Spectroscopy. New York: Plenum Press, 1983, 52–53.
23. L Li, LB McGown. Effects of gel material on fluorescence lifetime detection of yes and dye-labeled DNA primers in capillary electrophoresis. J Chromatogr A 841:95–103, 1999.
24. SH Mihindukulasuriya, TK Morcone, LB McGown. Characterization of acridone dyes for use in four-decay detection in DNA sequencing. Electrophoresis 24:20–25, 2003.

3 Microscale Sample Preparation for DNA Sequencing and Genotyping

Yonghua Zhang and Edward S. Yeung

CONTENTS

INTRODUCTION

The greatest achievement in molecular biology in the past decade is undoubtedly the sequencing of the human genome. The announcement of the completion of the Human Genome Project in June 2000 marks the availability of the rough draft about 3 years ahead of schedule. The remaining sequencing tasks should be completed in 2003. There is general agreement that one of the major developments that allowed such rapid progress is the availability of high-throughput DNA sequencers based on capillary electrophoresis (CE). These automated instruments eliminated most of the manual operations associated with the older generation of DNA sequencers, thereby substantially increasing the throughput. The completion of the Human Genome Project does not mean the end of technology development relevant to DNA sequencing. There are many more genomes to be sequenced. With each sequenced genome, there are many more individuals to be compared with the "standard" sequence. In this chapter, we review three different types of DNA technologies that may be the key to future applications of genomics.

Life sciences took center stage virtually around the world on June 26, 2000, when President Clinton announced the completion of the working draft of the entire human genome.[1] The working draft consisted of sequences of 85 to 90% of the 3 billion DNA bases. It is essentially the "blueprint" for the construction of humans, which holds great potential in the discovery of functional genes, the elucidation of disease-causing mutations within the genes, and the development of diagnostic and therapeutic procedures to detect, treat, and prevent diseases ranging from cancer to AIDS.[2]

The human genome is only one of almost 100 genomes currently being sequenced around the world. *De novo* sequencing, sequencing of other organisms, comparative genomics, single nucleotide polymorphism (SNP), and other genome-related issues are examples of applications that will continue to push the limits of DNA sequencing.[2] To realize the full potential that genomics holds, the current techniques for DNA sequencing need to undergo further reduction in cost and increase in throughput by integration, automation, and miniaturization of the sample preparation steps prior to sequencing.

The clinical use of genomic information is one of the main objectives of the Human Genome Project.[2] The entire human genome is estimated to comprise at least 100,000 genes, of which 4000 have already been identified as the causes of known heritable genetic diseases. Use of genomic information will also help pharmaceutical companies create drugs tailored to a patient's genetic profile, boosting effectiveness while drastically reducing side effects. It could even change our very concept of what a disease is, replacing broad descriptive categories with precise genetic definitions that make diagnosis confident and treatment swift. Advances in molecular biology have allowed the identification of genes and the specific mutations linked with a variety of human diseases. Rapid and cost-effective methods for the detection of such mutations are the basis of genetic diagnosis, which is expected to play an increasingly important role in the fields of molecular pathology and genetics.

A powerful technique for the detection of polymorphism is polymerase chain reaction (PCR), which allows the amplification of selected regions of DNA extracted from a variety of sample sources to a detectable level.[3] It is also possible to perform multiplexed amplification in cases where multiple mutations are present. Traditionally, DNA used for PCR-based diagnostic analysis has originated from blood, which involves labor-intensive sample preparation. Large-scale DNA polymorphism detection will require the development of new techniques that are fast, cost-effective, and easily automated.

Current DNA sequence analysis involves (1) DNA library construction and template preparation; (2) sequencing reaction, separation and detection; and (3) information analysis. The first two parts are the most challenging and the rate-limiting steps in the whole process.[4] Since the inception of the Human Genome

Project, the individual steps of DNA sequencing technology have improved dramatically. These include the development of bacterial artificial chromosomes (BACs) for longer insert clones, new thermostable enzymes, high-sensitivity dye sets, and perhaps most importantly, highly multiplexed high-speed capillary array electrophoresis (CAE) instruments.[5–8]

CHALLENGES IN GENOMIC ANALYSIS

PLASMID PREPARATION FOR DNA SEQUENCING

Many procedures have been developed over the years for isolation of bacterial plasmids.[8] Further, several proprietary methods also give satisfactory results. One aspect that nearly all of these methods share is that they involve three basic steps: growth of bacteria, harvesting and lysis of the cells, and purification of the plasmid. All include centrifugation or several treatments of the colonies that are labor intensive and difficult to automate and to interface with CAE. Some chromatographic columns (size exclusion, ion-exchange, high-performance membranes, hydrophobic interaction) have also been developed to avoid the use of centrifugation. All these methods, which may meet the stringent quality criteria for gene therapy, involve high cost and intensive labor. Another complication is that reagents used in the purification of plasmids, such as ethanol and SDS, might become interferences to the subsequent cycle-sequencing reaction.

Alternative methods, which require only heat-induced lysis of cells in bacterial colonies, also exist.[9,10] Centrifugation or vigorous vortex is still needed to isolate cell debris. The resulting lysate is then used as the template in cycle sequencing using labeled primers. Even though the performance and the ruggedness demonstrated so far by this method are still inferior to that of the standard protocol, the method shows promise for significant savings in time and cost. In fact, a microwave protocol similar to above procedure has already become the preferred method for purification of double-stranded DNA at the Washington University Genome Sequencing Center.[11] In its method, the growth of bacterial cultures and subsequent DNA isolation took place in the same 96-well block and no further purification of DNA, by precipitation or other means, was necessary. The 96-well blocks can be reused indefinitely provided they were cleaned between uses. Furthermore, the lysis solution was easily made and was stable at room temperature for a minimum of 3 months, allowing liter-quantity batches to be made and stored. The cost was estimated to be $0.03/sample. This compared very favorably with the cost of commercial preparation methods, which can be as high as $1/sample. Using this approach, the average high-quality sequence length was 427 bases, while 70.52% of the sequences had at least 400 bases of high-quality data.

DNA LADDER PURIFICATION AND SAMPLE INJECTION

Capillary gel electrophoresis (CGE) is an attractive technique for DNA analysis because the narrow-bore, gel-filled capillaries provide high-speed, high-resolution separations, as well as automated gel and sample loading. The use of CGE for DNA sequencing was first demonstrated in 1990, when sequencing separations of ~350 bases were obtained in cross-linked gels in ~80 min.[12] Much progress has been made in the past decade, and sequencing read-lengths of more than 1000 bases can now be obtained using replaceable gels.[13,14] In fact, CGE has eliminated the bottleneck involved in the separation of the DNA ladder produced by the Sanger reaction and has become the key technology for DNA sequencing.

Compared with slab-gel electrophoresis, the sample in CGE is injected into the separation capillary instead of being loaded into wells. Therefore, dye-labeled DNA fragments must compete with ions in the sample matrix when they enter the capillary. It is well documented that the performance and reliability of DNA sequencing by CE is sensitive to the quality of the DNA sample due to the employment of electrokinetic injection. Residual salt and dideoxynucleotides in the sequencing sample cause discrimination against DNA because they have higher mobilities. As a result, large variability in signal strength is often observed in CGE. Another problem associated with sample purification is the rehybridization of the single-stranded DNA injected. This causes the mobilities to change and makes base calling difficult.

The benefit of stringent sample purification has been demonstrated before.[15,16] A poly(ether sulfone) ultrafiltration membrane pretreated with linear polyacrylamide was first used to remove template DNA from the sequencing samples. Then, gel filtration in a spin column format (two columns per sample) was employed to decrease the concentration of salts to below 10 μM in the sample solution. The method was very reproducible and increased the injected amount of the sequencing fragments 10- to 50-fold compared to traditional cleanup protocols. Using M13mp18 as a template, the resulting purified single DNA sequencing fragments could routinely be separated to more than 1000 bases with a base-calling accuracy of at least 99% for 800 bases. A systematic study to determine the quantitative effects of the sample solution components, such as high-mobility ions (e.g., chloride and dideoxynucleotides) and template DNA, on the injected amount and separation efficiency revealed that, in the presence of only 0.1 μg of template in the sample (one third of the lowest quantity recommended in cycle sequencing) and at very low chloride concentrations (~5 μM), the separation efficiency decreased by 70%. The deleterious effect of template DNA on the separation of sequencing fragments was not observed in slab-gel electrophoresis because it was suppressed in the presence of salt at a concentration above 100 μM in the sample solution. The latest results showed that read length up to 1300 bases (average 1250) with 98.5% accuracy can be achieved in 2 h for single-stranded M13 template.[17] Thus, the purified DNA ladder dramatically improved the result but at the expense of high cost and manual manipulation.

Swerdlow et al.[18] first tried to perform DNA sequencing with unpurified DNA sequencing samples. They used a method called base stacking, which allowed direct injection of unpurified products of dye-primer sequencing reactions onto capillaries without any pretreatment. Briefly, on-column concentration of DNA fragments is achieved simply by electrokinetic injection of hydroxide ions. A neutralization reaction between these OH$^-$ ions and the cationic buffer component Tris$^+$ results in a zone of lower conductivity, within which field focusing occurs. Without base stacking, a drastic loss in signal was observed for the crude samples. This method can generate separation resolution of at least 0.5 up to 650 bp. The signal strength was excellent relative to conventional injection of highly purified samples. Furthermore, no significant degradation of the capillary performance was observed over at least 20 sequencing runs using this new sample injection method. One shortcoming of this method is that it did not yield satisfactory results with dye terminator chemistry due to the interference of unreacted dye terminators.

PREPARATION OF BIOLOGICAL SAMPLES FOR PCR

It is interesting that a new PCR method, FoLT (formamide low temperature) PCR, has been developed for reactions directly from whole blood. Formamide solubilizes blood cells and frees the DNA for amplification An important finding was that an alternative DNA polymerase, Tth polymerase, was less sensitive than Taq polymerase to the presence of proteins in blood. All these make PCR directly from blood possible.[19,20]

Although blood can be used directly in PCR, there is clearly a need for simpler, noninvasive, and more cost-effective means of sample collection, DNA extraction, and genetic diagnosis in general. There are several disadvantages of using blood. First, blood collection can be very inconvenient, because genetic testing often involves analysis of multiple family members. Furthermore, drawing blood can be uncomfortable for the patient and, most important, the handling of blood samples can increase the chances of infection by blood-borne pathogens such as HIV and hepatitis. To date, a variety of alternative sources of DNA have been used for genetic testing including finger-prick blood samples, hair roots, as well as the use of cheek scrapings and oral saline rinses as a means of collecting buccal epithelial cells. The oral saline rinse is perhaps the most extensively used nonblood-based sampling technique. However, it still involves liquid sample handling and requires an additional centrifugation step to spin down the cells, which is difficult to automate and interface with subsequent analysis.

By avoiding centrifugation, a simpler method has been developed and validated by using just swabs and brushes.[21] The buccal cells were collected on a sterile brush by twirling the brush on the inner cheek for 30 s. Although still requiring a neutralizing step later, this method is generally easy and very reliable. In a blind study comparing the analysis of 12 mutations responsible for cystic fibrosis in multiplex products amplified with DNA from both blood and buccal cell samples from 464 individuals, there was 100% correlation of the results for blood and cheek-cell DNA. The success rate of PCR amplification on DNA prepared from buccal cells was 99%. This method has also been used to analyze DNA for genetic polymorphism by matrix-assisted laser desorption/ionization mass spectrometry.[22]

TOTAL SYSTEM INTEGRATION

Numerous endeavors have been made in developing robotic workstations to perform sequencing reaction, purification, preconcentration, and sample loading. Although robotics has shown advantages in repetitive operation with high precision, the adaptation to highly multiplexed capillary array separation interface suffers from many incompatibilities in terms of the total reaction volume, purification by centrifugation, and sample injection. Online microfluidics systems based on either capillaries or microchips hold promise for the next generation of totally automated DNA sequencers.

Capillary Microfluidics Systems

In capillary microfluidics,[23] dye-labeled terminator cycle sequencing reactions are performed in a 250 μm i.d. fused-silica capillary, which was placed into a hot-air thermal cycler. After PCR was completed, the reaction mixture was transferred online to a size-exclusion column to purify the reaction product from the unreacted dye terminators. The purified product was then injected through a cross into a gel-filled capillary for size separation. This system was closed and the operation was reliable since no moving parts were involved. Cleaning of the system with 0.1 M NaOH was required to remove cross-contaminants before reuse. The sequence could be called from 36 to 360 bases with an accuracy of 96.5% using in-house software. By manual editing, the accuracy improved to 98% for 370 bases. Later on, a multiplexed system based on above scheme was developed in which eight DNA sequencing samples could be processed simultaneously starting from template to called bases.[24] The major achievement in the instrument was the use of freeze/thaw switching valves instead of rotary valves, which were unsuitable for multiplexed systems due to their size. For all eight processed samples, sequences could be called up to 400 bases with an accuracy of 98%. PCR analysis directly from blood was also demonstrated with a similar flow management concept.[25]

Another fully integrated single capillary instrument comparable in design has also been designed and prototyped.[26] The reaction was performed inside Teflon tubing. The purification and separation columns were interfaced through a simple T-connector instead of a cross. The instrument was reliable and fast, performing PCR reaction cycling, purification, and analysis all in 20 min. Adaptation of the instrument prototype for separation of DNA-sequencing reactions was described; cycle sequencing and electrophoresis of a single lane were complete in 90 min with base calling to beyond 600 bases.

Miniaturization of the online system will reduce the cost of DNA sequencing substantially below current levels because only 1/100 of the reagent is required for actual CE. Soper et al.[27] developed a miniaturized solid-phase cycle sequencing reactor coupled with CE. The nanoreactor consisted of a fused-silica column with a total volume of 62 nl. Biotinylated DNA template was bonded to the surface by biotinstrepavidin–biotin linkage. The main disadvantage of this scheme is the loss of template surface coverage. One solution for reactivation of the nanoreactor could be through adding fresh streptavidin and new biotinylated target DNA. The read length for a single-color run was approximately 450 bases. The system is considered amenable to automation even though there is still manual operation in the published design.

Online Microchip Systems

Microchips provide a new platform for integration with unique electro-osmotic pumping and nonmechanical valves. A true nano-total analysis device was developed by Burns et al.[28] who used microfabricated fluidic channels, heaters, temperature sensors, and fluorescence detectors to analyze nanoliter-size DNA samples. The device is capable of starting with 100 nl reagent and DNA solution, online mixing, amplifying or digesting the DNA to form discrete products, and separating and detecting those products in microfabricated channels. No external lenses, heaters, or mechanical pumps are necessary for complete sample processing and analysis. The components have the potential for assembly into complex, low-power, integrated analysis systems at low unit cost.

Microchips still need to overcome some technical difficulties before they can be routinely utilized. These include solvent evaporation and interface with the outside bioanalysis laboratory where the common working volume is µl. Litborn et al. described using a closed humidity chamber to address the problem of solvent evaporation.[29] Later, they reported an improved technique for performing parallel reactions in open, 15-nl volume, chip-based vials. The evaporation of solvent from the reaction fluid was continuously compensated for by addition of solvent via an array of microcapillaries. Their results showed that the concept for continuous compensation of solvent evaporation should be applicable to reaction volumes down to 30 pl.[30]

There are other approaches to avoid the problem of solvent evaporation. Soper developed a hybrid system that coupled nanoliter sample preparation to PMMA (polymethylmethacrylate)-based microchips.[31] Unlike standard sample preparations that are performed off-chip on a µl scale, true integration was demonstrated at nl volumes. An integrated system for rapid PCR-based analysis on a microchip has also been demonstrated recently.[32] The system coupled a compact thermal cycling assembly based on dual Peltier thermoelectric elements with a microchip gel electrophoresis platform. This configuration allowed fast (~1 min/cycle) and efficient DNA amplification on-chip follo- wed by electrophoretic sizing and detection on the same chip. An unique on-chip DNA concentration technique based on adsorption and desorption has been incorporated into the system to reduce analysis time further by decreasing the number of thermal cycles to 10 cycles, or 20 min for DNA amplification and subsequently detection.

CAPILLARY ARRAY INSTRUMENTS

In 1992, Mathies's group developed this approach to address the throughput requirement of genomic analysis.[6] They constructed a confocal fluorescence scanner and demonstrated DNA sequencing in 25 parallel capillaries. Since then, additional improvements in optical design and separation matrixes have made the commercialization of CAE sequencers possible. Today, there are four commercial versions of CAE instruments. PE Biosystems has developed the ABI PRISM 3700 DNA analyzer. This 96-capillary array instrument is based on the approaches of Kambara[8] and Dovichi.[7] In this instrument, DNA sequencing fragments are detected in a sheath flow and spectrally resolved using a concave spectrograph and a cooled charged coupled device (CCD) camera. Bare capillaries are used with dynamic coating, which is stable for more than 300 runs. The turnaround time is roughly 2.6 h with 600 bp in 120 min. The Molecular Dynamics instrument MegaBACE 1000 is based on confocal detection after Mathies's group.[6] A microscope objective is used to focus the laser light inside the capillaries and, at the same time, collect the emitted light from the center of the column. The lifetime of the instrument may be limited by mechanical stress of moving the scanner when fast sampling rates are required. The system uses linear polyacrylamide (LPA)-coated capillaries, which are stable for 200 runs. The average sequencing data is 500 bp and the turnaround time is less than 2 h. Beckman Coulter has entered the market with an 8-capillary array design, the CEQ 2000 DNA analysis system. The optical design of this instrument is similar to that of Molecular Dynamics except that separate excitation and emission paths are used. It features four-color IR dideoxy-terminator chemistry. On-column detection is the approach implemented in the SpectruMedix instrument that is based on our work.[33] The laser beam crosses all 96 capillaries after the laser is expanded by a cylindrical lens. The fluorescent light is collected at right angles from the laser axis and detected by a CCD camera. Because no moving parts are involved in detection, the optic design is very rugged. Bare fused-silica capillaries are used with dynamic coating with a hydrophobic polymer. The turnaround time is 2 h with average base calling of 500 bp. The SpectruMedix system is also employed in the new 16-capillary ABI 3100 and 96 capillary 3710 sequencers.

Alternatives in system design mainly concern illumination and detection. Kambara's group[34] further tested side illumination with detection on column. For this design, the number of capillaries in an array is generally limited by laser-power attenuation along the array due to reflection and divergence. They overcame these problems by placing the capillaries in water and adding glass-rod lenses between the capillaries. As a result, up to 45 capillaries could be simultaneously irradiated with a single laser beam and the fluorescence from all the capillaries could be detected with high sensitivity. Quesada and Zhang[35] took another approach for a multiple capillary instrument by the use of optical fibers for illumination and collection of the fluorescence in a 90° arrangement. A subsequent version of this instrument utilized cylindrical capillaries as optical elements in a waveguide, where refraction confined a focused laser beam to pass through 12 successive capillaries in a flat parallel array.[36] However, larger capillary arrays are limited by the refractive effects that spread the light along the length of the capillaries.

Handling more than 96 capillaries is very challenging and several groups have attempted to address this problem by modifying existing designs. Dovichi's group used sheath-flow detection and a novel two-dimensional arrangement that can hold up to 576 capillaries.[37] A prototype 384-capillary array electrophoresis instrument has also been developed for higher throughput analysis by SpectruMedix. Their instrument design is based on the 96 capillary platform with a redesign in the camera lens. Mathies and coworkers have also continued to push the limit of the confocal system. They have developed a system with the capillaries aligned in a circular array. The microscope objective spins inside a drum, illuminating the capillaries one at a time. They have shown sequencing data from 128 capillaries, but a larger number of capillaries could be easily accommodated in this geometry.[38,39]

A totally different platform to perform CAE is the microchip. The first demonstration of CAE in microchip was by Mathies's group for genotyping.[40] A microplate that can analyze 96 samples in less than 8 min was produced by bonding 10 cm diameter micromachined glass wafers to form a glass sandwich structure. The microplate had 96 sample wells and 48 separation channels with an injection unit that permitted the serial analysis of two different samples in each channel. An elastomer sheet with an 8 × 12 array of holes was placed on top of the glass sandwich structure to define the sample wells. Samples are addressed with an electrode array that makes up the third layer of the assembly. Detection of all lanes with high temporal resolution was achieved by using a laser-excited confocal fluorescence scanner as described above. An SNP typing assay has also been developed and evaluated in a microfabricated array electrophoresis system.[41] That study demonstrates the feasibility of using allele-specific PCR with covalently labeled primers for high-speed fluorescent SNP typing.

DNA sequencing on the microchip array is very interesting and challenging. Early studies of single channel on a chip required channel lengths comparable to capillaries. Making many turns in a chip also proved to be deleterious to separation performance. It also implies that a larger-diameter chip is needed for DNA sequencing.[42] Recent results show that there is still much room for improvement. Liu et al.[43] demonstrated DNA sequencing by 16 channel CAE in a microchip format. Samples are loaded into reservoirs by using an eight-tip pipetting device, and the chip is docked with an array of electrodes. Under computer control, high voltage is applied to the appropriate reservoirs in a programmed sequence that injects and separates the DNA samples. An integrated four-color confocal fluorescence detector automatically scans all 16 channels. The system routinely yields more than 450 bases in 15 min in all 16 channels. In the best case using an automated base-calling program, 543 bases have been called at an accuracy of >99%. Separations, including automated chip loading and sample injection, normally are completed in less than 18 min. This demonstrates the potential of the microchip as the next generation CAE platform.

EMERGING TECHNOLOGIES

SEQUENCING DIRECTLY FROM SINGLE BACTERIAL COLONIES

The small diameter of capillaries coupled with ultrasensitive detection associated with laser-induced fluorescence in CE can reduce the sample requirement from 1 to 10 fmol of DNA in a given band on the slab gel to 1 to 10 amol in capillary gel.[44] This means we can reduce the reaction volume from 20 μl to 20 nl if the reaction efficiency is maintained and an efficient sample-loading method can be found.[27] However, present sequencing technology still leaves much to be improved if the potential for genomic science is to be fully realized. The front end of the sequencing process, especially sample preparation, which is typically labor intensive, repetitive, and time-consuming, needs to interface smoothly with the back end.[45]

In a recent study,[46] we show that the implementation of colony sequencing in our multiplexed online reaction–purification-sequencing instrument is feasible. Bases can be called directly from a single colony without compromise in spite of the complex matrix of the cell lysate and the small size of the sample. This completes the long-sought-after goal of integration and automation of the front and back ends of sequencing in one instrument.

The experimental setup is similar to that described previously.[23,24] *Escherichia coli* DH5α competent cells and plasmid pUC19, Plasmid Bluescript, and pGEM were used. Cell transformation was according to the protocol provided by Clontech.

Transformed cells were plated on $2 \times$ TY agar plates and incubated at 37°C for 24 h. A colony was picked up by a wire loop and put into a vial with 15 µl autoclaved deionized water inside. After heating at 96°C for 11 min, sequencing premix (8 µl), BSA (2 µl, 2.5 mg/ml), and 3.2 pmol primer were added. Then, 20 µl of the above solution was aspirated into the reaction capillary. Cycle sequencing began immediately by holding at 96°C for 3 min, followed by 40 cycles (96°C for 10 s, 50°C for 30 s, and 60°C for 4 or 2.5 min). Three different sets of dyes, Rhodamine, dRhodamine, and Energy Transfer, and two polymerases, AmpliTaq FS from ABI and ThermoSequenase from Amersham Life Science, were used in the reaction. Performance in all cases is comparable. For the results presented here, dRhodamine terminators and AmpliTaq FS were used. After reaction, the sequencing ladders were purified by homemade size-exclusion columns (SEC) and injected online into the separation capillary, or purified by spin columns and injected off-line for comparison.

Three features associated with the combined lysis/reaction protocol may cause the difference in efficiency. First, cells in the colony are still viable. The enzyme they produced if not deactivated may interfere with the polymerase. In PCR reactions, the cells can be lysed in the reaction buffer because the exponential amplification effect can compensate for the loss caused by this kind of interference. This is not the case for the linear amplification character of cycle sequencing. Second, we suspect that the small inner diameter of the capillary reactor prevents uniform distribution of the plasmids. Cells in colonies tend to stick together even after being drawn into the capillary. After heating, the released plasmids localize in a confined region of the long capillary despite 10 min heating at 96°C. Convection here is small, so the molecular movement is dominated by diffusion. For a diffusion-controlled process, mixing may be taken as nearly complete for Dt/l^2 from 0.1 to 1. The diffusion constant for rhodamine-dGTP in water at room temperature is $D = (3 - 5) \times 10^{-6}$ cm^2/s.[47] Accounting for the increase in diffusion caused by temperature and viscosity at 96°C,[48] the effective mixing distance in $t = 10$ min even for small molecules is thus less than $l = 1$ cm. The aspiration speed used to load the mixture to the reactor was 2.4 cm/s, which corresponds to a Reynolds number of roughly 12. The flow will therefore be laminar and no extra mixing exists.[49] The third reason may relate to the thermal stability of the enzyme. Taq polymerase is thermostable with a half-life of 40 min at 95°C and 10 min at 97.5°C, respectively. The extra heating time for lysis may cause the enzyme to lose some activity.

Adjusting the parameters of cycle sequencing was vital to success. Compared to the standard recommended protocol, we use somewhat longer annealing times (5 to 30 s) to assure the correctness and completeness of annealing between the primer and the plasmid to minimize interference from chromosomal DNA and RNA. The total signal strength may also benefit from this, as there should be an increase in the amount of primed template.[50] The signals among the four bases were balanced, so no adjustment of ratio of ddNTP/dNTP was necessary. The sequencing reaction kinetics do not appear to be affected by the complex matrix. Reducing the extension time from 4 to 2.5 min was shown to be possible. Extension was also tested at temperatures higher than 60°C, where the Taq enzyme is supposed to have a faster polymerization rate. Similar results were obtained.

The lysate of *E. coli* is a very complicated matrix, which contains salts, proteins, and other biomolecules. Surprisingly, we did not find any extra problems in injection. Presumably these are greatly diluted during the elution process. The Sephadex beads can retain some of the biomolecules such as degraded RNA, lipids, polyamine, and lipopolysaccharide, as long as the molecular weight is less than 5000 Da. There are still some high-molecular-weight molecules that may elute with the DNA ladder. In fact, the carrier protein BSA (10^{-11} mol) that is added to the reagents may co-elute also. Each *E. coli* cell may contain 2,350,000 protein molecules so one colony with 10^7 to 10^8 cells will result in another 10^{-11} mol of protein being added to the reaction mixture. As with our earlier experiments, we did not observe any deleterious effect related to injection because of BSA. For BSA the isoelectric point is 4.7 and the molecular weight (MW) is 67,000, which is roughly the MW of 200-bp ss-DNA. But the net charge per molecule at pH 7.4 is only −17, much smaller than a 200-bp ss-DNA. This results in a mobility of BSA that is at least one order of magnitude smaller than DNA. More importantly, by long-time heating, a high fraction of BSA will be polymerized and aggregated and are thus further discriminated against in electrokinetic injection. We also note that the observed interference of BSA to DNA separation is buffer dependent.[51] Urea used in the sequencing buffer may also help to reduce the interaction between the protein and DNA.

In the reconstructed image plot of the sequencing results from six experiments in a 1-week period, the intensity was not uniform because different DNA samples were used. Generally, the intensities from purified DNA (40% of the channels) are two times higher than the intensities from the lysate. But even the raw data starting directly from the lysate provide adequate signal-to-noise ratios for base calling. Two clones, PGEM and M13mp18, were used in these experiments. The former has known problems with compressions, but the use of 7 *M* urea here provided excellent separation even at room temperature. Among all 48 runs, only two runs did not produce a useful sequence due to bubble formation at the cross and one run produced only a short sequence (150 bp). All three appeared in the same channel. These intensities are, however, still reasonable, indicating that injection rather than reaction was the culprit. Two runs provided sequence around 350 bp, and the electropherogram of all others provided resolution greater than 0.5 well beyond 430 bp.

The turnaround time for each run is 0.2 h for lysate preparation, 3.5 h for reaction, 1 h for regeneration of the system, and 2.5 h for separation. With staggered operation of these individual steps analogous to an assembly line, the turnaround time will be essentially the reaction time, which can be reduced to 2.5 h even with lysates as the template. During these experiments, the reactors and SEC columns were regenerated by washing between runs with 1 × TE and water. No memory effect was observed in the different channels. Figure 3.1 shows the base-calling results (raw data, day 6, bottom channel) up to 620 bp with 98% accuracy. The resolution after 620 bp is still above 0.5. Base calling was accomplished by software written in Labview using the two-color scheme developed before.[52] Longer reads can be expected with more sophisticated algorithms. Loss in the front part of the sequence may be due to the long injection time in diluted buffer. This can be solved by a better design of the cross assembly (to fully sweep the volume inside the cross) to allow more efficient injection. This is not a problem in shotgun sequencing because for the EcoR.I and

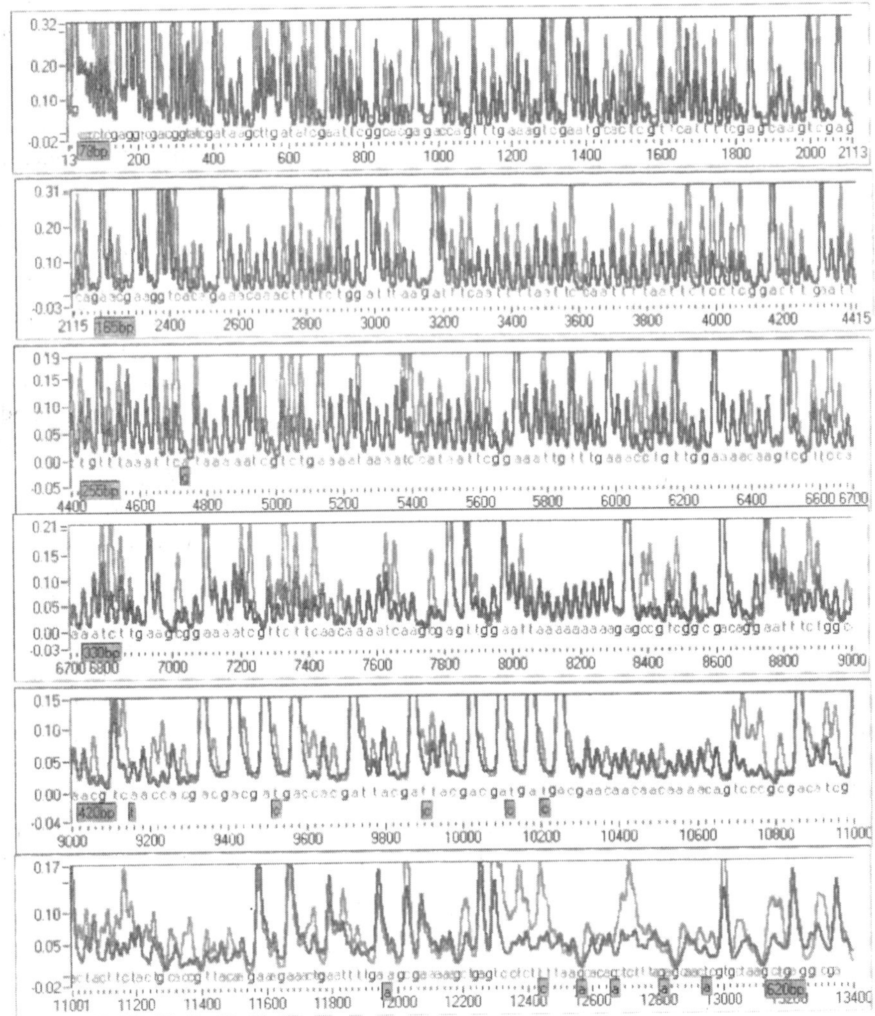

FIGURE 3.1 (**Color Figure 3.1 follows page 84.**) Sequencing of a DNA insert in the pBluescript vector using the cell lysate from one colony as the template. Primer: M13-40. The raw data from the blue and red channels are plotted. The miscalls are also corrected under the corresponding bases. The resolution is above 0.5 at 620 bp.

the universal primer, which are widely used, the vector sequence is around 100 bp. For primer walking, such a limitation will need to be considered.

A major advantage of the present system is the potential for further multiplexing and miniaturization. Multiplexing 100 channels using freeze–thaw valves can be manageable even in the present assembly. By reducing the inner diameter yet keeping the same reagent concentrations, even smaller volume samples could be injected using the current protocol. Our instrumentation therefore promises to further reduce the reagent cost and labor requirement in high-throughout DNA sequencing.

Sequencing Based on Nanoliter Reaction Volumes

Sample Manipulation

Miniaturization of cycle sequencing in a glass capillary[27,48,53,54] has been demonstrated at the microliter level. Moving the cycle-sequencing reaction into a capillary has the additional advantage of increasing the reaction speed due to the small heat capacity of a capillary vs. a heating block or a water bath. A capillary reactor is also compatible with highly multiplexed electrophoresis in a parallel capillary array. Several groups[23,26] have demonstrated online reaction in a capillary coupled with electrophoretic separation. Multiplexed systems also have been demonstrated.[24,25,46] However, they did not exploit the small amount of DNA sample required for CE separation. A microchip providing integrated operation from Sanger reaction to sequencing separation is a promising approach.[28,55,56] Evenson et al.[57] demonstrated that by using a piezoceramic actuator we can rapidly mix two 1-μl solutions in under 3 s inside a capillary. However, that system may pose a challenge for multiplexed operation.

The real issue is whether we can prepare small-volume solutions and deliver these to the reaction zone for the cycle-sequencing reaction, purify (if necessary), and then inject them for CE separation. If we need to premix the reagent with the template on a microliter scale before introducing into the reaction zone, large amounts (1 μl) of reagent would be wasted even when only a small volume (50 nl) is injected into the capillary for electrophoresis.

The work presented here[58] describes an automated nanoreactor for cycle-sequencing reaction with online SEC purification and capillary gel electrophoresis. As little as 25 nl reagent volume was required. A simple procedure allows the reagent solution to mix with the template solution inside the nanoreactor. By using this protocol, the bulk reagent solution can be reused without contamination. This provides real cost savings based on the amount of reagents used. A simple and inexpensive flexible heater design for the nanoreactor allows for future scale-up for capillary-array DNA sequencing.

Either the ABI PRISM dye terminator cycle-sequencing ready reaction kit with AmpliTaq DNA polymerase FS or the ThermoSequenase dye terminator cycle sequencing premix kit was used; 0.05 μg/μl M13mp18 ssDNA in 50 mM Tris, 2.5 mM HCl, 2 mM MgCl$_2$, and 1 $\times$ BSA as stock solution was used. The reaction mixture for AmpliTaq FS polymerase consisted of 8 μl of terminator ready reaction mix, 3.2 pmol universal –21 M13 primer, and 2 μl 10 $\times$ BSA. The reaction mixture for ThermoSequenase consisted of 8 μl of terminator ready reaction mix, 5 pmol universal –17 M13 primer, and 2 μl 10 $\times$ BSA. These are prepared in advance and a 0.5-μl aliquot was used for a series of reactions.

The temperature protocol for the AmpliTaq cycle-sequencing reaction was as follows: the reaction mixture was heated to 95°C and held for 2 min; 35 thermal cycles were performed with denaturation at 96°C for 10 s, annealing at 50°C for 15 s, and extension at 60°C for 4 min. Then the sample was ramped to 95°C and held for 2 min. When ThermoSequenase cycle-sequencing reaction was used, the annealing temperature was adjusted to 45°C.

Figure 3.2 shows the schematic diagram for the instrumental setup. The system consists of a nanoreactor system, an SEC system, and a CE electrophoretic system

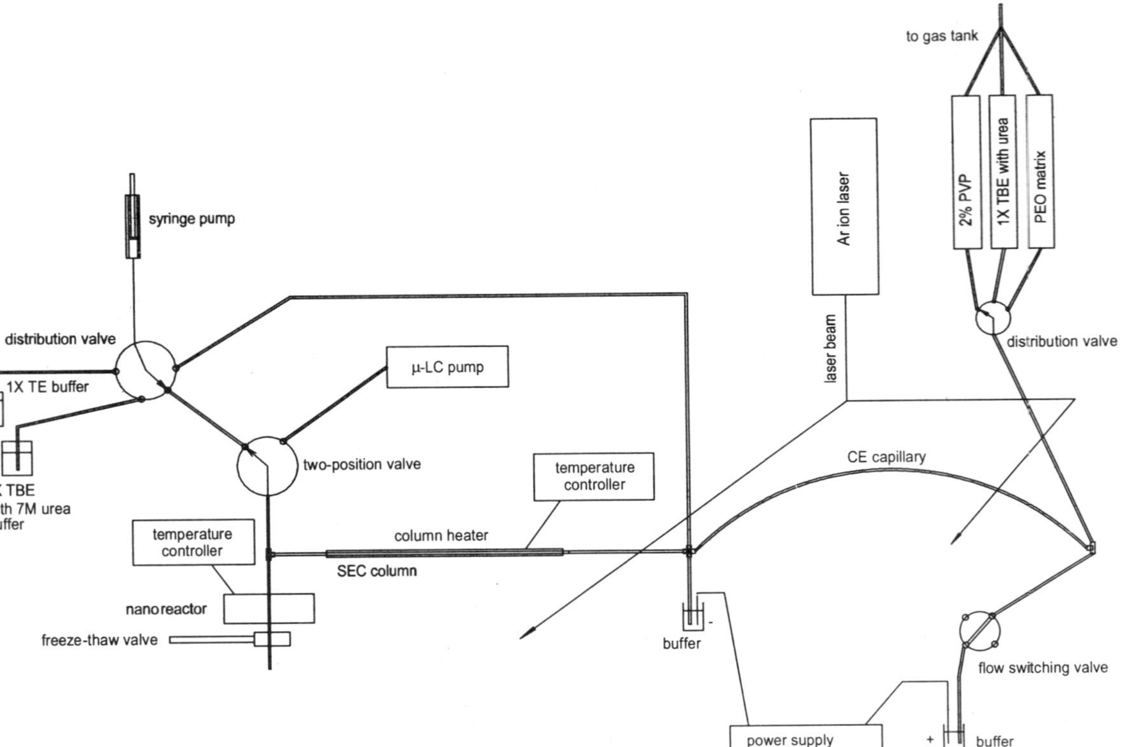

FIGURE 3.2 Schematic diagram of the experimental setup. Samples and reagents are introduced from the lower left and transported from left to right for reaction and then separation.

with gel filling. A microtee was used to connect the nanoreactor system, SEC system, and the pumping system, which consisted of a syringe pump and a µLC-500 pump. A two-position valve was used to selectively connect one of the pumps to the nanoreactor or the SEC column. A microcross was used to connect the SEC system, gel capillary electrophoretic system, and the syringe pump. The syringe pump was equipped with a 25 µl syringe with a resolution of 0.52 nl per step. An 8-position multiposition valve was used in conjunction with the syringe pump to allow the selection of different solutions to pump through the reactor capillary for cleaning or to the microcross for CE separation.

The nanoreactor was constructed with layers of brass sheet ($7.5 \times 2.5 \times 0.025$ cm) and a Kapton insulated flexible heater (7.5×2.5 cm, 2 W/cm^2, resistively heated). A thermal epoxy was used to bond the brass sheet and the flexible heater together. This nanoreactor has a very small thermal mass since the total thickness excluding the capillary reactor is ~300 µm so that it allows the temperature to change rapidly. The length (7.5 cm) of the heater allowed the use of 8 capillaries (9-mm spacing is the standard 8×12-tray format) or 16 capillaries (4.5 mm spacing for the 384-tray format) for simultaneous reactions. A 10-cm-long, 360 µm o.d. and 75 µm i.d. capillary was placed in between two brass sheets. A silicone heat sink compound was applied onto the capillary surface and in between the brass sheets to ensure proper heat transfer. For the 75 µm i.d. capillary, the 23 mm reaction length corresponds to ~110 nl maximum reaction volume. We could simply change to different inner diameters to accommodate different reaction volumes. A 0.08 cm diameter bare K type thermocouple was used to monitor the temperature of the nanoreactor. This thermocouple was inserted into a 250 µm i.d., 360 µm o.d. capillary in which water was filled and both ends of the capillary were sealed. The thermocouple was position ~0.5 cm away from the reaction capillary. A PID temperature controller was used to set the temperature profile for cycle sequencing. A computer was used to communicate with the temperature controller to effect the temperature change and duration. In this way, ±0.5°C accuracy can be obtained and the heating rate was ~3°C/s. A room-temperature nitrogen gas jet, which was controlled by a solenoid valve, was directed to the heater to lower the temperature quickly during the transition from denature to annealing conditions. When ~5 psi gas pressure was used, ~3°C/s cooling rate can be obtained. A faster cooling rate can be obtained by using a higher gas flow rate, but slight overshoot may occur.

For capillary cleaning, 250 µl of $1 \times$ TE buffer was first pumped through the reaction capillary by the syringe pump. Then 200 nl of 50 m*M* Tris, 2.5 m*M* HCl, and 2 m*M* MgCl$_2$ solution was aspirated into the capillary reactor followed by aspiration of 50 nl of cycle-sequencing reaction mixture at 21 nl/s (Figure 3.3, left). The Tris solution was used as a buffer zone to isolate the reaction solution from the TE separation buffer since EDTA will interfere the cycle-sequencing reaction. To add the sample, 500 nl template was placed in a microcentrifuge tube. The tip of the reaction capillary was placed into the bottom of the tube. The syringe pump first aspirated 100 nl of the template solution (Figure 3.3, middle) and then dispensed 100 nl (Figure 3.3, middle). This mixing procedure (last two steps) was repeated 12 times to allow complete mixing of the reaction mixture and the template solution. The bulk of the premix reagents was thus maintained inside the capillary throughout

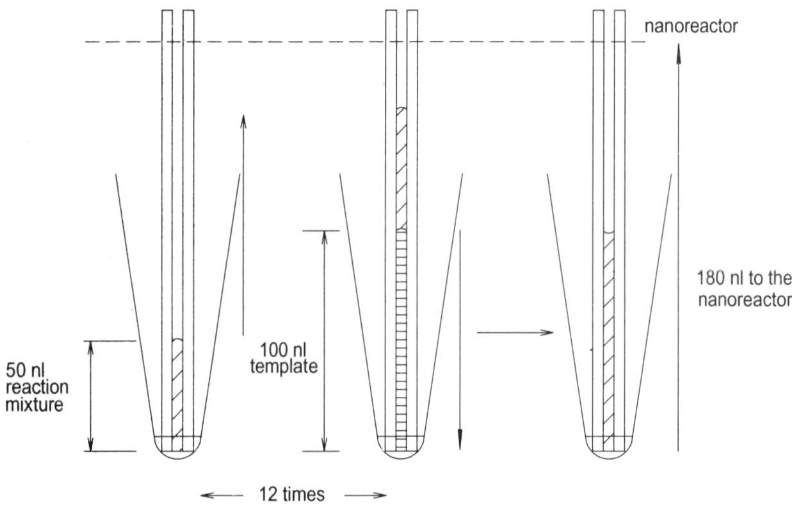

FIGURE 3.3 Schematic diagram of the solution mixing procedure. (Left) 50 nl of reagent mixture was introduced. (Middle) The pump was cycled 12 times to aspirate and then redispense 100 nl of template solution. (Right) After such mixing, 180 nl total volume was taken up for reaction.

this procedure. After the mixing, the reaction solution was moved up to the nanoreactor by aspirating an additional 180 nl of solution while the capillary tip was still positioned in the template tube (Figure 3.3, right).

By aspirating a dye solution into the capillary with a syringe pump for a fixed time period, we can visually determine how accurately the syringe pump could function. It was found that we can easily aspirate 25 nl of the solution into the 75 μm i.d. capillary with ~10% error. Although only a 10-cm length of capillary was used for the nanoreactor, there was a time delay on the fluid movement due to friction. Therefore, between each syringe pump action, a 6-s waiting period was added to ensure that the fluid completed the motion.

To evaluate the mixing procedure, 50 nl (corresponded to 1 cm in length) of concentrated Rhodamine 6G solution was aspirated into the capillary. After mixing with water following the above mixing protocol, the dye solution spread out to ~2 cm. Although this mixing procedure may not provide uniform distribution of the reagent, primer, and template across the entire reaction zone, the cycle-sequencing reaction is relatively robust and tolerates such a variation. For further testing, a series of 10, 12, 14, and 16 mixing cycles was used to determine the cycle-sequencing efficiency by observing the fluorescence signal after the SEC separation. It was found that 12 mixing cycles provided the largest DNA signal under these reaction conditions. The initial template concentration also played a role on the reaction efficiency. A range of concentrations from 0.2 μg/μl to 0.025 μg/μl was used to determine the cycle-sequencing reaction efficiency. It is found that 0.05 μg/μl template concentration provides the highest efficiency.

A freeze–thaw valve, which was described previously,[24] was positioned at the entrance end of the reaction capillary to close it during the cycle-sequencing reaction or SEC separation. During the reaction, TE buffer was flowed through the SEC system to condition the column and to pressurize (>60 psi) the nanoreactor to eliminate bubble formation.

After the reaction capillary was cleaned, the reagent mixture was aspirated into the capillary followed by mixing with the template solution with the procedure described above. The freeze–thaw valve was closed followed by switching the two-position valve so that $1 \times$ TE effluent can flow through the SEC column for conditioning during the cycle-sequencing reaction. Before the reaction was completed, TBE buffer, PVP coating solution, and PEO gel was filled into the electrophoretic capillary in preparation for the CE separation. After the cycle-sequencing reaction, the nanoreactor was heated up to 95°C for 2 min for denaturing the DNA products. Then the two-position valve was switched back to connect the syringe pump to the nanoreactor. The freeze–thaw valve was opened and allowed the aspiration of an additional 800 nl $1 \times$ TE solution to move the reaction products over the microtee. The freeze–thaw valve was closed and the two-position valve was switched back to connect the μ-LC pump to the SEC column to push the reaction products into the SEC column for purification. Fluorescence from the SEC column was monitored. A positive high voltage was applied at 60 V/cm field strength when the DNA signal appeared at the first detector (~10 s delay time due to the 38-cm-long connection capillary). Then, 30 s later, the field strength was increased to 120 V/cm for 60 s followed by a decrease in the field strength to 60 V/cm for the remaining injection period. The total injection time was ~3 min. Stacking occurs during the entire period. After injection, the two-position valve was switched to disconnect the μ-LC pump and the SEC column to stop the flow of the effluent into the microcross. Otherwise, the dye-labeled terminators will continuously pump through the microcross and subsequently become injected into the CE column. The syringe pump then delivered $1 \times$ TBE with $7\,M$ urea buffer to the microcross at a flow rate of 1.3 μl/min for the CE separation. The field strength for the CE separation was set at 120 V/cm.

Figure 3.4 shows the electropherogram of the cycle-sequencing products from M13mp18 DNA template after nanoreaction followed by SEC purification with one-wavelength excitation and dual wavelength detection. Here, 50 nl reaction mix was aspirated initially for cycle-sequencing reaction. No dye-labeled ddNTPs interference can be observed. The success rate is 100% over 15 consecutive runs. Good signal-to-noise ratio and separation resolution were obtained from the electropherogram with a DNA read length of over 450 bp using an in-house two-wavelength intensity ratio scheme.[52] Future implementation of multiwavelength base calling will significantly extend the read length. Miniaturization of the injection region will also provide better signal-to-noise ratio for base calling. It is interesting to see that no denaturing is necessary during the sample injection period. It is possible that when the reaction plug moved over the microtee into the SEC column, sufficient dilution of the reaction plug prevented the DNA from renaturing.

As little as 25 nl reaction mix can be used in this system with slightly compromised read length due to a lower signal-to-noise ratio. The mixing procedure was altered

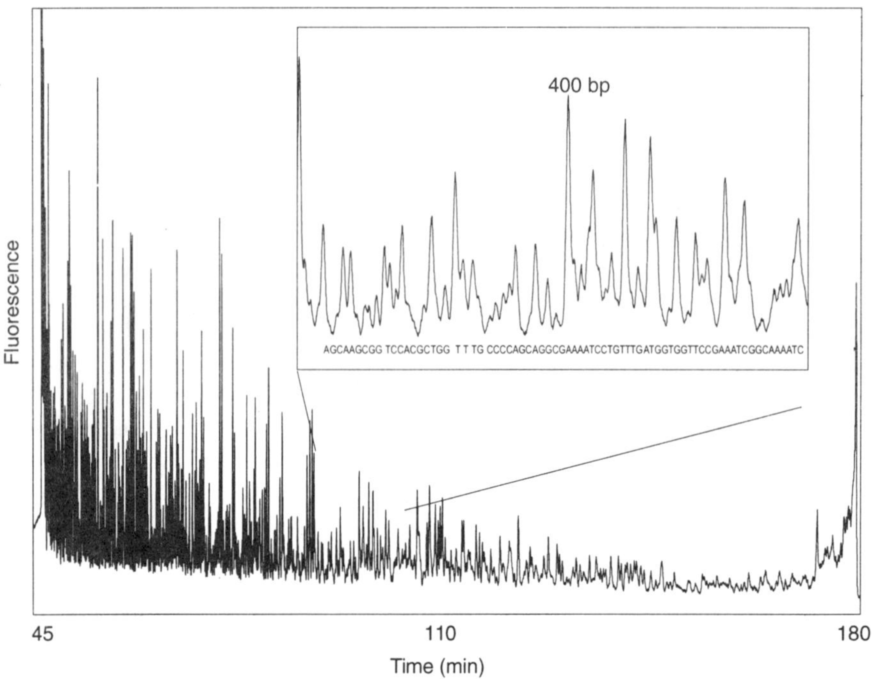

FIGURE 3.4 Electropherogram of DNA fragments after cycle-sequencing reaction for M13mp18 ss-DNA template amplified by AmpliTaq polymerase inside the nanoreactor with purification by SEC followed by online injection into the CE column. Only one wavelength channel was shown for clarity; 120 V/cm field strength with 50 cm effective length was used for CE separation.

slightly to accommodate the smaller uptake volume. Only 60 nl aspirate-dispense mixing cycle was used. The electropherogram still allowed base calling up to 380 bp.

To reduce the cost of the primers rather than the sequencing reagents, instead of putting the primer into the ready reaction mixture, we can premix the template with the reaction solution. In this case, 0.4 μg of M13mp18 template was added to the ready reaction mix. The primer stock solution was 0.6 μM. The primer was then mixed with the reaction solution later inside the nanoreactor (Figure 3.3, middle). Such a protocol will be desirable whenever the primer is the expensive component. In this way, we can simply use a different primer for multiple sequencing to implement, for example, the primer-walking scheme.[59] We found no apparent difference in the reaction efficiency and the sequencing separation up to 450 bp in this "primer-mixing" procedure. However, the signal is reduced significantly (3×) afterward and no compression peak can be observed.

The use of 500 nl template volume is convenient for standard microvials. If a smaller sample tube such as those used in the 384-tray format is used to hold the reagent solution and a 150 μm o.d. capillary is used as the nanoreactor, only ~150 to 200 nl solution volume is necessary for the operation in Figure 3.3. In the reaction solution, ~60% of the volume (30 of 50 nl) is the expensive ready reaction premix solution containing dye-labeled terminators and enzyme. In the standard reaction protocol, 4 μl of premix solution is used in a total of 10 μl reaction volume. More than 130-fold reduction in reagent used is thus achieved. Even compared to the state-of-the-art 1 μl reaction volume protocol,[48,53,54] more than 13-fold reduction is obtained.

The turnaround time for this integrated system was ~6 h including 3 h for cycle-sequencing reaction, 10 min for the SEC separation, and 3 h for CE separation. If an additional syringe pump is employed to provide the flow during CE separation, the next round of cycle-sequencing reaction can be performed while the first round of separation is taking place. This will cut the turnaround time by half. The cycle-sequencing time can in principle be reduced to 30 min.[26] Here we are limited by the heating rate and the cooling rate of our heater. The separation speed can also be increased to allow sequencing in 30 min.[60]

Online Sample Purification

The above system is also quite complicated as a result of the use of pressure-driven fluidics, which is required by the SEC purification column. The SEC column also introduces substantial dilution prior to injection into the CGE capillary. Here,[61] a simple, miniaturized, and integrated online sample preparation system is developed for DNA sequencing at the scale of nanoliters. The whole system features the use of CZE for purification of the cycle-sequencing products, the integration of nanoreactor and CZE in one capillary, and online coupling of CZE with CGE.

Modifications to the original cycle-sequencing reaction mixture developed for ABI Model 9600 thermocycler were made to fit the small-volume reaction in the capillary. A typical 20- μl reaction mixture was composed of 2 μl of 2.5 mg/ml BSA, 1 μl of 20 mM MgCl$_2$, 2 μl of 5 μM 40M13 (5′-GTTTTCCCAGTCACGAC 3′) universal primer, 3 μl of 0.2 μg/μl ss-DNA (M13mp18) or 5 μl of 0.2 μg/μl ds-DNA (pGEM) in 1 × TE buffer (pH ~ 7.5), 8 μl of sequencing reagent premix, and 4 μl of deionized water. The reagent premix consists of 125 mM Tris-HCl, pH 9.5, 5 mM MgCl$_2$, 1.25 mM dITP, 0.25 mM each dATP, dCTP, dTTP, ddATP (dye-labeled), ddCTP (dye-labeled), ddGTP (dye-labeled) and ddTTP (dye-labeled), Thermo-Sequenase DNA polymerase, Thermoplasma acidophilum thermostable inorganic pyrophosphatase (TAP), Nonidet P40, Tween 20 and 6.25% glycerol.

Figure 3.5 is a schematic diagram of the entire instrumental setup. The capillaries C$_1$ and C$_2$ were first conditioned by 1 M NaOH, deionized water, and THM buffer. Before loading cycle-sequencing reaction mixture, the liquid level in R$_1$ and R$_2$ were equilibrated for 20 min by using a wide-bore plastic tube (30 cm × 3 mm i.d.) filled with THM buffer to connect R$_1$ and R$_2$. R$_1$ and R$_2$ remain connected during cycle sequencing. In addition, the liquid level in R$_3$ was also manually adjusted to the same level as that in R$_2$. This was to ensure that hydrodynamic flow in the three capillaries was nearly zero, and the small-volume reaction mixture stayed in the same location in the capillary during the entire period (~2.5 h) of reaction.

A small volume of reaction mixture was introduced into the reaction region in the thermocycler by hydrodynamic injection followed by a plug of THM buffer. The

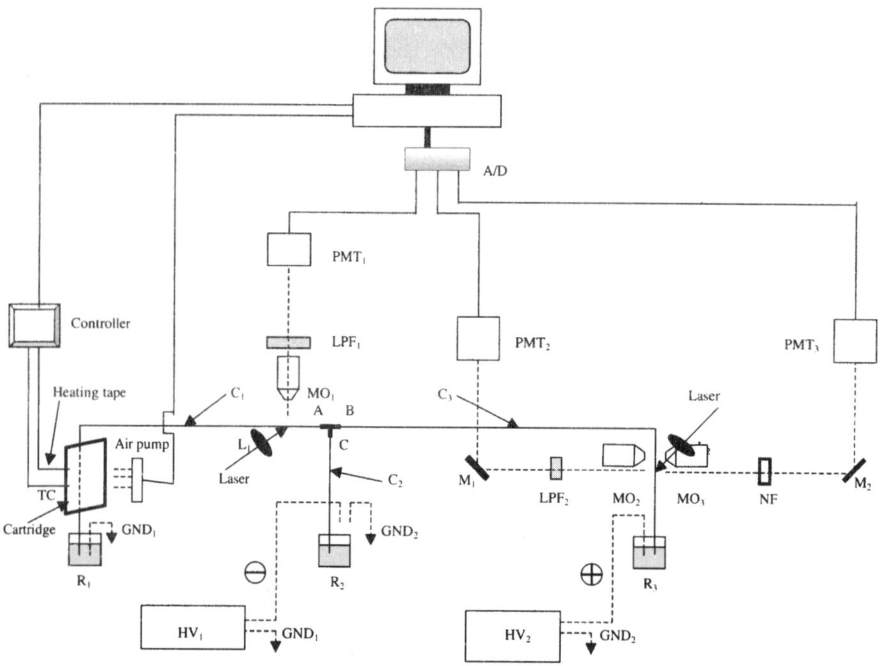

FIGURE 3.5 Schematic of the integrated online cycle-sequencing-CZE-CGE system. TC, thermocouple; C_1 and C_2, CZE capillaries; C_3 CGE capillary; R_1, R_2, and R_3, buffer reservoirs; HV_1, negative-voltage power supply; HV_2, positive-voltage power supply; GND_1 and GND_2, grounded lines for HV_1 and HV_2; L_1 and L_2, lenses; MO_1, MO_2, and MO_3, microscope objectives; LPF_1, 550-nm long-pass filter; LPF_2, 610-nm long-pass filter; NF, 543-nm notch filter; M_1 and M_2, mirrors; PMT_1, PMT_2, and PMT_3, photomultiplier tubes; A/D, data acquisition board.

distance between the inlet tip and the reaction region was ca. ~3 cm. The reaction volume was in the range of 90 to 120 nl with corresponding length in the range of 2 to 3 cm. The temperature protocol for the on-column cycle-sequencing reaction was adjusted to the following: the sample mixture was heated to 96°C and held for 1 min; 35 cycles were performed with denaturation at 96°C for 10 s, annealing at 45°C for 5 s, and extension at 60°C for 3 min; then the temperature was ramped to 96°C and held for 2 min.

After the cycle-sequencing reaction, the wide-bore plastic tube was removed from R_1 and R_2. A negative high-voltage power supply was used to drive electrophoresis for purification from the anode (R_1) to the cathode (R_2). Note that the anode in R_1 should be grounded to avoid electrical arcing in the μ-thermocycler when the high voltage is applied. Also, the electrode in R_3 needed to be removed from the buffer during the CZE separation. Otherwise, the current in the CZE capillary will be seriously affected even if HV_2 was not applied. During CZE separation, dye-terminators passed the detection window first. Once the DNA fragments were detected by PMT_1, timing was initiated. When the peak of the DNA fragments was expected to go past the tee and enter capillary C_2 (based on observed time to the detection window extrapolated to include the extra distance), HV_1 was turned off. The electrode at the cathode from HV_1 was removed from the buffer in R_2. The ground electrode from a positive high-voltage power supply (HV_2) was immersed in the buffer in R_2, and the anode was placed in the $1 \times$ TBE buffer in R_3. Then, a positive voltage (9 kV) was applied between R_2 and R_3 for 40 to 50 s to inject negatively charged DNA fragments into the CGE capillary. A steel wire with its two ends immersed in the buffer in R_1 and R_2 should be used to equalize the potential of R_1 and R_2 during injection and subsequent CGE separation. After injection of the DNA ladder into C_3, HV_2 was turned off. The THM buffer solutions in capillaries C_1 and C_2, and in R_1 and R_2 were replaced by $1 \times$ TBE buffer. Then, HV_2 was turned on to apply 12 kV between R_2 and R_3 to start the separation of the DNA ladder in CGE.

In the analysis of DNA fragments produced from the dye-labeled terminator-sequencing reaction, the major interference comes from the unincorporated dye-labeled terminators (d-ddNTPs) and the high concentration of salt. The d-ddNTPs will interfere with the detection of the dye-labeled DNA ladder (d-DNAs) in CGE, and seriously reduce the accuracy of base calling. The high concentration of salt will significantly reduce the amount of DNA injected electrokinetically into the CGE capillary (lower effective potential), and will also degrade the separation of the DNA ladder (perturbation of the separation buffer).

In previous sample-cleanup methods, the separation of d-ddNTPs and d-DNAs was based on their solubility differences in organic solvent (e.g., ethanol precipitation) or size differences (e.g., SEC). In the new approach investigated here, the separation of d-ddNTPs and d-DNAs is attained on the basis of their mobility differences in CZE. In CZE, the electrophoretic mobility of a molecule is primarily determined by its charge-to-mass ratio (z/m). It is well known that DNA fragments of different sizes (greater than ~20 bp) have almost equal electrophoretic mobilities (μ_{ele}) in free solution electrophoresis due to their very similar z/m ratios.[50,62,63] When a dye is attached to the DNA fragments, μ_{ele} should decrease because the dye has much smaller z/m ratio than the DNA fragments. However, the decrease of μ_{ele} should be very small for DNA fragments with more than 20 bp, i.e., all fragments larger than the primer, as the z/m ratio of the whole fragment is primarily determined by the nucleotides. So, various d-DNAs are expected to migrate close to each other in free-solution CZE. On the other hand, the attachment of rhodamine dye to the ddTTP reduces its z/m by about half from approximately $-3/500$ to $-3/1000$. Note that both ddTTP and d-ddTTP have net -3 charge in weakly basic solutions, and ddTTP and rhodamine dye have similar molecular weights (around 500). Additionally, the d-ddTTP will form a stable 1:1 complex with Mg^{2+} present in the reaction mixture,[64,65] reducing its net charge from -3 to -1. Hence, the z/m for ddATP-Mg^{2+} complex is only about $-1/1000$. On the other hand, the z/m ratio of d-DNA_{20} is only slightly reduced by the attachment of dye and the presence of Mg^{2+}, and is thus slightly lower than $-1/500$. Therefore, it can be expected that the μ_{ele} of d-DNAs should be higher than that of the ddNTPs in basic solution, which forms the basis of their separation by CZE.

In previous work on on-line injection of d-DNAs from SEC to CGE, injection was initiated when the top of the d-DNAs peak reached the center of the cross junction.[23] In the present system, however, it was observed that small amounts of d-ddNTPs would be injected together with the d-DNAs if the same protocol was utilized. Moreover, the reproducibility of the signal intensity in CGE was poor. This confirms that electro-osmotic flow is less reliable than pressure-driven flow. To circumvent this problem, voltage switching was performed after the d-DNAs peak entered C_2 and reached a point about 1.5 cm beyond the center of the tee junction. This was to ensure that the d-DNAs were totally inside C_2 despite the run-to-run variations in migration time. Since the electro-osmotic flow from C_3 to C_2 was largely suppressed by the PEO gel in C_3, the d-DNAs moved faster than d-ddNTPs during injection. Therefore, it became easier to control the injection of d-DNAs while avoiding the injection of d-ddNTPs.

The performance of the optimized integrated online system was investigated in actual DNA sequencing. In the electropherograms of M13mp18 recorded by using one-wavelength excitation and dual-wavelength detection, data in both channels show high signal-to-noise ratios, and adequate resolution for base calling from 5 to 460 bp with an accuracy of 97%. The majority of miscalled bases were from small G peaks following high T peaks. This implies that the on-column reaction generated more DNA fragments terminated by ddTTP as compared to off-line reaction, e.g., in Reference 52. Note that a minimal amount of dye was present around 20 bp, which, however, did not interfere with base calling. Related experiments indicated that the small amount of dye present in this region was not introduced during injection, but might be caused by diffusion of the concentrated dye labels as it initially passed the center of the μ-tee.

Because the excess old-dye terminators comigrate with ~60 bp and ~110 bp DNA fragments in CGE, they will mask several base pairs around those regions in the sequencing separation and introduce errors in base calling. Also the high salt content of the reaction mixture makes electrokinetic injection to the CGE capillary extremely difficult. Proper purification of the cycle-sequencing product to eliminate these interferences must be incorporated into the online system. CZE separation is especially attractive because it does not require additional instrumental components such as a high-pressure pump. However, in bare fused-silica capillaries, because electro-osmotic flow (EOF) dominates the direction of migration, the dye terminators, which have smaller electrophoretic mobilities, migrate faster than the DNA fragments. The strong tailing of the terminator peaks, which is probably due to their strong hydrophobicity, makes the separation very difficult. Furthermore, the uncertainty in EOF due to variable surface conditions of the capillary inner wall makes the migration times in a capillary array very irreproducible.[26] This makes multiplexed CZE purification, where very precise timing and very short injection time to the CGE capillary is required, almost impossible.

We also use CZE with suppressed EOF to separate the dye terminators from the DNA fragments.[66] Several separation systems were studied to optimize the online system. The most obvious idea would be to use the $1 \times$ PCR buffer as the separation buffer (pH = 8.9), which is completely compatible with the cycle-sequencing reaction. Because of its high ionic strength, the cations would shield the deprotonated silanol groups on the inner wall of the capillary to suppress EOF. A very good separation was achieved (Figure 3.6a). The DNA fragments migrated as a narrow

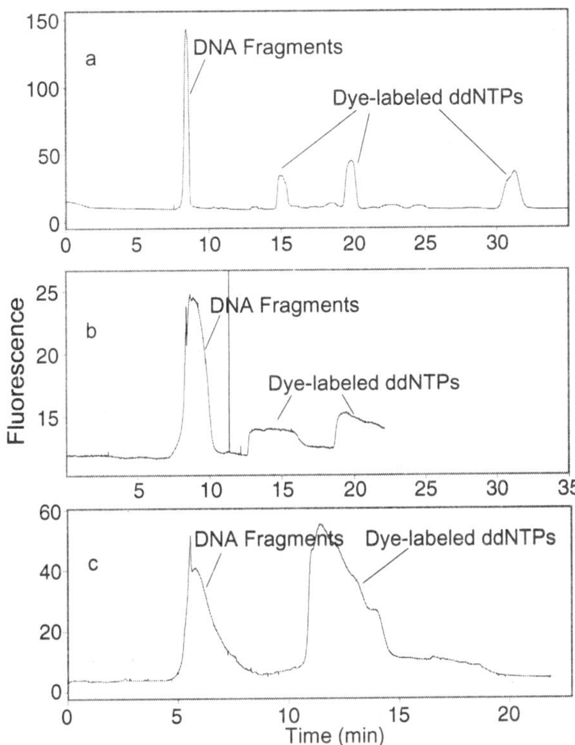

FIGURE 3.6 CZE purification of cycle-sequencing product. (a) Bare fused-silica capillary, 1 × PCR buffer with 50 mM KCl. (b) PVA-coated capillary, 1 × PCR buffer with 4 mM KCl. (c) Bare fused-silica capillary, 1 × PCR buffer with 2 mM KCl, 0.3% 1,000,000 MW PVP.

band because of their similar mass-to-charge ratios. However, when high voltage is applied across the CZE capillary and the CGE capillary during electrokinetic injection, most of the voltage will drop along the CGE capillary. Thus, the field strength in the CZE capillary will be much smaller than that in the CGE capillary, which is just the opposite scenario as in electrostacking. Therefore, electrokinetic injection turned out to be very inefficient.

One of the other choices would be using coated capillaries. Different coated capillaries, such as fluorocarbon (FC), polyethleneglycol (PEG), polyvinylalcohol (PVA), DB-WAX coated capillaries, were tested as the integrated reaction vessel and separation column. 1 × PCR buffer with low concentrations of KCl (pH = 8.9) was tested as the separation medium. The PVA-coated capillary proved to be the best choice in terms of suppressing EOF. Good separation from the dye terminators was achieved (Figure 3.6b), although the DNA fragments appeared as a broader band. Different concentrations of KCl were tested for the CZE separation and electrokinetic injection efficiency. From 2 to 4 mM KCl proved to be best separation condition as a trade-off between DNA bandwidth and online injection efficiency, which was confirmed by good DNA signals and sequencing separation with base calling up to 500 bases. No dye-terminator interference was observed. However, there are two inevitable problems inherent to coated capillaries. One is the cost in a multiplexed system and the other is degradation of the coating. After about 20 reaction and CZE separations, a noticeable decrease in the reaction efficiency was observed. Also, EOF gradually increased. Attempts to regenerate the capillary by washing it with methanol and D.I. water were not successful.

In the previous studies, we demonstrated that EOF of bare fused-silica capillary could be substantially suppressed by the dynamic coating of PVP solution due to the strong hydrogen bonding between the hydrophilic carbonyl group of PVP and the residual hydroxyl group on the capillary wall.[67] Compatibility of the PVP coating with the cycle-sequencing reaction was therefore studied by flushing the long capillary loop with 2% PVP followed by placing the capillary in the commercial air thermocycler for offline cycle-sequencing reaction. After reaction, about ~2 μl of reaction product was collected and purified by spin column. CGE separation of the purified DNA product showed similar signal strength as that without PVP coating, which confirmed that PVP did no harm to the reaction.

During CZE separation, PVP was added to the separation buffer to achieve more even coating. Different concentrations of PVP were investigated for efficiency for suppressing EOF and for separation. At pH 8.9, 1.0% PVP was required to suppress EOF consistently. However, because the entanglement limit of 1,000,000 MW PVP is about 0.7%, at such a high PVP concentration the DNA fragment peak was substantially broadened by the sieving effect. An alternative approach was to lower the pH of the separation buffer. It was found that at pH 8.2, only 0.3% PVP was needed to suppress EOF. Different sized DNA fragments comigrated as a narrow band with half peak width of ~1.5 min (Figure 3.6c). More important was that there was no compromise in the reaction efficiency in such a low pH buffer. The surfactant, Triton X-100, was taken out from the buffer to avoid bubble formation around the microtee connection during CZE separation.

Very reproducible separation (±2% in migration times) was achieved with PVP dynamic coating. The bare fused-capillary could be reused again and again by simply flushing the capillary with the separation buffer between runs. For over 1 month of experiments, more than 40 reactions and separations were performed on the same capillary. No decay in the online reaction efficiency was observed and the DNA fragments showed very reproducible migration times.

High-Throughput PCR Analysis of Clinical Samples

For laser-induced fluorescence (LIF) detection in CE, as few as 10^5 molecules are typically detectable in most laboratories and with state-of-the-art equipment. Even single-molecule detection is attainable.[68] However, the DNA fragments need to be fluorescently labeled, which involves expensive reagents and increases the concern for waste disposal because of the toxic nature of the reagents. Instrumentation for LIF detection in CAE is also costly. Here,[69] we show a new PCR sample preparation protocol starting directly from cheek cells or from blood that can be used in multiplexed CE with UV detection. No purification of the PCR products was necessary by simply applying base stacking.[18] The basis for this approach is that for every

PCR reaction, the concentration of the product will eventually reach a plateau that approaches the μM range.[3] This is sufficient for UV detection because each DNA fragment has more than 100 absorbing units. The complete integration of sample preparation and detection provides a very cost effective scheme to be used in the clinical and forensic laboratories for PCR-based DNA analysis.

In the approach, 5-ml blood samples were collected from volunteers into the containers, which have 0.057 ml, 0.34 M K_3EDTA as a preservative. The final concentration of EDTA in blood is 3.9 μM, which has no effect on the PCR reaction. The blood sample was stored in a refrigerator until use to prevent hemolysis. For analysis, a 6- μl blood sample was mixed thoroughly with 90 μl formamide. The mixture was incubated at 95°C for 10 min before PCR reaction. Then, 3 μl of the above sample was used in a 20- μl reaction mixture to amplify a 110-bp fragment of the β-globin gene with specific primers. The 20- μl reaction mixture had the following final composition: standard 1 × PCR buffer from Promega, 50 mM KCl and 10 mM Tris/HCl, pH 8.3; 3.5 mM MgCl$_2$; 250 μM each of dNTP, 1 μM of each primer; 0.25 $\mu g/\mu l$ BSA; 0.5 $\mu g/\mu l$ T4 gene 32 protein; 0.25 unit/μl Tth enzyme. The reaction was performed in a 360 μm o.d., 250 μm i.d. capillary using a Rapid Cycler. The PCR protocol is listed in Table 3.1. The fully automated, integrated online setup and operation are similar to what we used before.[25] Briefly, PCR solution was aspirated by a syringe pump to the reaction capillary and sealed by freeze–thaw valves. After reaction, the reaction mixture was online-transferred to the injection cross and injected at 50 V/cm for 6 min. Then injection of 0.1 M NaOH followed to ensure the stacking of the DNA sample. Finally, 1 ×TE buffer was employed to run the electrophoresis.

A 115-bp fragment from the gag region of the HIV-1 DNA was amplified using the HIV test kit from Perkin-Elmer. The individual components were added according to the manufacturer's suggestion, except that 0.25 $\mu g/\mu l$ BSA was added to prevent the adsorption of the Taq enzyme on the capillary. The protocol is 94°C for 2 min followed by 40 cycles of denaturation at 95°C for 15 s, annealing and extension at 60°C for 1 min. The annealing and extension temperatures were the same for this protocol.

Buccal epithelial cells were collected by twirling a sterile swab on the inner cheek for 30 s. The swab was immersed into 400 μl of formamide in a microfuge tube.

TABLE 3.1
PCR Protocol Using Blood Directly as the Template

Operation	Temp. (°C)	Time (s)	Cycle No.
Incubation	85	120	1
Denature	80	60	3
Anneal	45	60	
Extension	60	60	
Denature	80	15	40
Anneal	45	30	
Extension	60	60	
Hold	60	600	1

The tube with the swab was heated at 95°C for 15 min and the swab was removed. A 20-µl reaction consists of 3 µl of the buccal cell DNA solution, 1 μM of each of the primers, 200 μM of dNTP; 3 mM of $MgCl_2$, 2.5 units of Taq DNA polymerase and 1 ×Tris/HCl buffer with 0.25 µg/µl BSA. The cheek cell can also be transferred into the reaction mixture directly using a plastic toothpick, in which case lysing of the cheek cell was achieved by the incubation step of the PCR reaction. Cheek cells from five individuals were tested and PCR reactions were performed in a Perkin-Elmer GeneAmp PCR system 2400. The temperature protocol was incubation at 80°C for 2 min, followed by 35 cycles of denaturation at 80°C for 20 s, annealing at 53°C for 1 min and extension at 60°C for 1 min. When using cheek cells directly in the PCR reaction mixture, the incubation step is 7 min at 80°C.

The concentration of DNA after reaction generally approaches μM and should be sufficient for UV detection. However, for CE with UV detection, excess salts in the sample matrix interfere with effective sample injection.[70,71] A possible solution is to use a size-exclusion column to purify the sample and online inject the sample into the capillary. As we already demonstrated in online DNA sequencing, samples in low-ionic-strength buffers can be injected up to 3 min at the running voltage without sacrificing the resolution.[46] However, a simpler way exists for the analysis of crude PCR mixtures. There are several stacking methods that can be used to overcome problems associated with inefficient sample utilization.[72,73] Swerdlow[18] described a base-stacking method for DNA sequencing from unpurified products. Briefly, the hydroxide ions neutralize the Tris ions to generate a zone of low conductivity. On-column concentration of the DNA fragments was achieved by electrokinetic injection of hydroxide ions. A low-conductivity zone was produced by the neutralization reaction between OH^- ions with the cationic buffer. The higher electric field in the zone causes DNA to be concentrated. We use this stacking method to inject PCR reaction mixtures which contain 50 mM KCl, although the KCl is not always necessary.[73] The signal is higher in the case of base stacking than for ordinary injection. We find there is no interference from dNTP because they move faster than the PCR products. To achieve base stacking in these small volumes, the 0.1 M NaOH vials should be kept closed after stacking. Otherwise CO_2 will neutralize some OH^- and ruin the stacking.

In genetic analysis, it is always desirable to amplify DNA directly from clinical materials to avoid sample manipulation. Although blood is a very complicated mixture and using blood is prone to contamination, the amount of DNA in blood is relatively constant: 1 µl of blood generally contains 4.1 to 10.3×10^3 white blood cells. It was found that the major problem for PCR direct from blood was the inability of the DNA polymerase to access the target DNA.[19] The solution is to find conditions that will release DNA from cells in a form suitable for PCR while preserving the activity of Taq DNA polymerase. We tried to use water to lyse the cells, but PCR was not successful. Based on the observed cell debris, the DNA most likely was trapped by coprecipitation with proteins. So, FoLT (formamide low temperature) PCR, which was demonstrated before for slab-gel electrophoresis, was used here for UV-CE analysis.[20]

For cheek cells, the matrix is cleaner than blood. Cheek cells lysed in water have been used in PCR. The common method is to use 15 ml NaCl solution to wash the mouth. This procedure produces a yield of total genomic DNA of about 2 to 5 µg and is sufficient for many PCR reactions. One inconvenient feature is that centrifugation is needed to precipitate the cells. The use of cytology brushes and swabs as a more efficient means to cell collection/DNA extraction has also been demonstrated and validated in several research groups.[21,22] This method is easily performed in a 96-well format and is compatible with high-throughput testing. In previous reports, cheek cells were lysed in 0.1 M NaOH and later neutralized with Tris/HCl. Here, we show that we can avoid the neutralization step by heating in formamide to lyse the cells either before or during the PCR reaction. Compared with normal PCR starting from pure DNA, PCR from cheek cells or from blood gave good results.

In Figure 3.7, PCR amplification of β-globin directly from blood and HIV analysis were demonstrated. First, 1 × TE buffer was injected into capillary 13 and was used for base-line subtraction to cancel out the flicker noise from the mercury lamp, as reported before.[74] In all, 12 different samples using cheek cells from five individuals with two different primer sets were amplified in a commercial thermal cycler and analyzed by the array. The electropherograms from two capillaries show negative PCR results, for which no cheek cell was added but formamide and all other reagents for PCR reaction were present. This shows the lack of DNA contamination in the system. A longer-term study with actual clinical samples will be necessary to verify that memory effects are absent from repeated use of the system.

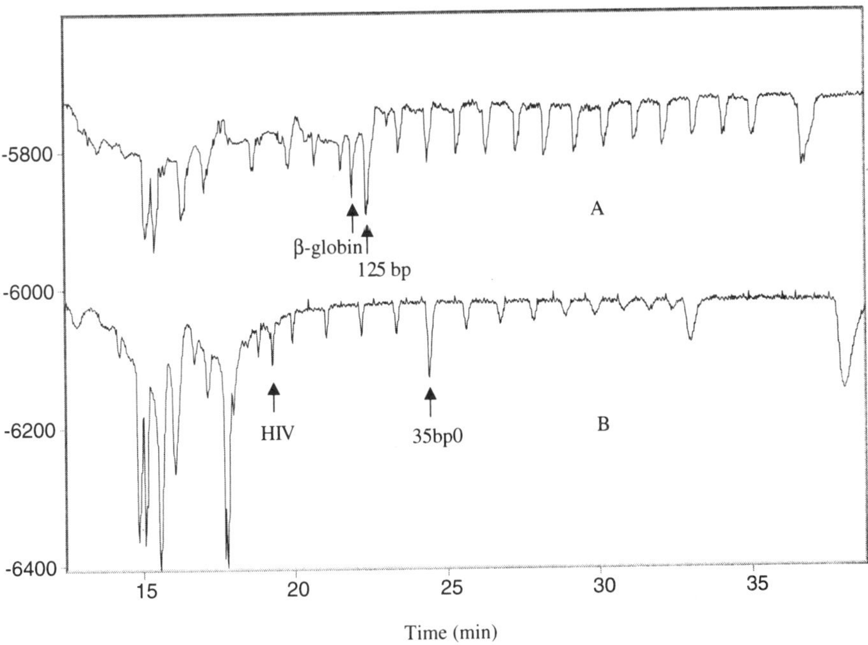

FIGURE 3.7 Online PCR analysis of (A) a 110-bp fragment of β-globin gene amplified from blood and spiked with 25-bp DNA ladder; (B) a 115-bp fragment of HIV gag DNA spiked with 50-bp DNA ladder.

Amplification from mitochondrial DNA generates more DNA products because hundreds of copies of mitochondrial DNA are present in each human cell.[75] To increase the confidence level for identification, the PCR products were coinjected with a 100-bp ladder for some of the capillaries. The electropherograms show the two products, which are the 241-bp fragment for the cyto primer and the 256-bp fragment for the M69 primer. The capillary array was flushed with water between runs and did not show degradation over months.

Because the sample preparation step and sample injection step can be easily integrated and automated by this protocol, this has the potential of becoming a cost-efficient and high-throughput approach to perform genetic analysis or disease diagnosis at a low cost. For example, one can envision equipping a doctor's office with a CAE instrument and a thermal cycler. Test kits in 96-well microtiter plates can be factory-prepared for such analyses. Each vial in the microtiter plate would contain the premix solution as specified here plus a distinct primer pair. PCR analysis can thus be performed 96 at a time for multiple genetic regions or for multiple patients directly from cheek cell swabs or microliter volumes of blood. No solution preparation or sample preparation will be required at the test site. Such a system will allow doctors and nurses, as opposed to highly skilled laboratory technicians, to gain immediate access to genetic dispositions relevant to disease in 1 or 2 h. The protocol described here is applicable to both fluorescence-based and absorption-based capillary array instruments, although the latter is less expensive to operate (no lasers and no dyes). Because such instruments are already commercially available, practical applications should be imminent.

ACKNOWLEDGMENT

The authors thank the many coworkers whose work laid the foundation of this chapter. The Ames Laboratory is operated for the U.S. Department of Energy by Iowa State University under Contract No. W-7405-Eng-82. This work was supported by the Director of Science, the Office of Biological and Environmental Research, and by the National Institutes of Health.

REFERENCES

1. E Pennesi. Human genome: Finally, the book of life and instructions for navigating it. Science 288:2304–2307, 2000.
2. F Collins, A Patrinos, E Jordan, A Chakravarti, R Gesteland, L Walters. New goals for the U.S. human genome project: 1998–2003. Science 282:682–689, 1998.
3. W Bloch. A biochemical perspective of the polymerase chain reaction. Biochemistry 30:2735–2747, 1991.
4. Y Hu, JL Glass, AE Griffith. Observation and simulation of electrohydrodynamic instabilities in aqueous colloidal suspensions. J Chem Phys 100:4674–4682, 1994.
5. K Ueno, ES Yeung. Simultaneous monitoring of DNA fragments separated by capillary electrophoresis in a multiplexed array of 100 capillaries. Anal Chem 66:1424–1431, 1994.

6. XC Huang, MA Quesada, RA Mathies. DNA sequencing using capillary array electrophoresis. Anal Chem 64:2149–2154, 1992.

7. NJ Dovichi, H Swerdlow, JZ Zhang, DY Chen, HR Harke. Three DNA sequencing methods using capillary gel electrophoresis and laser-induced fluorescence. Anal Chem 63:2835–2841, 1991.

8. H Kambara, S Takahashi. Multiple-sheathflow capillary array DNA analyser. Nature 361:565–566, 1993.

9. C Kilger, M Krings, H Poinar, S Pääbo. "Colony sequencing": Direct sequencing of plasmid DNA from bacterial sources. BioTechniques 22:412–418, 1997.

10. Q Chen, C Neville, A MacKenzie, RG Korneluk. Automated DNA sequencing requiring no DNA template purification. BioTechniques 21:453–457, 1996.

11. MA Marra, TA Kucaba, LW Hillier, RH Waterstion. High-throughput plasmid DNA purification for 3 cents per sample. Nucleic Acids Res 27:e37, i–vi, 1999.

12. AS Cohen, DR Najarian, BL Karger. Separation and analysis of DNA sequence reaction products by capillary gel electrophoresis. J Chromatogr 516:49–60, 1990.

13. O Salas-Solano, E Carrilho, L Kotler, AW Miller, W Goetzinger, Z Sosic, BL Karger. Routine DNA sequencing of 1000 bases in less than one hour by capillary electrophoresis with replaceable lineary polyacrylamide solutions. Anal Chem 70:3996–4003, 1998.

14. W Wei, ES Yeung. Improvements in DNA sequencing by capillary electrophoresis at elevated temperature using polyethylene oxide) as a sieving matrix. J Chromatogr A 745:221–230, 2000.

15. MC Ruiz-Martinez, O Salas-Solano, E Carrilho, L Kotler, BL Karger. A sample purification method for rugged and high-performance DNA sequencing by capillary electrophoresis using replaceable polymer solutions. A. Development of the cleanup protocol. Anal Chem 70:1516–1527, 1998.

16. O Salas-Solano, MC Ruiz-Martinez, E Carrilho, L Kotler, BL Karger. A sample purification method for rugged and high-performance DNA sequencing by capillary electrophoresis using replaceable polymer solutions. B. Quantitative determination of the role of sample matrix components on sequencing analysis. Anal Chem 70:1528–1535, 1998.

17. H Zhou, AW Miller, Z Sosic, B Buchholz, AE Barron, L Kotler, BL Karger. DNA sequencing up to 1300 bases in two hours by capillary electrophoresis with mixed replaceable linear polyacrylamide solutions. Anal Chem 72:1045–1052, 2000.

18. Y Xiong, SR Park, H Swerdlow. Base stacking: pH-mediated on-column sample concentration for capillary DNA sequencing. Anal Chem 70:3605–3611, 1998.

19. M Panaccio, AM Lew. PCR based diagnosis in the presence of 8% (v/v) blood. Nucleic Acids Res 19:1151, 1991.

20. M Panaccio, AM Lew. Direct PCR from whole blood using formamide and low temperatures. In: HG Griffin, AM Griffin, eds. PCR Technology: Current Innovations. New York: CRC Press, 1994, 151–157.

21. B Richards, J Skoletsky, AP Shuber, R Balfour, RC Stern, HL Dorkin, RB Parad, D Witt, KW Klinger. Multiplex PCR amplification from the CFTR gene using DNA prepared from buccal brushes/swabs. Hum Mol Genet 2:159–163, 1993.

22. Y Liu, J Bai, Y Zhu, X Liang, D Siemieniak, PJ Venta, DM Lubman. Rapid screening of genetic polymorphisms using buccal cell DNA with detection by matrix-assisted laser desorption/ionization mass spectrometry. Rapid Commun Mass Spectrometry 9:735–743, 1995.

23. H Tan, ES Yeung. Integrated on-line system for DNA sequencing by capillary electrophoresis: From template to called bases. Anal Chem 69:664–674, 1997.

24. H Tan, ES Yeung. Automation and integration of multiplexed on-line sample preparation with capillary electrophoresis for high-throughput DNA sequencing. Anal Chem 70:4044–4053, 1998.

25. N Zhang, H Tan, ES Yeung. Automated and integrated system for high-throughput DNA genotyping directly from blood. Anal Chem 71:1138–1145, 1999.

26. H Swerdlow, BJ Jones, CT Wittwer. Fully-automated PCR and DNA sequencing: Reaction and analysis in a fluidic capillary instrument. Anal Chem 69:848–855, 1997.

27. SA Soper, DC Williams, Y Xu, SJ Lassiter, Y Zhang, SM Ford, RC Bruch. Sanger DNA sequencing reactions performed in a solid-phase nano-reactor directly coupled to capillary gel electrophoresis. Anal Chem 70:4036–4043, 1998.

28. MA Burns, BN Johnson, SN Brahmasandra, K Handique, J Webster, M Krishnan, TS Sammarco, PM Man, D Jones, D Heldsinger, CH Mastrangelo, DT Burke. An integrated nanoliter DNA analysis device. Science 282:484–487, 1998.

29. E Litborn, A Emmer, J Roeraade. Chip-based nanovials for tryptic digest and capillary electrophoresis. Anal Chim Acta 401:11–19, 1999.

30. E Litborn, A Emmer, J Roeraade. Parallel reactions in open chip-based nanovials with continuous compensation for solvent evaporation. Electrophoresis 21:91–99, 2000.

31. SA Soper, SM Ford, Y Xu, S Qi, S McWhorter, S Lassiter, D Patterson, RC Bruch. Nanoliter-scale sample preparation methods directly coupled to polymethylmethacrylate-based microchips and gel-filled capillaries for the analysis of oligonucleotides. J Chromatogr A 853:107–120, 1999.

32. J Khandurina, TE McKnight, SC Jacobson, LC Waters, RS Foote, JM Ramsey. Integrated system for rapid PCR-based DNA analysis in microfluidic devices. Anal Chem 72:2995–3000, 2000.

33. Q Li, T Kane, C Liu, H Zhao, R Fields, J Kernan. Fully automated DNA sequencing with a commercial 96-capillary array instrument. HPCE '99, Palm Springs, CA, 1999, Abstract 32.

34. T Anazawa, S Takahashi, H Kambara. A capillary-array electrophoresis system using side-entry on-column laser irradiation combined with glass rod lenses. Electrophoresis 20:539–546, 1999.

35. M Quesada, S Zhang. Multiple capillary DNA sequencer that uses fiber-optic illumination and detection. Electrophoresis 17:1841–1851, 1996.

36. M Quesada, H Dhadwal, D Fisk, F Studier. Multi-capillary optical wavelengths for DNA sequencing. Electrophoresis 19:1415–1427, 1998.

37. NJ Dovichi. DNA sequencing by capillary electrophoresis. Electrophoresis 18:2393–2399, 1997.

38. JR Scherer, I Kheterpal, A Radhakrishnan, WW Ja, RA Mathies. Ultra-high throughput rotary capillary array electrophoresis scanner for fluorescent DNA sequencing and analysis. Electrophoresis 20:1508–1517, 1999.

39. I Kheterpal, RA Mathies. Capillary array electrophoresis DNA sequencing. Anal Chem 71:31A–37A, 1999.

40. PC Simpson, D Roach, AT Woolley, T Thorsen, R Johnston, GF Sensabaugh, RA Mathies. High-throughput genetic analysis using microfabricated 96-sample capillary array electrophoresis microplates. Proc Natl Acad Sci USA 95:2256–2261, 1998.

41. I Medintz, W Wong, G Sensabaugh, RA Mathies. High speed single nucleotide polymorphism typing of a hereditary haemochromatosis mutation with capillary array electrophoresis microplates. Electrophoresis 21:2352–2358, 2000.

42. S Liu, Y Shi, WW Ja, RA Mathies. Optimization of high-speed DNA sequencing on microfabricated capillary electrophoresis channels. Anal Chem 71:566–573, 1999.

43. S Liu, H Ren, Q Gao, D Roach, R Loder, TM Armstrong, Q Mao, L Blaga, D Barker, S Jovanovich. Automated parallel DNA sequencing on multiple channel microchips. Proc Natl Acad Sci USA 97:5369–5374, 2000.

44. H Drossman, JA Luckey, AJ Kostichka, J D'Cunha, LM Smith. High speed separations of DNA sequencing reactions by capillary electrophoresis. Anal Chem 62:900–903, 1990.

45. JC Venter, HO Smith, L Hood. A new strategy for genome sequencing. Nature 381:364–366, 1996.

46. Y Zhang, H Tan, ES Yeung. Multiplexed and automated DNA sequencing directly from single bacterial colonies. Anal Chem 71:5018–5025, 1999.

47. X Xu, ES Yeung. Direct measurement of single-molecule diffusion and photodecomposition in free solution. Science 276:1106–1109, 1997.

48. O Kalinina, I Lebedeva, J Brown, J Silver. Nanoliter scale PCR with TaqMan detection. Nucleic Acids Res 25:1999–2004, 1997.

49. LA Kolmodin, JF Williams. PCR Cloning Protocols. Totowa, NJ: Humana Press, 1997, 3–15.

50. LG Lee, CR Connell, SL Woo, RD Cheng, BF McArdle, CW Fuller, ND Halloran, RK Wilson. DNA sequencing with dye-labeled terminators and T7 DNA polymerase: Effect of dNTPs on incorporation of dye-terminators and probability analysis of termination fragments. Nucleic Acids Res 20:2471–2483, 1992.

51. JP Landers, RP Oda, TC Spelsberg, JA Nolan, KJ Ulfelder. Capillary electrophoresis: A powerful microanalytical technique for biologically active molecules. BioTechniques 14:98–111, 1993.

52. Q Li, ES Yeung. Simple two-color base-calling schemes for DNA sequencing based on standard four-label Sanger chemistry. Appl Spectrosc 49:1528–1533, 1995.

53. DR Meldrum. A biomechatronic fluid-sample-handling system for DNA processing. IEEE/ASME Trans Mechatronics 2:99–109, 1997.

54. DR Meldrum, HT Evensen, WH Pence, SE Moody, DL Cunningham, PJ Wiktor. ACAPELLA-1K, a capillary-based submicroliter automated fluid handling system for genome analysis. Genome Res 10:95–104, 2000.

55. MU Kopp, AJ deMello, A Manz. Chemical amplification: Continuous-flow PCR on a chip. Science 280:1046–1048, 1998.

56. LC Waters, SC Jacobson, N Kroutchinina, J Khandurina, R Foote, JM Ramsey. Multiple sample PCR amplification and electrophoretic analysis on a microchip. Anal Chem 70:5172–5176, 1998.

57. HT Evenson, DR Meldrum, DL Cunningham. Automated fluid mixing in glass capillaries. Rev Sci Instr 69:519–526, 1998.

58. H-M Pang, ES Yeung. Automated one-step DNA sequencing based on nanoliter reaction volumes and capillary electrophoresis. Nucleic Acids Res 28:e73, i–viii, 2000.

59. MC Raja, D Zevin-Sonkin, J Shwartzburd, TA Rozovskaya, IA Sobolev, O Chertkov, V Ramanathan, L Lvovsky, LE Ulanovsky. DNA sequencing using differential extension with nucleotide subsets (DENS). Nucleic Acids Res 25:800–805, 1997.

60. EN Fung, H-M Pang, ES Yeung. Fast DNA separations by using poly(ethylene oxide) in non-denaturing medium with temperature programming. J Chromatogr A 806:157–164, 1998.

61. Y He, H-M Pang, ES Yeung. Integrated electroosmotically-driven on-line sample purification system for nanoliter DNA sequencing by capillary electrophoresis. J Chromatogr A 894:179–190, 2000.

62. DM Goodall, SJ Williams, DK Lloyd. Quantitative aspects of capillary electrophoresis. TrAC 10:272–279, 1991.

63. D Rickwood, BD Hames, Eds. Gel Electrophoresis of Nucleic Acids: A Practical Approach. Washington, DC: IRL Press, 1983.

64. H Sigel, Ed. Metal Ions in Biological Systems. New York: Marcel Dekker, 1979, Vol. 9.

65. F Sanger, S Nicklen, AR Coulson. DNA sequencing with chain-terminating inhibitors. Proc Natl Acad Sci USA 74:5463–5467, 1977.

66. G Xue, H-M Pang, ES Yeung. On-line nanoliter cycle sequencing reaction with capillary zone electrophoresis purification for DNA sequencing. J Chromatogr A 914:245–256, 2001.

67. Q Gao, ES Yeung. A matrix for DNA separation: Genotyping and sequencing using poly(vinylpyrrolidone) solution in uncoated capillaries. Anal Chem 70:1382–1388, 1998.

68. Y-H Lee, RG Maus, BW Smith, JD Winefordner. Laser-induced fluorescence detection of a single molecule in a capillary. Anal Chem 66:4142–4149, 1994.

69. Y Zhang, Y He, ES Yeung. High throuput PCR analysis of clinical samples by capillary electrophoresis with UV detection. Electrophoresis 22:2296, 2001.

70. XC Huang, SG Stuart, PF Bente, TM Brennan. Capillary gel electrophoresis of single-stranded DNA fragments with UV detection. J Chromatogr A 600:289–295, 1992.

71. PE Williams, MA Marino, SA Del Rio, LA Turni, JM Devaney. Analysis of DNA restriction fragments and polymerase chain reaction products by capillary electrophoresis. J Chromatogr A 680:525–540, 1994.

72. R-L Chien, DS Burgi. On-column sample concentration using field amplification in CZE. Anal Chem 64:489A–496A, 1992.

73. JP Quirino, S Terabe. Exceeding 5000-fold concentration of dilute analytes in micellar electrokinetic chromatography. Science 282:465–468, 1998.

74. X Gong, ES Yeung. An absorption detection approach for multiplexed capillary electrophoresis using a linear photodiode array. Anal Chem 71:4989–4996, 1999.

75. JS Hanekamp, WG Thilly, MA Chaudhry. Screening for human mitochondrial DNA polymorphisms with denaturing gradient gel electrophoresis. Hum Genet 98:243–245, 1996.

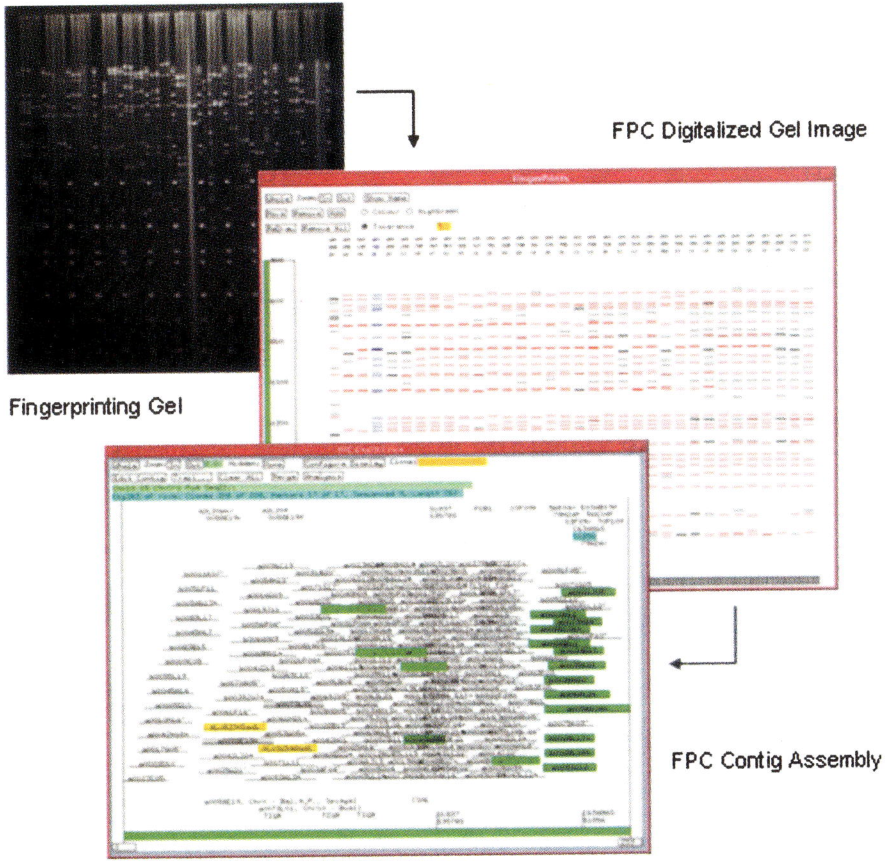

FIGURE 7.1 Physical mapping (high-resolution agarose system).

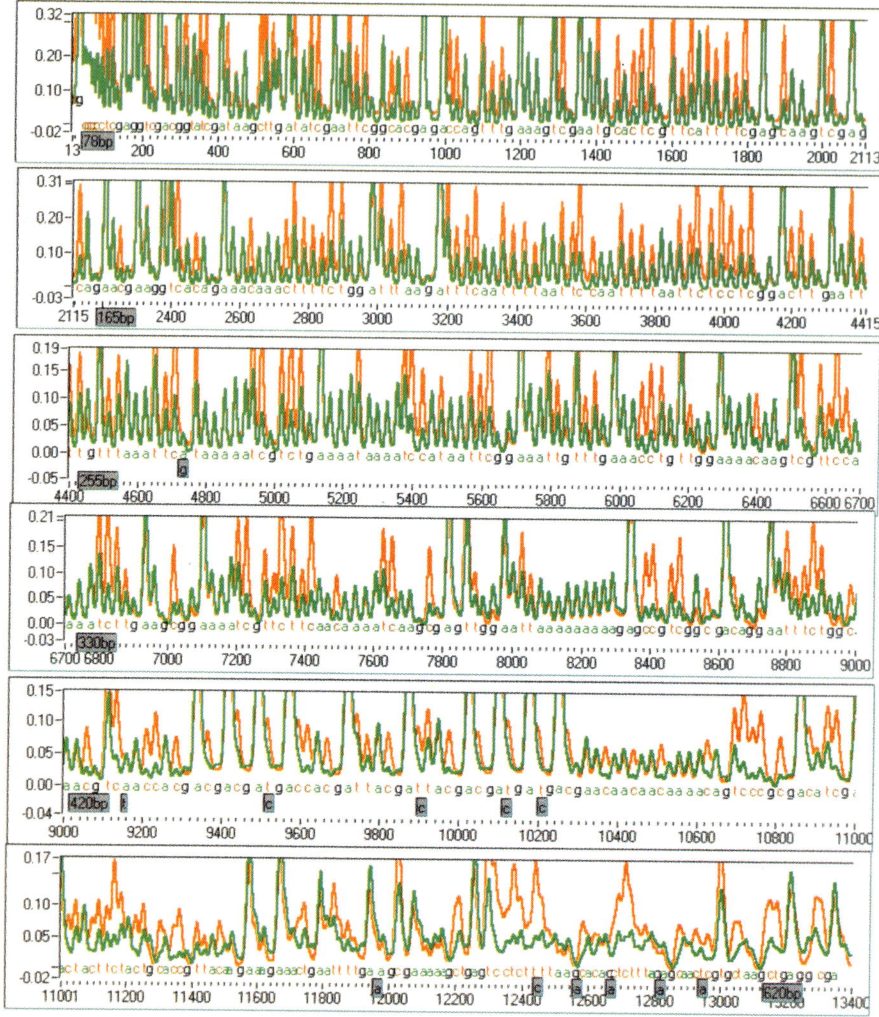

FIGURE 3.1 Sequencing of a DNA insert in the pBluescript vector using the cell lysate from one colony as the template. Primer: M13-40. The raw data from the blue and red channels are plotted. The miscalls are also corrected under the corresponding bases. The resolution is above 0.5 at 620 bp.

*Brian M. Paegel, Robert G. Blazej,
and Richard A. Mathies*

CONTENTS

INTRODUCTION

The evolution of analytical technology is crucial for the advancement of any scientific discipline, but perhaps never in history has this fact been more poignantly illustrated than in the rapid and efficient completion of the human genome sequence.[1] Over the course of 10 years, the costs, complexity, and analysis time were systematically reduced and the sample throughput increased through the development of fluorescence-based methods,[2,3] engineered polymerases,[4,5] efficient fluorescent dye-labeling constructs,[6] and high-throughput capillary array electrophoresis (CAE) instrumentation.[7–9] Now that the Human Genome Project is in its finishing stage, our thirst for genomic knowledge is only intensifying, with genome sequencing projects targeting myriad model and industrially important organisms,[10,11] as well as the emergence of interspecies studies of genetic variation.[12] The current CAE paradigm was instrumental in providing the sample throughput required to tackle the massive, repeat-rich human genome, but the dependence of this format on large, expensive robotic systems that are incapable of manipulating submicroliter volumes make it an interim solution to a much larger analytical challenge: How do we exploit the nanoliter sampling capabilities of capillary electrophoresis (CE) analysis, eliminate robotics and difficult-to-automate sample preparation steps, increase data quality and throughput, and further reduce costs?

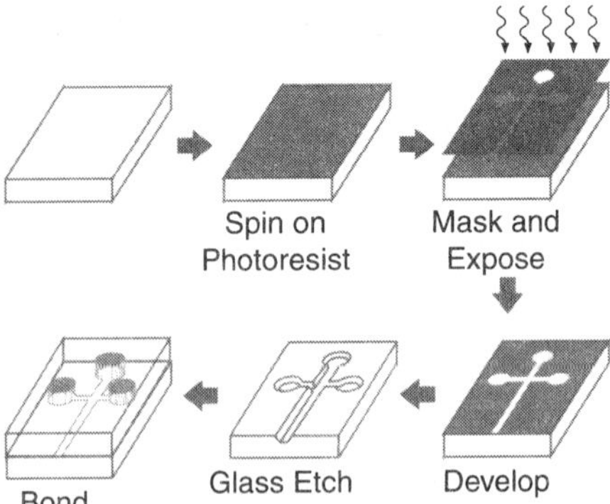

FIGURE 4.1 Schematic of the microfabrication process for making CE devices. A blank glass wafer is spin-coated with PR and the desired pattern is transferred to the PR film by exposure to UV through a photomask. The PR is developed and the exposed glass regions are etched in HF to produce trenches 20 to 50 μm deep. After etching, the remaining PR is stripped, the etched wafer is drilled through at reservoir locations, and thermally bonded to a blank glass wafer to form the capillary channels.

In 1992, CE analysis in planar silicon and glass chips was introduced by Manz and Harrison.[13,14] The planar glass wafer fabrication method is schematically outlined in Figure 4.1. A blank glass wafer is spin-coated with a photoreactive resin called photoresist (PR). The desired pattern of microcapillary channels described on a master photomask is then transferred to the PR by contact exposure with ultraviolet (UV) light. Exposed regions of PR are washed away in a developer bath, and the substrate is immersed in HF to etch the glass isotropically. Microchannels for DNA analysis range from 10 to 50 μm deep, and from 30 to 200 μm wide. The remaining photoresist is stripped away, holes are drilled through the etched substrate at the locations of access reservoirs, and the channels are completed by thermal compression bonding with a blank glass substrate of similar size. The entire fabrication procedure is compatible with the microelectronics industry very large scale integration (VLSI) processing techniques and instrumentation.[15]

The key advantage of planar CE chip technology lies in the photo-patterning step. Modern contact printing permits the control of patterns to the UV diffraction limit, facilitating the micron-scale control of such features as capillary geometry, reactor shape and placement, and array interconnectivity. Alternative conventional capillary systems rely on cumbersome and unreliable physical joints to connect sections of capillaries and are geometrically confined to the cross section of a drawn capillary.[16] Photolithography permits monolithic construction of massively parallel fluidic networks and micron-scale control of local channel geometry. The quintessential demonstration of these advantages is the cross-injector, a ubiquitous feature of CE microchips. In 1993, Harrison[17] showed that an intersection of channels used as an injector allows the formation of extremely narrow sample plugs, making possible the separation of complex mixtures in a fraction of the column length required in conventional capillary systems. This concept is diagrammed in Figure 4.2.

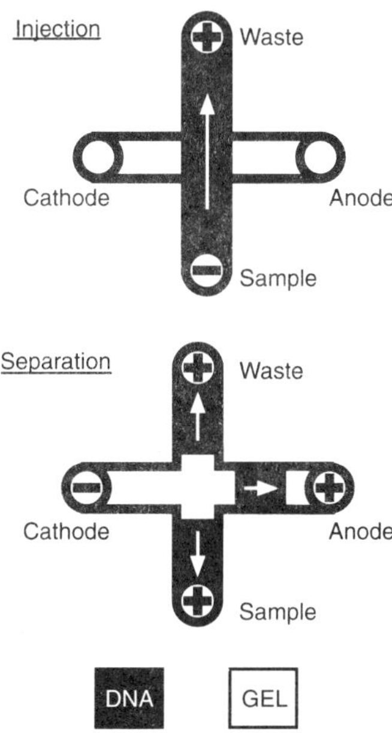

FIGURE 4.2 Schematic of the cross-injection process. The sample reservoir is filled with DNA solution and the remaining reservoirs and channels are filled with gel buffer. During injection, sample is electrophoresed to the waste, thus filling the intersection of channels with sample. The anode and cathode reservoirs are poised at a slightly negative potential such that the sample is confined to the intersection (top). During separation, the sample contained in the intersection is driven to the high-voltage anode, while a positive potential is applied to the sample and waste reservoirs to avoid bleeding (bottom).

A cross-injector is an intersection of channels addressed by four reservoirs, typically labeled sample, waste, cathode, and anode. The running buffer is introduced through the anode, filling the entire system. The sample reservoir is filled with the analyte mixture and the rest are filled with the run buffer. A negatively charged analyte, such as DNA, is driven through the intersection by applying a positive potential at the waste reservoir while grounding the sample reservoir during "injection." Switching to "separation" mode entails applying a large positive potential to the anode reservoir while grounding the cathode. Excess analyte in the sample and waste arms of the intersection is cleared by applying a small, positive potential to the sample and waste reservoirs. This process, called back-biasing, ensures that no analyte will leak or bleed from the reservoirs onto the separation column during the run. Processes for additional control of analyte flow during injection were refined by Jacobson et al.[18] For example, the initial plug size can be further reduced through the application of slightly negative potentials at the anode and cathode, thus "pinching" the analyte stream and confining it to less than the volume defined by the intersection. This mode of injection allowed separations of dye-labeled amino acids in ~2 cm of column in *less than 10 s.*[17]

The implications of these early amino acid analysis experiments were profound because analysis times were reduced by more than an order of magnitude and the possibilities for controlling channel density, interconnectivity, and geometry were seemingly limitless. This original work nucleated a variety of experiments exploring simple online labeling,[19] characterizing electroosmotic pumping and valving,[20] and porting classical CE applications such as electrochromatography to the microchip CE platform.[21] The most dramatic reductions in analysis time, however, would come in the separation of DNA. In 1994, Woolley and Mathies[22] reported separations of restriction enzyme-digested DNA using only a 3.5-cm-long channel, requiring less than 120 s for completion. The DNA separation experiment amplified the major advantage of miniaturized and integrated injection microfluidics: a small initial plug length requires less column length and therefore less time to achieve separation. With a comparable separation length and time to the original amino acid analysis device, separations of DNA are limited solely by the length of the initially injected plug because the diffusion coefficient of DNA in a gel is much lower than that of small molecules in free solution. Therefore, the demanding DNA sequencing separation requiring high-efficiency, single-base resolution is also an ideal candidate application for microchip CE.

DNA SEQUENCING BIOPROCESSORS

SINGLE-CHANNEL BEGINNINGS

The first sequencing results from a microfluidic CE device were described by Woolley and Mathies in 1995.[23] In these seminal experiments, a cross-shaped channel was filled with acrylamide (9% T) and polymerized *in situ*. With an effective separation distance of 3.5 cm, ~200 bases (≥97% accuracy) were sequenced in 10 min. These results were not wholly unexpected in the context of previous DNA fragment sizing separations,[22,24] but they strikingly presented the potential of the microchip CE paradigm. However, the work did leave open some very important questions. For example, the use of *in situ* polymerization of high-percentage linear polyacrylamide (LPA) made device reuse impossible because the gel matrix was too viscous to be removed from the channels. Furthermore, 200-base read lengths were relatively low compared to concurrently evolving CAE instrumentation. A solution to these issues was to be found in the replaceable, low-viscosity, large-chain LPA[25] and longer effective length channels.

Research focused on implementing replaceable polymers for DNA sequencing has advanced with the goal of allowing device reuse and increasing the read length through extended channel lengths. Schmalzing and coworkers[26] reported single-color sequencing with an effective separation distance of 11.5 cm, generating 400 bases in 20 min, while the Liu and coworkers,[27] utilizing the replaceable LPA formulation and optimized energy transfer (ET) primer cycle sequencing chemistry,[6] generated 500 bases of four-color sequence (≥99% accuracy) in 20 min on a chip with an effective separation distance of only 7 cm. Early work with production sequencing samples indicated that the microchip paradigm could yield comparable, if not superior, results to commercially available CAE systems for "real-world" samples.[28] Further optimization of sequencing on 11.5-cm-long channels was achieved by implementing a mixed molecular weight blend of emulsion-polymerized LPAs,[29] resulting in 580-base read lengths

(≥98.5% accuracy) in 18 min.[30] Paralleling this work, sequencing was also presented on a 40-cm-long channel that produced an average of 800 bases in 80 min. However, in this format, the analysis time was close to that required by conventional CAE sequencers.[31] From these four collected studies, we may conclude that enhanced DNA sequencing in microfluidic devices requires a combination of superior sequencing chemistry, optimized polymer matrix, and extended separation distance, all of which play a pivotal role in determining a system's performance.

CAPILLARY ARRAY ELECTROPHORESIS MICROCHIPS

The single-channel experiments laid the groundwork for the development of microfabricated DNA sequencing bioprocessors: massively parallel arrays of sequencing channels. Microfabricated devices, in addition to providing reductions in analysis time, also offer the unique advantage of monolithic array construction. Dense arrays of precisely arranged channels are fabricated simply by increasing the complexity of the master photomask. An elegant demonstration of this principle may be found in the rapid scale-up of array complexity and density from 12, 48, 96, and finally to 384 lanes over a 5-year period.[32–36] These devices, consisting of short channel lengths, showcased the benefits of microfabrication in constructing intricate and high-density arrays coupled to novel, compact injection microfluidics for genotyping, but were incompatible with the requirements of a successful sequencer device.

The first prototype microfabricated CAE (μCAE) DNA sequencing array of 16 lanes was presented by Liu et al.[37] The schematic for the device is presented in Figure 4.3A. The 16 lanes (effective separation distance of 7 cm) are fanned out on a 100-mm-diameter glass wafer, each lane containing an individual cross-injector with three reservoirs. The lanes converge on the detection area at the high-voltage bottom end of the device, which is scanned by a galvanometer-based rectilinear scanner. However, this channel design does not utilize the wafer surface area effectively and requires distance compensation at the anode end to equalize the electric field in all channels. Nonetheless, the array was an excellent preview of the throughput increases possible for μCAE, generating an average of 457 bases called to ≥99% accuracy across all lanes (effective separation distances ranging from 7 to 7.6 cm) in 15 min. With 16 channels, the device has a throughput of ~0.5 kbp/min, greater than the conventional, commercial 96-lane CAE instrumentation, which operates at ~0.4 kpb/min.

An alternative design was accomplished by abandoning the constraints of VLSI wafer processing standards and adopting a custom plate size that permits the layout of long, straight channels (Figure 4.3B). The array of 48 lanes was fabricated on a large, rectangular glass plate measuring 53 × 13 cm. Substrates of this size are difficult to work with, requiring nonstandard processing equipment and large baths of dangerous glass etchants. Furthermore, reproducible thermal compression bonding becomes increasingly problematic as the device surface area increases. Of greater note here is the individual channel design, in which a 50-cm-long channel is terminated with a single well at each end with an effective separation distance of 46.5 cm. In lieu of a cross-injector, this device operates identically to a standard capillary. The penalty for abandoning the cross-injector is manifested in the 2.5-h

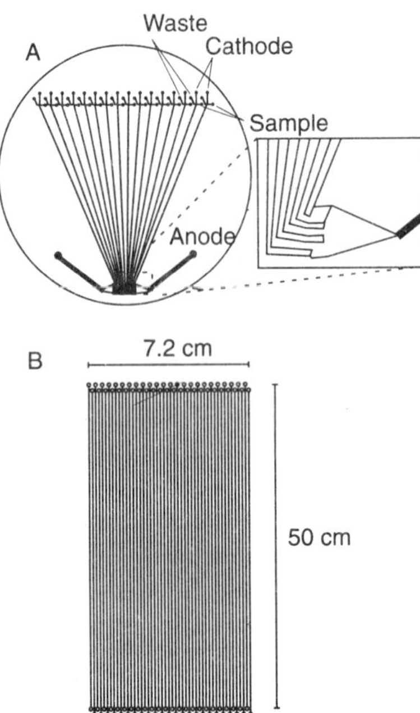

FIGURE 4.3 Prototype µCAE sequencing devices. (A) A 16-lane array on a 4-in.-diameter glass wafer. The 16 standard, individually addressed cross-injectors line the top of the array and the channels converge at the scanning region near the bottom of the device. The magnified view shows the structure used to make the total column length identical for each channel. (From SR Liu et al., Proc Natl Acad Sci USA 97:5369–5374, 2000. With permission.) (B) A 48-lane device on a rectangular 21 × 5 in. glass plate. Two rows of 24 channels are staggered. Each 50-cm-long channel is comprised of a sample inlet and outlet. (From C Backhouse et al., Electrophoresis 21:150–156, 2000. With permission.)

run time necessary to acquire 640 bases with accuracy ≥98%.[38] Array data for this device have yet to be presented.

The only alternative method for increasing effective separation distance while maintaining a compact, VLSI-compatible wafer size is to fold the channel into a serpentine geometry. Serpentine channel geometries maximize space utilization on the wafer; however, they necessitate the introduction of turns in the separation path. Early in the microchip CE literature, turns incorporated in the separation portion of the device were identified as a potential source of geometric dispersion.[18] The source of this geometric dispersion is the path length difference between the inner and outer radii of the turn, and was dubbed the "racetrack effect." Given a set of cars traveling at equal velocity around a racetrack, those vehicles closest to the inside corner of the turn have a shorter path length to traverse, thus completing the turn more quickly than those traversing the outer radius. Electrophoresing molecules behave in an identical fashion, and turns tend to tilt separating bands of analyte. Culbertson and coworkers[39] described the dependence of this phenomenon in terms of the width added to a band, Δl, by a turn with a channel width, w, and turning an angle, θ, as the product of these components. This effect was analyzed in detail by Paegel et al.,[40] who used a novel rotary scanner to interrogate the same separation at multiple points. The dramatic effect of turns on separations of DNA digest ladders is presented in the left frame of Figure 4.4. Here, fragments of the ϕX174 *Hae*III digest are cross-injected and separated in a serpentine channel containing two U-shaped turns. Three detection points along the course of the channel (before a turn, x, after a left turn, y, and after a right turn, z) illustrate the effect of the turn on band morphology. At x, the bands are substantially orthogonal to the direction of electromigration, while at y, the bands have been tilted by the racetrack effect. After the right turn, the bands are restored to their original orientation, but signal intensity and resolution have decreased due to transverse diffusion while the bands traveled in a tilted orientation. Electropherograms in the lower left frame of Figure 4.4 demonstrate the peak shape as seen by detecting at points x, y, and z. The peaks at y are extremely broad and unresolved due to the tilted band orientation. Because the severity of band tilting was expected to be proportional to the channel width, w, Paegel et al.[40] introduced the concept of a tapered turn, or "hyper-turn," in which the channel width is constricted in the region of the turn. The effect on the separation is presented in the right frame of Figure 4.4. Here, we see that the bands remain substantially orthogonal to the direction of electromigration at y, and the peaks in the corresponding electropherogram remain sharp and well resolved. Utilizing the optimal tapering geometry, the serpentine channels operate at 91% the theoretical efficiency of a comparable straight-channel device. Numerical optimizations of the hyper-turn design have suggested asymmetrically tapered structures that will potentially further diminish geometry dispersion introduced by turns.[41,42]

With the optimal hyper-turn geometry in hand, the first high-throughput μCAE DNA sequencing bioprocessor was fabricated on a 150-mm-diameter glass wafer. The channel layout schematic, presented in Figure 4.5, is based on the radial array design developed by Mathies' group in 1999.[34] In the design, 48 "doublet" structures are arrayed around a common, central anode reservoir. Each doublet, shown in the bottom portion of Figure 4.5, comprises two sample wells addressed by common waste and cathode reservoirs, halving the number of required reservoirs for cross-injection. As before, analyte is injected through the intersection of channels by driving it toward the waste reservoir, and the plug is separated by driving it down the 15.9 cm serpentine separation channel. As the plug approaches the center of the device, fluorescence is detected by the Berkeley rotary scanning confocal microscope, which is presented in Figure 4.6. Developed by Mathies' group,[34] the "chip scanner" directs 488-nm laser excitation through the hollow shaft of a stepper motor. A rhomb prism displaces the beam 1 cm from the axis of rotation into a 0.7 NA 60× microscope objective. The objective collects multispectral fluorescence from the migrating DNA sequencing fragments, passing the light back through the same optical path and through the dichroic beamsplitter to a four-color detector housing. Here, fluorescence is sequentially sorted into four color channels, spatially filtered, and detected by four PMTs. The μCAE bioprocessor sits on a heated stage directly above the rotary scanning objective. An elastomeric buffering loop fixed to the top of the bioprocessor creates two 3-ml concentric and electrically isolated buffer moats for fluidically

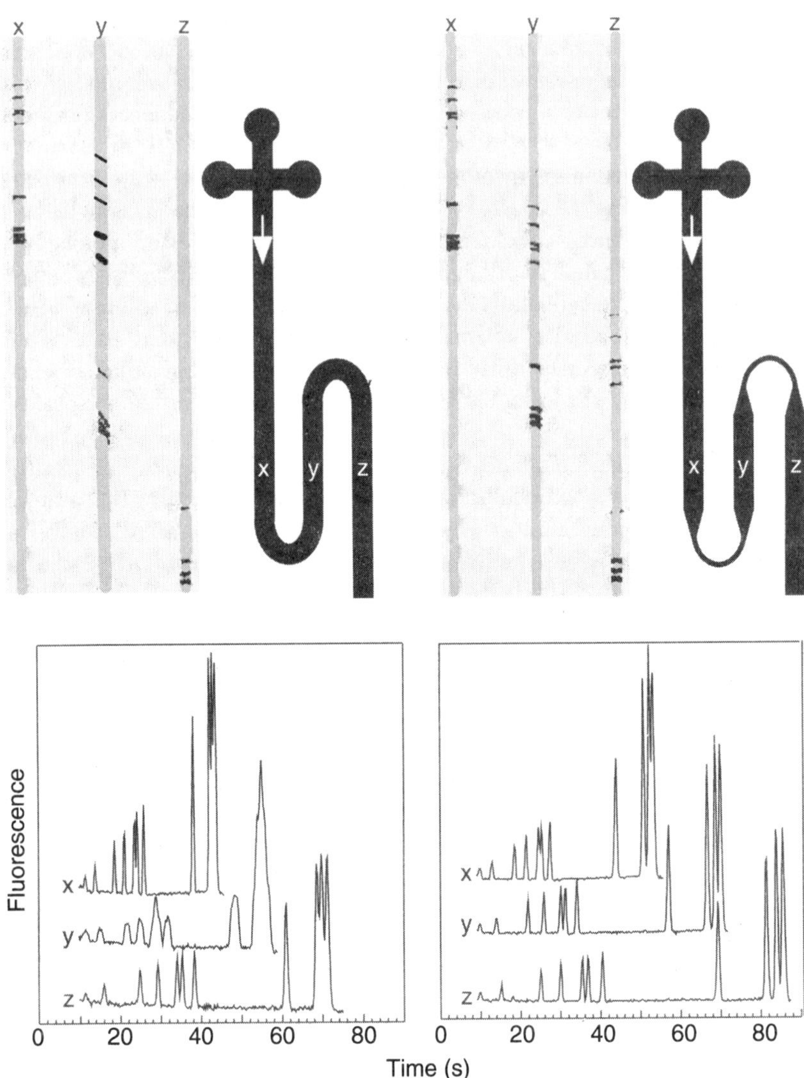

FIGURE 4.4 The effect of turns on separation quality in folded channels. U-shaped turns introduce dispersion and disrupt separation quality (left side). Bands of the same separation imaged before a turn (x), after a left turn (y), and after a second right turn (z), show a tilting distortion at y due to the racetrack effect. The opposing turn rectifies the bands, but signal and resolution are lost due to lateral diffusion. Hyper-turns mitigate the racetrack effect, maintaining separation performance and signal intensity (right side). Peaks in the corresponding electropherogram retain resolution and signal intensity at all positions. (From BM Paegel et al., Anal Chem 72:3030–3037, 2000. With permission.)

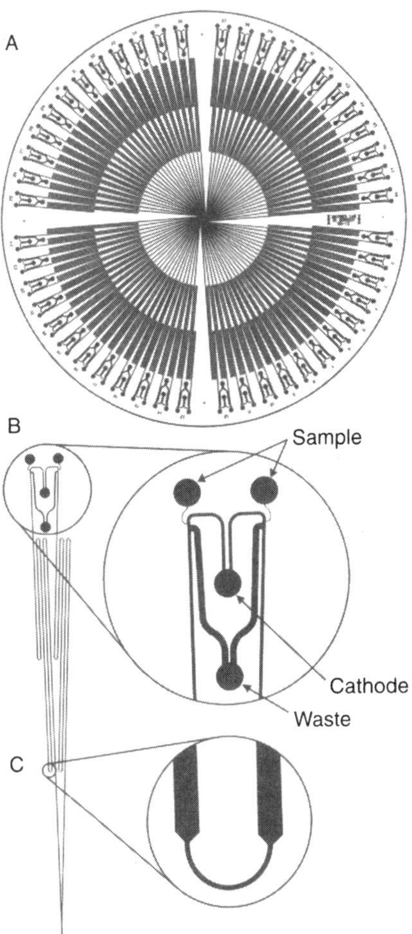

FIGURE 4.5 The 96-lane μCAE DNA sequencing bioprocessor. (A) The 48 identical doublet structures are arrayed around a common anode at the center of the device. (B) Expanded view of a doublet. Each injector contains two different sample reservoirs addressed by a common cathode and waste reservoir located in the center of the injector. Channels leading away from the injection intersection are fluidically balanced by using different channel widths. (C) The serpentine channel contains four hyper-turns for an effective channel length of 15.9 cm. (From BM Paegel et al., Proc Natl Acad Sci USA 99:574–579, 2002. With permission.)

addressing all cathode and waste ports with enough buffer to sustain electrophoresis for 30 min. An acrylic 96-pin electrode ring electrically addresses the sample wells on the periphery of the bioprocessor. Filling the array with viscous LPA sequencing matrix, as well as high-pressure water-mediated removal of the matrix, is accomplished using a loader device developed by Scherer et al.[43]

Sequencing products of single-stranded M13mp18 run in all 96 lanes are shown in a gel image format in Figure 4.7. Each vertical section of the image corresponds to one lane of output from the bioprocessor, and each band within a lane represents

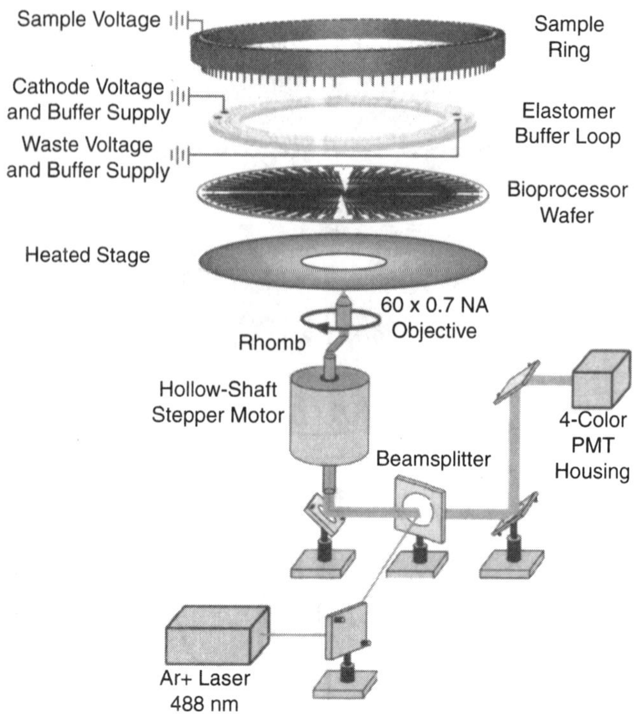

FIGURE 4.6 The Berkeley four-color rotary chip scanner and an exploded view of the assembled bioprocessor. An acrylic ring electrode array individually addresses the 96 peripheral sample reservoirs. An elastomeric buffer loop creates two annular, concentric, electrically isolated troughs (3-ml capacity) that separately address the cathode and waste wells. This assembly is placed on the heated stage of the four-color chip scanner. Excitation from an Ar$^+$ laser (488 nm) is reflected from a dichroic beamsplitter through the hollow shaft of a stepper motor. The beam is translated 1 cm from the axis of rotation by a rhomb prism and focused onto the channels through a 60× 0.7 NA microscope objective. Fluorescence is collected by the same objective and passed through the beamsplitter to the four-color PMT (photomultiplier) housing for spectral sorting and spatial filtering prior to detection. (Adapted from Emrich et al., 2002.)

one called base. Of the 96 lanes, only one lane failed to produce sequence of any appreciable length due to a defect in the lithography. An average of 430 bases with ≥99% accuracy were called per channel and the entire analysis required only 24 min. A plot of the average phred quality value as a function of base position in a 10-base moving window is shown in Figure 4.8. This plot indicates that the average accuracy equals or exceeds 99.9% for the majority of the run.[44] The rapid decline in quality at the end of the run was due to the properties of the LPA matrix used in this particular experiment. Using an optimized, blended LPA as described in previous single-channel experiments, the bioprocessor produces on average 621 bases to ≥99% accuracy in 28 min. This level of throughput corresponds to ~2 kbp/min, fivefold higher than currently deployed 96-lane CAE instruments, and even surpassing the 384-lane CAE instruments currently in testing.

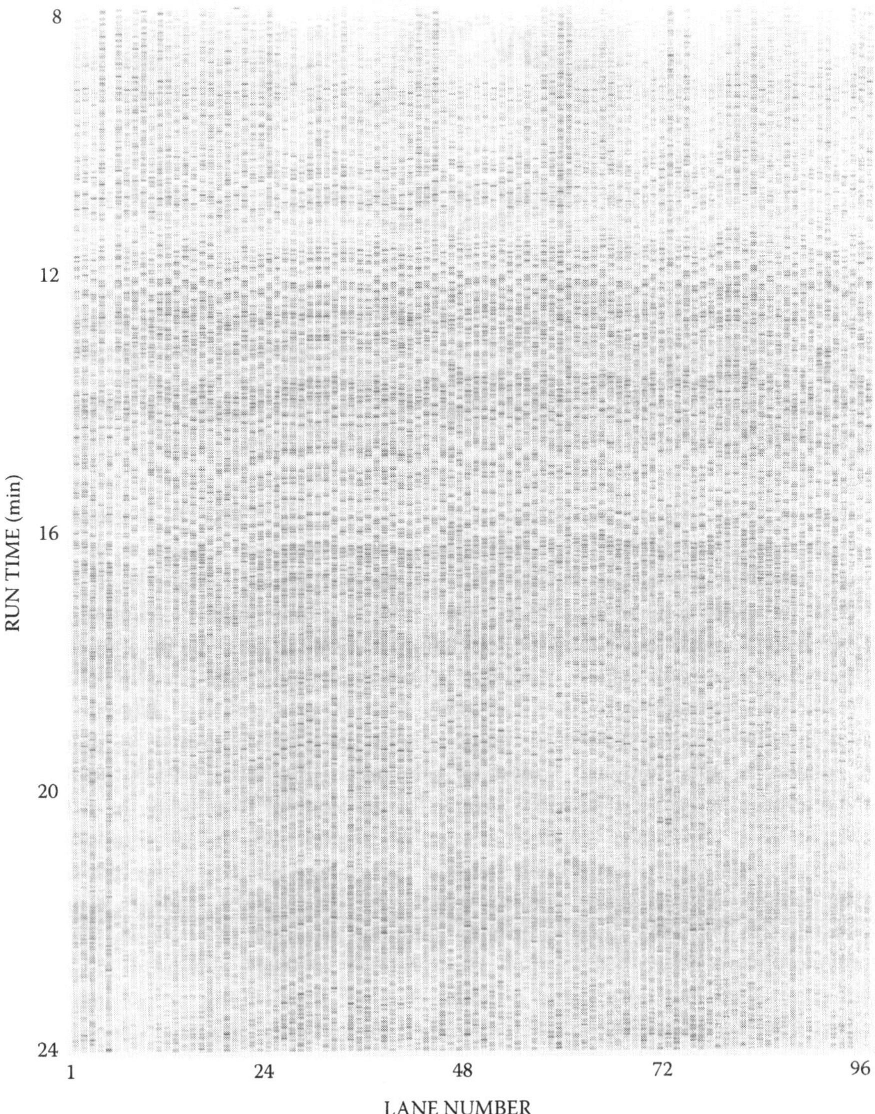

FIGURE 4.7 Gel image from the 96-lane µCAE DNA sequencing bioprocessor. Lanes are represented in the vertical dimension. Each horizontal band corresponds to one called base. The image contains 41,000 bases called with ≥99% accuracy acquired in only 24 min.

APPLICATIONS OF THE µCAE BIOPROCESSOR

With the expanded sequencing capacity provided by µCAE, polymorphism detection and screening based on DNA sequencing becomes a tractable and, in fact, very attractive alternative to modern genotyping techniques. To demonstrate this concept, Blazej et al.[45] developed a novel, sequencing-based polymorphism screening and

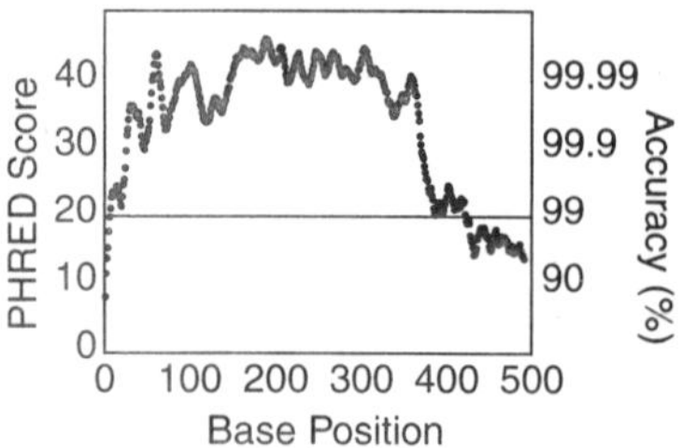

FIGURE 4.8 Average phred quality value in 10-base windows as a function of base position for the data in Figure 4.7. The average read length over 96 lanes was 430 bp, defined as the number of phred $q \geq 20$ (accuracy $\geq 99\%$) bases. In-house synthesized linear polyacrylamide (4.5% T, 7 M urea, 1× TTE), 50°C, ET dye-primer chemistry, M13mp18 ssDNA template, 200 V/cm, 24-min operation. (From BM Paegel et al., Proc Natl Acad Sci USA 99:574–579, 2002. With permission.)

detection assay called Polymorphism Ratio Sequencing (PRS). In this technique, two sequences are compared by creating single base extension ladders via the Sanger cycle-sequencing reaction, but utilizing a different pooling scheme from conventional four-color sequencing. DNA from a reference individual (or population) and DNA from a sample individual (or population) are used to generate the four standard single-base extension ladders using dye-primer sequencing chemistry. In the case of the reference individual for the A ladder, the ET-R6G primer is used, whereas for the sample individual, the A ladder is prepared with the ET-FAM primer. Similarly, the C ladder for the reference would be prepared using the ET-ROX primer for the reference and the ET-TAMRA primer for the sample. To prepare the A/C PRS sample, these four ladder reactions are pooled, precipitated, and electrophoretically resolved on the bioprocessor in an identical fashion to a standard four-color sequencing sample. An analogous pooling scheme for the T and G ladders is used to generate the T/G PRS trace to provide complete coverage of all bases. To detect polymorphisms between the sample and reference sequence, extension ladders are color-corrected and normalized utilizing standard base-calling processing procedures. The corrected output is then examined for variations in signal intensity for a given base identity. The plot of the squared difference of intensities, for example, from the T ladder for the sample and reference individual is the PRS plot (Δ^2 plot), and peaks in this plot indicate the presence of a polymorphic base.

Examples of PRS output from the bioprocessor are presented in Figure 4.9. At top, box A shows the A/C and T/G PRS traces as well as the Δ^2 plot. Along the length of the A/C trace, almost all A and C terminations are shown to overlap exactly, and the PRS plot in these locations is essentially null. In the middle of the traces, however, a polymorphism shows an A $\rightarrow$ G transition, and is manifested as a peak disappearing from the blue (sample, A) channel of the A/C trace and appearing as a blue G (sample, G) in the T/G trace. The polymorphic base is clearly identified in the Δ^2 plot. Box B presents a transversion polymorphism, A $\rightarrow$ C, in which a peak has shifted from the blue channel (sample, A) to the green channel (sample, C). Although the polymorphic peaks overlap, the relative reference-to-sample signal ratio for the A and C traces has

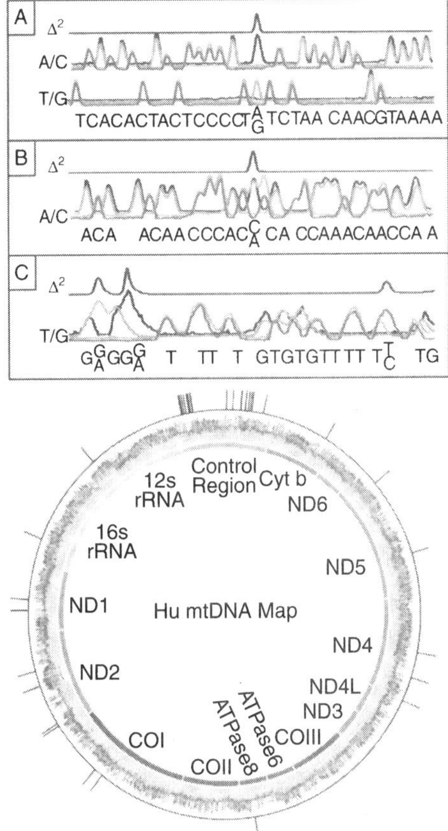

FIGURE 4.9 High-speed PRS analysis of human mtDNA. (Top) Examples of different polymorphism types. The PRS (Δ^2) plot is shown above the PRS traces. (A) Transition polymorphisms result in a peak disappearing from one PRS trace (in this case, the A/C trace) and appearing in its companion trace (G/T trace). (B) Transversion polymorphisms result in peaks changing color in the same PRS trace. (C) Due to fewer interfering peaks, extended PRS read lengths allow polymorphism detection in the classically low-quality region at the end of the run. (Bottom) Tiled PRS traces provide complete coverage of the human mitochondrial genome. The plots are overlaid on an mtDNA map showing the polymorphism positions relative to mitochondrial genes. (From RG Blazej et al., Genome Res 13:287–293, 2003. With permission.)

changed and is reflected in the Δ^2 plot. As a final demonstration of the utility of this technique, box C presents polymorphism detection in the classically low-quality region at the end of the sequencing run. Because any given PRS trace contains fewer interferences than a standard sequencing sample (containing all four base terminations) and because relative, not absolute signals are compared, greatly increased effective read lengths are possible, permitting polymorphism detection in regions that would otherwise contain no useful sequence information.[45]

The PRS method was applied to the analysis of human mitochondrial DNA (mtDNA) from two individuals to screen for polymorphic bases. The PRS traces and Δ^2 plot are overlaid on a genetic map of the mtDNA heavy strand in the bottom frame of Figure 4.9. Polymorphic bases are indicated as peaks in the Δ^2 plot shown in black on the outside of the map. Highly condensed PRS traces spanning the entire mtDNA genome are shown as the overlapping, unresolved peaks immediately adjacent to the map for scale comparison. The PRS analysis of the two individuals uncovered 30 previously published polymorphisms as well as *6 novel* polymorphisms, an effective demonstration of the advantages of adopting a global screening technique. An entire comparative analysis of the two genomes was completed in *one operation* of the bioprocessor, requiring only 29 min for separation.[45]

PROSPECTS: BIOPROCESSORS WITH INTEGRATED CHEMISTRIES

Although the absolute sequencing throughput of μCAE devices is impressive in the context of separation analysis alone, the fivefold decrease in analysis time pales in comparison to the total time required to process a sample from a BAC clone library to the front end of the electrophoresis device. For example, clone propagation in cell culture requires incubation and growth steps that may exceed 18 *hours*. Thus, the holy grail of bioanalytical instrumentation is the development of a process that eliminates lengthy cell growth steps, reduces the standard reaction volume from the present day values of 10 μl to 100 nl or lower, integrates all sample handling and transfer steps to eliminate tracking errors and sample contamination, and facilitates submicroliter fluid manipulation. To this end, research emphasis is shifting toward devices that will expedite and miniaturize the time-consuming chemistry and molecular biology steps that dominate the modern sequencing sample preparation process.

INTEGRATED DNA AMPLIFICATION

The DNA sequencing process universally relies on thermal cycling for the creation of dideoxy-terminated extension fragments, and in some cases also makes use of PCR to generate template DNA from subclones of a large-insert (e.g. BAC, YAC) library. These reactions are problematic in high- and low-throughput settings for three key reasons: (1) the reagents (polymerase cocktails, primers, etc.) used for thermally cycled reactions are costly; (2) large-volume reactions (5 to 20 μl) occur in plastic tubes that conduct heat poorly; (3) thermal cycling blocks typically have large thermal masses with low heating rates (3°C/s) and even lower cooling rates (1°C/s). Of the microfabricated separation devices we have considered thus far, injected sample volumes are on the order of 1 nl, and thus an integrated thermal cycling system that takes advantage of the microchip's native nanoliter-scale handling capabilities could address reagent cost, preparation time, and sample transfer problems.

Woolley and coworkers[46] first demonstrated hybrid integrated PCR-CE in a silicon microreactor attached to a microfabricated CE device in 1996. As a consequence of the low thermal mass of the reactor system, thermal cycling time was 25 s, compared to cycle times of 2 to 6 min in a commercial thermal cycler. Once cycling was complete, product was directly injected into the microfabricated CE system for high-speed electrophoretic resolution. This work highlighted the potential for microchips to eliminate sample transfer between processing steps, instead relying solely on electromigration to move the sample from reactor to CE column.[46] These experiments spurred subsequent studies of online PCR and cell lysis on a microfabricated device.[47,48] but the reaction volume was never reduced to the nanoliter scale. However, in 2000, Lagally et al.[49] described the low-volume amplification of M13 template DNA in which a 280-nl reactor was rapidly thermally cycled (20 cycles in 10 min), and the resulting product was directly injected onto a microfabricated CE column for electrophoretic analysis. Subsequent integration of heating and temperature-sensing elements with the nanoliter reactor further advanced the capabilities of this miniaturized analytical instrumentation.[50] Although these nanoliter-scale reactions provide significant improvements in thermal cycling speed and reagent consumption, the synthesis and analysis of DNA sequencing products presents some unique challenges. PCR generates a monodisperse, double-stranded product population, which allows nonspecific, multiple-site labeling by intercalating fluorescent dyes. The presence of literally hundreds of fluorophores per molecule allows detection of these species down to the single-molecule limit.[51] In contrast, the polydisperse DNA sequencing sample comprises approximately 1000 different single-stranded species, each of which is singly labeled. Purification and preconcentration of this sample is necessary to reach the requisite detection sensitivity, but conventional macroscale purification methodologies are incompatible with microchip-scale synthesis, thus necessitating a nanovolume sample preparation step prior to CE analysis.

INTEGRATED DNA PURIFICATION

In 1998, Ruiz-Martinez and Salas-Solano[52,53] published studies on the effects of the sequencing reaction composition (specifically, template and chloride concentration) on sequencing read length. They found that the longest read lengths were obtained from samples that contained a minimal concentrations of chloride ions and template molecules. These interferences were efficiently removed in a two-step process involving serial treatment with ion exchange followed by size exclusion spin columns.[52,53] A cheaper, more commonly adopted alternative to this protocol is ethanol precipitation. Ethanol precipitation, however, is extremely difficult to automate efficiently due to multiple sample transfer and centrifugation steps, suffers from poor (~65%) yield, and requires ~1 h to complete. Importantly, precipitation does not remove template, which can have the most detrimental effect on separation quality. A miniaturized purification technology is needed that will address the automation, speed, and cost issues associated with sample purification while providing the requisite sample purity.

A variety of methods for sample purification and preconcentration have been presented in the microchip electrophoresis literature such as solid phase extraction (SPE) and preconcentration using bead-based chromatography reagents,[54] and a porous glass frit allowing continuous preconcentration of DNA prior to injection.[55] None of these strategies accomplished all three objectives of desalting the sample, removing template, and selectively preconcentrating the sequencing fragments. A solution to this problem was to be found in exploiting the self-complementarity of DNA itself.

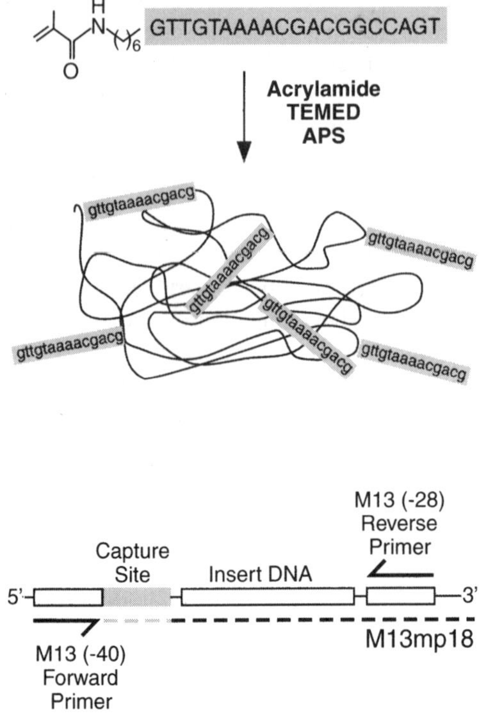

FIGURE 4.10 (Middle) Addition of the 5′-methacryl-modified capture oligo to a solution of acrylamide with radical-generating catalysts APS and TEMED yields a DNA-acrylamide capture copolymer. (Bottom) The M13mp18 cloning vector map. The capture sequence (gray) is held between the insert DNA site and the universal -40 forward M13 sequencing primer site. Extension products from the primer contain the capture sequence followed by the insert DNA sequence of interest.

In 1992, Baba and coworkers[56] at Tokushima introduced the concept of covalently linking DNA oligos of a chosen sequence to a polymeric gel matrix. An example structure of such an oligo is presented in Figure 4.10. The oligo is synthesized with a hexyl spacer separating the 5′-end from a methacryl group, which allows participation in the radical-initiated polymerization of acrylamide. When mixed with a low percentage solution of acrylamide and catalyzed with APS and TEMED, a DNA-acrylamide copolymer is formed, as shown in the middle frame of Figure 4.10, where the solid line represents the C–C backbone of the polymer. The gray oligos, or capture oligos, appear randomly along the backbone (~0.001% compared to acrylamide). This DNA-acrylamide copolymer matrix is capable of sequence-specific capture of DNA that is electrophoresed through it, providing a one-step selective capture, preconcentration, and cleanup. Sequencing fragments are selectively captured by choosing a capture sequence that is specific only to enzymatic extension products. The bottom panel of Figure 4.10 schematically presents a map of the M13mp18 cloning vector in the vicinity of the polylinker cloning site containing the insert DNA to be sequenced. Between the universal −40 forward M13 sequencing primer site and the beginning of the polylinker cloning site lies a region of constant vector sequence. Sequencing products of all insert DNA cloned into this vector will contain the capture sequence. The length of the capture oligo should be chosen such that the probability of the complementary sequence appearing in the target genome is minimized, and ideally should be short enough to allow full denaturation at the sequencing process temperature (a 19-mer satisfies these requirements for a process temperature of 65°C and a target genome of 1 Gbp). Driving an unpurified DNA sequencing sample through this capture matrix will cause the sequencing products to bind selectively to the matrix and preconcentrate to ~10 μM. Contaminating ionic species such as unspent nucleotide triphosphates, chloride, primer, and buffer components have no specificity for the capture matrix and will electrophorese through the capture matrix to the waste. Template DNA is too large to enter the polyacrylamide matrix appreciably (M13mp18 clones will be ~8 kbp) and will be excluded. This method of sample purification is advantageous for two primary reasons: (1) it combines solid-phase immobilization strategies with the high activity intrinsic of a solution phase reactions, and (2) it is entirely electrokinetically driven, obviating the need for robotic sample transfer and hydrodynamic driving apparatus.

An integrated bioprocessor for DNA sequencing sample purification followed by high-speed microchip electrophoresis was presented by Paegel et al. in 2002. The serpentine doublet separation channels and injector geometry were derived from the first-generation 96-lane bioprocessor, with a capture chamber placed in line with the cross-injector. A schematic of the injection microfluidics and device operation is shown in Figure 4.11. The 60-nl capture chamber has a 1-mm-wide body and 100-μm-wide inlet and outlet arms. The sample inlet side of the chamber is split between the sample input arm leading to the sample reservoir and a coupling arm leading to the cross-injector. Prior to operation, the entire device is filled with sequencing matrix and then the capture chamber is filled with capture matrix. All wells are filled with run buffer unless otherwise noted. In product binding mode, the device is heated to 50°C to maximize DNA hybridization kinetics (the oligo-DNA duplex $T_M = 60.4°C$), and DNA sequencing sample taken from the thermal cycler is driven electrophoretically through the capture chamber from well "S." DNA binding selectively occurs at the front taper of the chamber where the local electrophoretic flow rate is decreased due to the rapid increase in local conductivity. Product binding is usually complete in 90 s. Product washing is accomplished by evacuating the sample well and replacing it with fresh run buffer, followed by electrophoresis for an additional 30 s to remove residual low-mobility species such as primer. The bound product is released from the capture matrix by elevating the device to the DNA sequencing process temperature (67°C) and performing a standard cross-injection as previously discussed.

Sequence output from the integrated DNA sequencing purification processor is presented in Figure 4.12. In the top panel, the C terminations of three four-color sequencing experiments are shown. Trace A shows terminations obtained from the integrated purification process, while traces B and C were obtained by running the same sequencing sample on the first-generation 96-lane bioprocessor. Trace B is obtained from ethanol-precipitated sequencing sample while trace C is the result

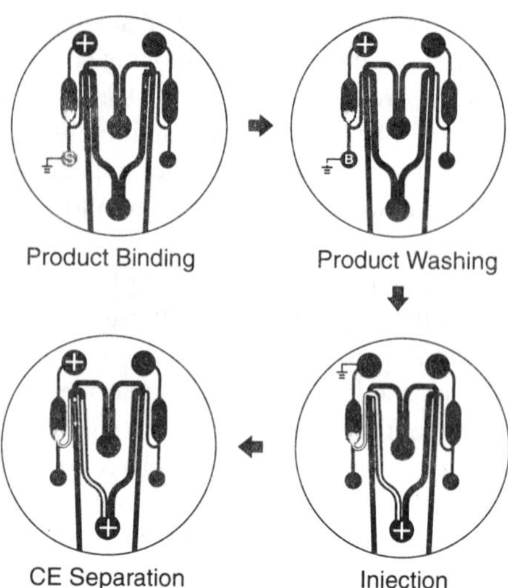

FIGURE 4.11 Schematic operation of a microfabricated device for integrated sequencing sample purification and CE analysis. Unpurified sequencing reaction mix (S, white) is electrophoretically driven through the capture matrix in the chamber and concentrated in the tapering region of the capture chamber. After 90 s, the input reservoir is cleaned and fresh buffer (B) is used to wash the bound product electrophoretically. During injection, the capture matrix–product duplex is thermally denatured and free, purified product is electrophoretically driven through the injection cross to the waste. Separation is initiated by applying a large positive potential to the central anode, drawing ~1 nl purified sample into the CE separation column. A small positive back bias is applied to the capture chamber and waste reservoirs to prevent leakage during the run.

of injecting unpurified sample. The signal intensity of the integrated purification device is on average 85% of the corresponding peak from the ethanol-precipitated sample, whereas the unpurified sample yielded peaks that were ≤10%. An examination of the early region of these electropherograms shows the specificity of the capture matrix exclusively for the sequencing extension fragments. In traces B and C, the unincorporated primer peak and first four terminations are shaded in black. These five peaks are missing from the capture matrix-purified sample because the fluorescent primers are not captured by the matrix, and the first four terminations are all contained within the capture sequence itself. These fragments have melting temperatures lower than the binding process temperature, and therefore are not captured. High-speed microchip electrophoresis of the bound product yielded 560 bases of ≥99% accuracy. This microfluidic circuitry reduces the time for sample purification 10-fold, reduces the process volume 100-fold, and eliminates all associated robotic transfer steps. This result provides dramatic evidence that microchip-level integration is evolving beyond merely expediting separations and toward a completely independent system for transferring, reacting, purifying, as well as analyzing samples.[57]

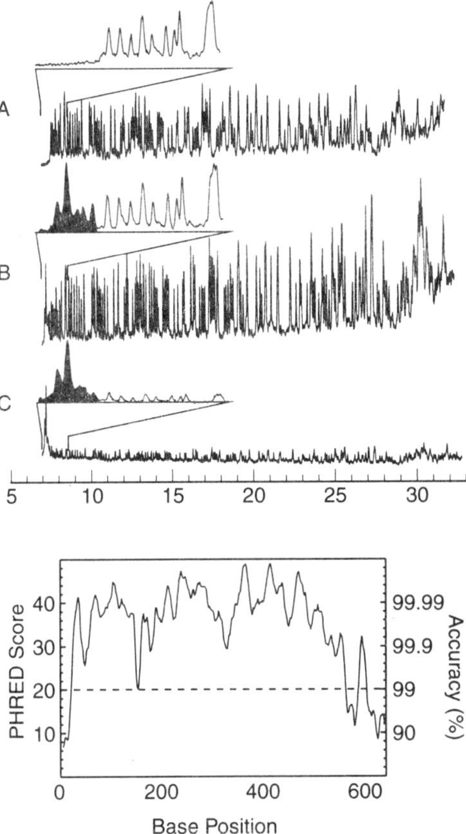

FIGURE 4.12 Integrated, high-speed oligonucleotide-capture purification of DNA sequencing products. (Top) C-terminations of four-color microchip CE sequencing runs for comparison of different purification methods. (A) Integrated oligonucleotide-capture purification followed by microchip CE. (B) Ethanol-precipitated sequencing sample that was resuspended in 50% formamide solution in DI H_2O to 1× manufacturer's specification. (C) Sample that was injected without purification. The magnified traces show the first 60 bases from each trace. Darkened peaks in traces B and C show the primer peak and the first four C terminations contained in the capture sequence, which are not present in the capture-purified sample. (Bottom) Phred analysis of the integrated capture purification microchip CE DNA sequencing run, showing 560 phred $q \geq 20$ bases called. Integrated purification required 2 min, and microchip DNA sequencing was complete in 30 min. (From BM Paegel et al., Anal Chem 74:5092–5098, 2002. With permission.)

The miniaturization of the current DNA sequencing processes in an integrated µCAE format will result in dramatic decreases in reagent consumption, analysis time, and reliance on robotics, centrifuges, and other large, expensive, and unreliable laboratory equipment. The next generation of commercial microfluidic DNA sequencers will contain entirely integrated nanoliter fluidic networks for sorting and distributing individual subclones, lysing and high-speed amplification of the template DNA molecules, isolating template amplicons for delivery to integrated cycle-sequencing reactors, sequencing product preconcentration and purification, and finally electrophoretic resolution.[58] We envision a sequencing lab of the future where microtiter plates, pipettors, test tubes, and macro robotics have been relegated to museums, replaced by stacks of highly efficient microchip bioprocessors.

ACKNOWLEDGMENTS

This work was supported by grants from the National Institutes of Health (HG01399) and from the Director, Office of Science, Office of Biological and Environmental Research of the U.S. Department of Energy under Contract DEFG91ER61125. B.M.P. acknowledges NIH fellowship support from the Berkeley Program in Genomics (T32 HG00047). Charles Emrich provided key insight on microfabrication techniques and circuit design, James Scherer designed the Berkeley four-color rotary confocal scanner, and Eric Lagally provided expertise on PCR. Additionally, we thank the entire Mathies Microchip Group for providing valuable input in the conception and execution of the μCAE DNA sequencing system.

REFERENCES

1. ES Lander, LM Linton, B Birren et al. Initial sequencing and analysis of the human genome. Nature 409:860–921, 2001.

2. LM Smith, JZ Sanders, RJ Kaiser, P Hughes, C Dodd, CR Connell, C Heiner, SBH Kent, LE Hood. Fluorescence detection in automated DNA-sequence analysis. Nature 321:674–679, 1986.

3. JM Prober, GL Trainor, RJ Dam, FW Hobbs, CW Robertson, RJ Zagursky, AJ Cocuzza, MA Jensen, K Baumeister. A system for rapid DNA sequencing with fluorescent chain-terminating dideoxynucleotides. Science 238:336–341, 1987.

4. S Tabor, CC Richardson. DNA-sequence analysis with a modified bacteriophage-T7 DNA-polymerase. Proc Natl Acad Sci USA 84:4767–4771, 1987.

5. S Tabor, CC Richardson. A single residue in DNA-polymerases of the *Escherichia-coli* DNA-polymerase-I family is critical for distinguishing between deoxyribonu-cleotides and dideoxyribonucleotides. Proc Natl Acad Sci USA 92:6339–6343, 1995.

6. JY Ju, CC Ruan, CW Fuller, AN Glazer, RA Mathies. Fluorescence energy-transfer dye-labeled primers for DNA sequencing and analysis. Proc Natl Acad Sci USA 92:4347–4351, 1995.

7. RA Mathies, XC Huang. Capillary array electrophoresis — an approach to high-speed, high-throughput DNA sequencing. Nature 359:167–169, 1992.

8. H Kambara, S Takahashi. Multiple-sheathflow capillary array DNA analyzer. Nature 361:565–566, 1993.

9. HJ Crabtree, SJ Bay, DF Lewis, JZ Zhang, LD Coulson, GA Fitzpatrick, SL Delinger, DJ Harrison, NJ Dovichi. Construction and evaluation of a capillary array DNA sequencer based on a micromachined sheath-flow cuvette. Electrophoresis 21:1329–1335, 2000.

10. T Sasaki. The rice genome project in Japan. Proc Natl Acad Sci USA 95:2027–2028, 1998.

11. MD Adams, SE Celniker, RA Holt, et al. The genome sequence of *Drosophila melanogaster*. Science 287:2185–2195, 2000.

12. P Dehal, P Predki, AS Olsen, A Kobayashi, P Folta, S Lucas, M Land, A Terry, CLE Zhou, S Rash, Q Zhang, L Gordon, J Kim, C Elkin, MJ Pollard, P Richardson, D Rokhsar, E Uberbacher, T Hawkins, E Branscomb, L Stubbs. Human chromosome 19 and related regions in mouse: Conservative and lineage-specific evolution. Science 293:104–111, 2001.

13. A Manz, DJ Harrison, EMJ Verpoorte, JC Fettinger, A Paulus, H Ludi, HM Widmer. Planar chips technology for miniaturization and integration of separation techniques into monitoring systems — capillary electrophoresis on a chip. J Chromatogr 593:253–258, 1992.

14. DJ Harrison, A Manz, ZH Fan, H Ludi, HM Widmer. Capillary electrophoresis and sample injection systems integrated on a planar glass chip. Anal Chem 64:1926–1932, 1992.

15. PC Simpson, AT Woolley, RA Mathies. Microfabrication technology for the production of capillary array electrophoresis chips. Biomed Microdevices 1:7–26, 1998.

16. HD Tan, ES Yeung. Integrated on-line system for DNA sequencing by capillary electrophoresis: From template to called bases. Anal Chem 69:664–674, 1997.

17. DJ Harrison, K Fluri, K Seiler, ZH Fan, CS Effenhauser, A Manz. Micromachining a miniaturized capillary electrophoresis-based chemical-analysis system on a chip. Science 261:895–897, 1993.

18. SC Jacobson, R Hergenroder, LB Koutny, RJ Warmack, JM Ramsey. Effects of injection schemes and column geometry on the performance of microchip electrophoresis devices. Anal Chem 66:1107–1113, 1994.

19. SC Jacobson, R Hergenroder, AW Moore, JM Ramsey. Precolumn reactions with electrophoretic analysis integrated on a microchip. Anal Chem 66:4127–4132, 1994.

20. K Seiler, ZHH Fan, K Fluri, DJ Harrison. Electroosmotic pumping and valveless control of fluid-flow within a manifold of capillaries on a glass chip. Anal Chem 66:3485–3491, 1994.

21. SC Jacobson, R Hergenroder, LB Koutny, JM Ramsey. Open-channel electrochromatography on a microchip. Anal Chem 66:2369–2373, 1994.

22. AT Woolley, RA Mathies. Ultra-high-speed DNA fragment separations using microfabricated capillary array electrophoresis chips. Proc Natl Acad Sci USA 91:11348–11352, 1994.

23. AT Woolley, RA Mathies. Ultra-high-speed DNA sequencing using capillary electrophoresis chips. Anal Chem 67:3676–3680, 1995.

24. CS Effenhauser, A Paulus, A Manz, HM Widmer. High-speed separation of antisense oligonucleotides on a micromachined capillary electrophoresis device. Anal Chem 66:2949–2953, 1994.

25. MC Ruiz-Martinez, J Berka, A Belenkii, F Foret, AW Miller, BL Karger. DNA-sequencing by capillary electrophoresis with replaceable linear polyacrylamide and laser-induced fluorescence detection. Anal Chem 65:2851–2858, 1993.

26. D Schmalzing, A Adourian, L Koutny, L Ziaugra, P Matsudaira, D Ehrlich. DNA sequencing on microfabricated electrophoretic devices. Anal Chem 70:2303–2310, 1998.

27. SR Liu, YN Shi, WW Ja, RA Mathies. Optimization of high-speed DNA sequencing on microfabricated capillary electrophoresis channels. Anal Chem 71:566–573, 1999.

28. D Schmalzing, N Tsao, L Koutny, D Chisholm, A Srivastava, A Adourian, L Linton, P McEwan, P Matsudaira, D Ehrlich. Toward real-world sequencing by microdevice electrophoresis. Genome Res 9:853–858, 1999.

29. W Goetzinger, L Kotler, E Carrilho, MC Ruiz-Martinez, O Salas-Solano, BL Karger. Characterization of high molecular mass linear polyacrylamide powder prepared by emulsion polymerization as a replaceable polymer matrix for DNA sequencing by capillary electrophoresis. Electrophoresis 19:242–248, 1998.

30. O Salas-Solano, D Schmalzing, L Koutny, S Buonocore, A Adourian, P Matsudaira, D Ehrlich. Optimization of high-performance DNA sequencing on short microfabricated electrophoretic devices. Anal Chem 72:3129–3137, 2000.

31. L Koutny, D Schmalzing, O Salas-Solano, S El-Difrawy, A Adourian, S Buonocore, K Abbey, P McEwan, P Matsudaira, D Ehrlich. Eight hundred base sequencing in a microfabricated electrophoretic device. Anal Chem 72:3388–3391, 2000.

32. AT Woolley, GF Sensabaugh, RA Mathies. High-speed DNA genotyping using micro-fabricated capillary array electrophoresis chips. Anal Chem 69:2181–2186, 1997.

33. PC Simpson, D Roach, AT Woolley, T Thorsen, R Johnston, GF Sensabaugh, RA Mathies. High-throughput genetic analysis using microfabricated 96-sample capillary array electrophoresis microplates. Proc Natl Acad Sci USA 95:2256–2261, 1998.

34. Y Shi, PC Simpson, JR Scherer, D Wexler, C Skibola, MT Smith, RA Mathies. Radial capillary array electrophoresis microplate and scanner for high-performance nucleic acid analysis. Anal Chem 71:5354–5361, 1999.

35. IL Medintz, CCR Lee, WW Wong, K Pirkola, D Sidransky, RA Mathies. Loss of heterozygosity assay for molecular detection of cancer using energy-transfer primers and capillary array electrophoresis. Genome Res 10:1211–1218, 2000.

36. CA Emrich, H Tian, IL Medintz, RA Mathies. Microfabricated 384-lane capillary array electrophoresis bioanalyzer for ultrahigh-throughput genetic analysis. Anal Chem 74:5076–5083, 2002.

37. SR Liu, HJ Ren, QF Gao, DJ Roach, RT Loder, TM Armstrong, QL Mao, I Blaga, DL Barker, SB Jovanovich. Automated parallel DNA sequencing on multiple channel microchips. Proc Natl Acad Sci USA 97:5369–5374, 2000.

38. C Backhouse, M Caamano, F Oaks, E Nordman, A Carrillo, B Johnson, S Bay. DNA sequencing in a monolithic microchannel device. Electrophoresis 21:150–156, 2000.

39. CT Culbertson, SC Jacobson, JM Ramsey. Dispersion sources for compact geometries on microchips. Anal Chem 70:3781–3789, 1998.

40. BM Paegel, LD Hutt, PC Simpson, RA Mathies. Turn geometry for minimizing band broadening in microfabricated capillary electrophoresis channels. Anal Chem 72:3030–3037, 2000.

41. SK Griffiths, RH Nilson. Low dispersion turns and junctions for microchannel systems. Anal Chem 73:272–278, 2001.

42. JI Molho, AE Herr, BP Mosier, JG Santiago, TW Kenny, RA Brennen, GB Gordon, B Mohammadi. Optimization of turn geometries for microchip electrophoresis. Anal Chem 73:1350–1360, 2001.

43. JR Scherer, BM Paegel, GJ Wedemayer, CA Emrich, J Lo, IL Medintz, RA Mathies. High-pressure gel loader for capillary array electrophoresis microchannel plates. BioTechniques 31:1150–1156, 2001.

44. BM Paegel, CA Emrich, GJ Wedemayer, JR Scherer, RA Mathies. High throughput DNA sequencing with a microfabricated 96-lane capillary array electrophoresis bioprocessor. Proc Natl Acad Sci USA 99:574–579, 2002.

45. RG Blazej, BM Paegel, RA Mathies. Polymorphism ratio sequencing: a new approach for single nucleotide polymorphism discovery and genotyping. Genome Res 13:287–293, 2003.

46. AT Woolley, D Hadley, P Landre, AJ deMello, RA Mathies, MA Northrup. Functional integration of PCR amplification and capillary electrophoresis in a microfabricated DNA analysis device. Anal Chem 68:4081–4086, 1996.

47. LC Waters, SC Jacobson, N Kroutchinina, J Khandurina, RS Foote, JM Ramsey. Multiple sample PCR amplification and electrophoretic analysis on a microchip. Anal Chem 70:5172–5176, 1998.

48. LC Waters, SC Jacobson, N Kroutchinina, J Khandurina, RS Foote, JM Ramsey. Microchip device for cell lysis, multiplex PCR amplification, and electrophoretic sizing. Anal Chem 70:158–162, 1998.

49. ET Lagally, PC Simpson, RA Mathies. Monolithic integrated microfluidic DNA amplification and capillary electrophoresis analysis system. Sens Actuator B-Chem 63:138–146, 2000.

50. ET Lagally, CA Emrich, RA Mathies. Fully integrated PCR-capillary electrophoresis microsystem for DNA analysis. Lab Chip 1:102–107, 2001.

51. BB Haab, RA Mathies. Single-molecule detection of DNA separations in microfabricated capillary electrophoresis chips employing focused molecular streams. Anal Chem 71:5137–5145, 1999.

52. MC Ruiz-Martinez, O Salas-Solano, E Carrilho, L Kotler, BL Karger. A sample purification method for rugged and high-performance DNA sequencing by capillary electrophoresis using replaceable polymer solutions. A. Development of the cleanup protocol. Anal Chem 70:1516–1527, 1998.

53. O Salas-Solano, MC Ruiz-Martinez, E Carrilho, L Kotler, BL Karger. A sample purification method for rugged and high-performance DNA sequencing by capillary electrophoresis using replaceable polymer solutions. B. Quantitative determination of the role of sample matrix components on sequencing analysis. Anal Chem 70:1528–1535, 1998.

54. RD Oleschuk, LL Shultz-Lockyear, YB Ning, DJ Harrison. Trapping of bead-based reagents within microfluidic systems: On-chip solid-phase extraction and electrochromatography. Anal Chem 72:585–590, 2000.

55. J Khandurina, SC Jacobson, LC Waters, RS Foote, JM Ramsey. Microfabricated porous membrane structure for sample concentration and electrophoretic analysis. Anal Chem 71:1815–1819, 1999.

56. Y Baba, M Tsuhako, T Sawa, M Akashi, E Yashima. Specific base recognition of oligodeoxynucleotides by capillary affinity gel electrophoresis using polyacrylamide poly(9-vinyladenine) conjugated gel. Anal Chem 64:1920–1925, 1992.

57. BM Paegel, SHI Yeung, RA Mathies. Microchip bioprocessor for integrated nano-volume sample purification and DNA sequencing. Anal Chem 74:5092–5098, 2002.

58. BM Paegel, RG Blazej, RA Mathies. Microfluidic devices for DNA sequencing sample preparation and electrophoresis analysis. Curr Opin Biotech 14:42–50, 2003.

5 Analysis of Nucleic Acids by Mass Spectrometry

Dirk van den Boom and Franz Hillenkamp

CONTENTS

INTRODUCTION

With the advent of the soft ionization techniques electrospray ionization (ESI) and matrix-assisted laser desorption/ionization (MALDI), mass spectrometry (MS) has developed over the last decade into a key analytical method for the analysis of biological macromolecules. ESI and MALDI mass spectrometers are today indispensable tools in all aspects of protein research and identification. For a detailed discussion of the techniques as such and their application to the analysis of peptides and proteins, the reader is referred to Reference 1. The routine application of these techniques for the analysis of nucleic acids is only emerging at this point in time and is still much less widespread than that of proteins. The main reason for this slow introduction is the relative lability of gas-phase ions of oligonucleotides, which limits the accessible size of these ions, particularly for MALDI-MS. The availability of several well-established alternative analytical techniques also contributes to the somewhat slow penetration.

The applications of mass spectrometry to the large variety of different fields in oligonucleotide research and analysis have been described in a recent review article, which also contains a list of 180 references.[2] The reader is directed to this review for a more complete coverage of the field, including such applications as identification of modified RNAs, DNAs, and of RNA–drug interaction, analysis of DNA–protein non-covalent complexes, to name a few. This chapter concentrates on the application of MALDI-MS to industrial-scale applications where automation and high throughput are the prime goals, requirements that are more difficult to meet by ESI and most of the mass analyzers commonly used with this ion source.

In this chapter "sequencing of DNA" is taken in a somewhat loose sense, encompassing resequencing, proofreading, detection of polymorphisms, and similar applications. At least at this point in time MS is not used for *de novo* sequencing in competition to gel-based sequencing.

THE CHALLENGE OF MALDI MASS SPECTROMETRY FOR OLIGONUCLEOTIDE ANALYSIS

A number of problems had to be solved before a routine successful MS analysis of oligonucleotides was established. First, careful sample purification is a prerequisite for MALDI and even more so for ESI-MS. Choice of the matrix and optimized preparation protocols for the samples as they are introduced into the mass spectrometer is a second, and last, but not least, molecular assays had to be developed, which take the specific requirements of the mass spectrometric analysis into account. The scientific basis for these requirements and the technical solutions are discussed in the first part of this chapter. The second part presents specific application examples.

SAMPLE PURIFICATION

Oligonucleotides as analytes in conjunction with the MALDI process introduce very specific requirements for a stringent sample purification. While all phosphate groups of the oligonucleotides are dissociated in solution, they get neutralized by a proton or any of the ubiquitous cations, such as Na^+ or K^+, upon incorporation into the solid MALDI matrix. Following desorption of the usually singly charged oligonucleotide ions, these heterogeneous salts lead to a multiplicity of signals in the spectrum, spaced by the mass of Na or K. The larger the sequence, the more extensive the mass heterogeneity. For small polymers, these different peaks are still resolved in the spectrum, but the total number of ions for a given oligomer is distributed over many peaks, which will decrease the signal-to-noise ratio and thereby limit the sensitivity. For larger oligomers the cluster of signals will not be resolved, which again limits sensitivity, but, more importantly, will prevent a correct mass assignment. Desalting of the analytes and the matrix is commonly achieved by adding ammonium salts such as di-ammoniumcitrate,[3] or by suitable treatment with ion exchange beads.[4] Upon desorption, ammonia is quantitatively lost from all phosphate groups leaving the free acid ion for detection. Reversed phase purification, commonly used in protein and peptide analysis (for example, with ZipTips), has not been similarly successful for oligonucleotides. The easy availability of negative charges on the backbone and the high proton affinity of the A, C, and G-bases make oligonucleotides amenable to an analysis in the positive as well as the negative ion mode, usually with comparable signal intensity. The choice of ion polarity depends somewhat on the choice of the matrix. For 3-hpa (see below) the positive ion mode is usually preferred because doubly charged ions are less abundant in the positive ion mode.

Successful UV-MALDI analysis with solid-state matrices requires an undisturbed crystallization of the sample on the target. Buffers, detergents, and other additives, such as glycerol, interfere with this crystallization even at low concentrations. Therefore, ammonium buffers should replace phosphate and similar buffers in the molecular biological procedures for sample generation. Detergents should be avoided, or non-ionic detergents should be used, if necessary. Proteins, such as polymerases, exo- or endonucleases or restriction enzymes, in the final sample can partially or fully suppress the oligonucleotide signals in the spectrum because of their higher proton affinity. Templates from polymerase chain reaction (PCR) or other enzymatic reactions may also cause problems. All these components need to be removed by precipitation or other suitable methods. Several companies market commercial purification kits. In so-called "homogeneous assays" (see below) a sample dilution may also decrease their concentration below a critical value.

Sample purification is even more critical in ESI-MS. A comparison of the efficacy of different purification procedures has recently been published by Null et al.[5]

SAMPLE PREPARATION FOR MALDI MASS SPECTROMETRY

The proper choice of matrix is of utmost importance for a successful MALDI-MS of oligonucleotides. 3-Hydroxypicolinicacid (3-hpa) has proved to be the matrix of choice for the UV-MALDI-MS of DNA at the commonly used 337 nm wavelength of the nitrogen laser or the 353 nm of the frequency tripled Nd:YAG laser, mainly because it induces the least fragmentation of analyte ions. Unless very special precautions are taken, only signals of the single strands are recorded in MALDI spectra. It is assumed that the acidity of this matrix leads to a separation of double-stranded species already in solution. Interestingly, acidic base hydrolysis is observed only rarely, provided that the solvent is evaporated off immediately after mixing analyte and matrix. Once in the solid state, the samples remain stable for prolonged times if kept in a strictly dry atmosphere.

Compared to the α-cyano-4-hydroxy-cinnamic acid matrix, frequently used for peptide analysis, 3-hpa has a less favorable crystallization behavior, forming rather heterogeneous samples with a pronounced statistical morphology and a separation of the analytes into so called sweet spots. Such sweet spots describe areas of a given sample, which turn out to give good spectra. The cause for this heterogeneity and why some areas give much better results than others are largely unknown. Fractionation of analytes into different areas may contribute, but is certainly not the only reason. The former limits mass accuracy in linear time-of-flight (lin-TOF) analyzers to typically a few mass units. The latter requires probing of a fair number of different areas of the sample in order to obtain spectra that truthfully reflect the analyte content and have a good signal-to-noise ratio. This is particularly worrisome in mixture analysis where a truthful representation of all components in the mixture is required (see "quantification" below). A very sizable improvement in the influence of this sample morphology has been achieved through the preparation ot microsample from a few nanoliters of matrix and analyte solutions, yielding samples only a few hundred micrometers in size;[6] the standard dried droplet method, in comparison, uses a few microliters of matrix and sample, resulting in millimeter size preparations. For high-throughput analysis such nanoliter samples are best prepared on chips a few centimeters in size, which accommodate 96 or 384 samples in a microtiter plate format.[7] These chips are surface-modified with hydrophilic plaques in a hydrophobic surrounding to reproducibly anchor the samples in predetermined locations, as has been described by Schürenberg et al.[8] for microtiterplate size sample plates. Several other matrices have been tested for the analysis of oligonucleotides with mixed success. Exception may be a mixture of 2,3,4-, 2,4,6-trihydroxyacetophenone (THAP) (both 0.2 M) and diammoniumcitrate (0.3 M) in a 1:1:2 ratio (v:v) for the analysis of RNA-oligomers and 6-aza-2-thiothymine (ATT) for the detection of ds-DNA[9] and of DNA/protein non-covalent complexes in research applications.

ION FRAGMENTATION

Quite in contrast to the situation in a physiological solution, ions of DNA in the gas phase are much more prone to fragmentation than proteins/peptides. This abundant fragmentation severely limits the accessible mass range and size of analytes at least in UV-MALDI-MS. Routine application of MALDI-MS for oligonucleotide analysis has therefore required and induced the development of suitable molecular assays, which allow us to obtain the information of interest from oligomers of limited size, such as the analytes described below. High-throughput analysis is currently limited to oligomers of less than approximately 50 bases in length. Fragmentation is another reason, which has so far prevented applications of MALDI-MS to *de novo* sequencing by the Sanger or similar approaches.

The cause for this lability of DNA has been discussed controversially in the literature over the years. Two recent publications showed by H/D-exchange, post-source-decay (PSD) experiments that the fragmentation is initiated by facile loss of any of the high-proton-affinity A, C, and G bases, induced by a weakening of the N-glycosidic bond after protonation of the base.[10,11] Surprisingly, this is even the case in negative ion mode where the base protonation results from a zwitterion formation of these bases with their 5′-neighboring phosphate groups. Following base loss, the DNA-strand becomes destabilized, which results in a series of zipper-type fragmentations of the backbone and a total loss of information on the original strand. Whereas base losses are also observed "prompt," i.e., on a timescale below ~100 ns, most of the following fragmentation occurs metastably on a micro- to millisecond timescale.

Many attempts have been reported to limit the fragmentation by chemical modifications. These modifications have involved the bases (e.g., 7-deaza-A and G), the ribose (e.g., 2′-F) and the phosphate backbone. None of these modifications has found large-scale applications mostly because they all more or less strongly compromise the molecular biology of the common sample generation procedures. The only exception is alkylation of the backbone as the very last step of sample generation, which is usually combined with charge tagging of the oligomers.[12] Although very successful in stabilizing the ions, a reproducibly complete alkylation is restricted to oligomers of at most 15 bases in length.

An interesting aspect of the fragmentation mechanism is the inherently higher stability of RNA relative to DNA: the 2'-OH group stabilizes the N-glycosidic bond and thereby essentially prevents base loss. Recently, assays comprising transcription and analysis of RNA strands have been reported[13] (see below).

IR-MALDI with infrared wavelengths in the 3-μm region exhibits a dramatically reduced fragmentation. Accordingly, ions of single-stranded restriction enzyme fragments up to 2 kb in length have been recorded with a glycerol matrix.[14] Intense matrix adduct formation unfortunately limits the mass resolution in these spectra to values of 100 or less, which has, so far, prevented routine application of this technique. ESI is also inherently softer than MALDI. Highly charged ions of plasmid and phage DNA of more than 10 kb have been generated by ESI. Exact determination of the charge state is difficult in these cases, limiting the accuracy of mass determination to less than ±10%. This work is reviewed in Reference 2.

INSTRUMENTATION

Fragmentation also strongly influences the choice of mass analyzers suitable for DNA analysis. Lin-TOF analyzers are almost exclusively used in conjunction with UV-MALDI sources. Ions decaying in the field free drift tube after the ~1 μs time of (delayed extraction) acceleration in the source will continue to travel at essentially the same speed as the parent ion and arrive at the detector concurrently with it, thus contributing to the signal of interest. Mass resolutions of 500 to 1000 are routinely achieved in such instruments. Reflectron-TOF (ref-TOF) analyzers are not suitable, because of the mass dispersion of the fragments in the reflector field, except for the identification of specific fragments by PSD analysis in a research mode. All other standard mass analyzers such as ion traps (IT) and Fourier-transform ion-cyclotron-resonance (FT-ICR) analyzers require milliseconds to seconds for the analysis, which interfere with the ion lifetimes. Recently, hybrid quadrupole-TOF analyzers with an orthogonal ion extraction into the ref-TOF analyzer have been marketed by several companies (Q-TOF: Micromass Ltd, Manchester, U.K. or Q-star: MDS Sciex, Inc., Toronto, Canada). These instruments offer the option of injecting the MALDI ions into a region of elevated pressure at typically 1 to 100 mbar, where they become thermalized by collisions with the neutral gas. This thermalization dramatically reduces metastable fragmentation and permits the analysis even of oligonucleotide ions in the following ref-TOF. The first results of the analysis of DNA fragments with such instruments were very promising.[15] A combination of such thermalization MALDI-ion sources with ion traps and FT-ICR also holds some promise and has been successfully tested for other analytes of similar lability.[16]

FT-ICR analyzers have been used very successfully for the analysis of oligonucleotides in combination with ESI sources. The very high mass resolution of these analyzers permits the unambiguous identification of the signals of different charge state even for rather complex mixtures of analytes in a given sample. A good example for such an application has recently been described by Null et al.[17] The stringent sample purification requirements and the complexity and analytical speed of FT-ICR instruments somewhat limit this approach in routine and high-throughput analysis of oligonucleotides.

QUANTIFICATION AND MIXTURE ANALYSIS

Absolute quantification is not an option for most MS applications and even relative quantification requires great care. Fortunately, even oligonucleotides of different mixed sequence desorb and ionize with at least comparable efficiencies in a MALDI source, except for a general decrease of signal intensity with increasing size of the oligomer. This decrease can be attributed to the decreasing yield of ion formation as well as to the increasing fragmentation. Signals of oligonucleotides of nearby masses in the same spectrum can, therefore, be compared quantitatively with a precision of a few percent. Algorithms, which account for the decrease of signal with size and the minor influence of the base content, can also be developed. The main limitation for quantification is the dispersion and disproportionation of analytes in the matrix preparation discussed above. Quantification of 3-hpa UV-MALDI spectra, therefore, strictly requires accumulation of a large enough number of spectra from different locations of a given preparation. Improvements in sample preparation toward more homogeneous samples would be highly desirable for analyses requiring quantification.

Sample disproportionation can also be a problem in mixture analysis. Again, spectra obtained from one given location of the preparation may not be representative for the full content of analytes in the sample. This can be particularly worrisome for the analysis of primer extension products for the analysis of single nucleotide polymorphisms (SNPs), where a heterozygote may be classified as homozygous because of analyte disproportionation. Analysis of mixtures with a very large number of different oligonucleotides such as would be encountered in a Sanger ladder or an exo- or endonuclease digest may give rise to another problem. It is generally assumed that the total charge available for the formation of ions in a given desorption event is limited and largely independent of the number of different analytes in the mixture. If this charge is distributed among an increasing number of ions of different mass, the signal-to-noise ratio of each single signal will decrease. The limited dynamic range of typically 8 bit of the ion detection may also introduce a limitation, if signals of a much wider range of intensities occur in the spectrum.

ROLE OF MALDI-TOF-MS IN HIGH-THROUGHPUT SEQUENCE ANALYSIS APPLICATIONS

Efficient sequence analysis of whole genomes of individual species and organisms as well as efficient sequence comparison of the information contained therein is still one of the major tasks to advance our understanding in fields such as biology, medicine, or evolution. The fulfillment of these tasks demanded technological developments in the field of DNA sequence analysis: first, methods for large-scale *de novo* sequencing were (and still are) required, which allow for timely and cost-effective analysis of megabase to gigabase DNA stretches (generation of reference sequences); second, methods for large-scale comparative sequence analysis, which allow for fast but accurate elucidation of inter- and intraspecies genetic variation and heterogeneity (exploration of genetic markers); and third, methods that can efficiently score very large numbers of genetic markers in selected populations to determine genotype-phenotype correlations.

MALDI-TOF-MS–Based DNA Sequencing and Resequencing

In the early phase of the Human Genome Project, MS, and MALDI-TOF-MS in particular, received attention as an alternative method for separation and detection of Sanger sequencing ladders.[18] The basic underlying idea was to improve the speed and accuracy of the sequence analysis process. MALDI-TOF-MS offered an unparalleled speed in signal acquisition (microseconds in TOF systems opposed to hours for conventional gel-electrophoresis-based systems), a higher accuracy due to analysis of an intrinsic molecule property, the molecular mass, and the potential for a high degree of automation.

Several biochemical schemes had then been developed, which generated DNA sequencing ladders of sufficient yield and purity to suit the specific requirements for the analysis by MALDI-TOF-MS.[19–22] Following the concept of conventional dideoxy sequencing, the nested set of truncated sequences originating from a primer can, in principle, be analyzed by MALDI-TOF-MS and the mass difference between the DNA fragments can be used to calculate the nucleotide sequence. However, because of the nearly exponential decay in sensitivity of MALDI-TOF-MS with increasing mass of the DNA fragments, as described in earlier paragraphs of this chapter, the read length of this approach was rather limited and hardly competitive with fluorescence-based dideoxy sequencing. Despite very promising results for solid-phase-based sequencing and cycle sequencing, the 100 bp barrier was never overcome on a routine basis. In addition to sensitivity issues limiting the overall read length, the mass resolution of conventional axial-TOF instruments was in some cases insufficient for very accurate sequence determination. Insufficient discrimination between polymerase pausing signals generated by secondary structures of the template and "real" termination signals, moreover, significantly hamper sequence analysis in the higher mass range. Sensitivity as well as mass resolution and mass accuracy issues in the higher mass range have all contributed to the fact that analysis of dideoxy sequencing ladders by MALDI-TOF-MS has not yet been implemented in high-throughput sequencing applications.

For all these reasons, early advances in the use of MS for the analysis of oligonucleotides have, therefore, occurred for the analysis of single nucleotide polymorphisms rather than sequencing. These applications are discussed further below.

Recently, however, a new scheme for sequence analysis by MALDI-TOF-MS has been introduced.[13] Rather than using a primer extension–based method, which yields a ladder of DNA fragments with increasing sizes starting from the primer length and thus rapidly interferes with the current capabilities of MALDI-MS, this scheme relies on the generation of rather short base-specific fragments from a given nucleic acid amplificate, in principle similar to peptide mapping in protein identification. It uses base-specific RNases for template digestion, followed by an analysis of the resulting cleavage products by MS.[23]

Base-specific cleavage of nucleic acids represents a paradigm shift in sequencing by MS. The principle resembles more closely the original approach of Maxam and Gilbert for DNA sequencing.[24] It is, however, not a *de novo* sequencing method and rather represents identification or resequencing, where an experimentally determined sequence is cross-compared to a known reference sequence.

In base-specific cleavage, a single-stranded copy of a target sequence is generated and cleaved to completion in four separate base-specific reactions. Each reaction reduces the original sequence to a set of oligonucleotides, which is readily separated and analyzed by MALDI-TOF-MS. To each mass signal of a base-specific cleavage reaction, a set of compomers can be assigned. The sequence can be reconstructed from this set of compomers by combining the information of all four cleavage reactions and comparing it to a predicted set of mass signals as provided for an *in silico* digest of the reference sequence. Sequence changes between the reference sequence and the experimentally analyzed sample sequence have a profound impact on the mass signal pattern. A heterozygous sequence change can generate up to five observations in the mass signal patterns: it can add or remove a cleavage site in one or more cleavage reactions as well as shift the mass of single products by the mass difference between exchanged nucleotides. For heterozygous samples, additional mass signals appear in the spectra. A homozygous sequence change might even provide up to 10 observations, because not only additional but also missing signals can be used to detect, identify, and locate the sequence change.

Figure 5.1 illustrates how the principle would relate to the detection of a G-C sequence change in a target region. Assuming that the region of interest is analyzed after a C- and a T-specific cleavage of both, the forward and the reverse strand (equivalent to four base-specific cleavages), the sequence change from G (as the reference nucleotide) to C would generate five additional mass signals for a heterozygous sample. In the T-specific cleavage (forward reaction), one oligonucleotide resulting from the cleavage would shift by 40 Da (as the mass difference between C and G). In C-specific cleavage (forward), the sequence change would generate a new cleavage site and would thus split the original 8-mer fragment into a 3-mer and a 5-mer fragment. Supporting information would then be generated in the analysis from the reverse strand. Here again, the T-specific cleavage would show one fragment shifted by the mass difference between C and G, and for the C-specific reaction, a cleavage site would be removed, generating a 12-mer fragment instead of the original 10-mer fragment. In most cases, the combined observations of the four reactions allows for an unambiguous identification and localization of the sequence change.

This scheme has the limitation that some cleavage information is lost in either the low or the high mass range or by overlapping cleavage products. With increasing length of the target region (from 200 bp upward to 1 kb), the mono-, di-, tri-, and even sometimes tetra-nucleotides are completely non-informative due to many fragments of nominally equal mass. Additionally, their detection is often hindered by strong matrix signals in the low mass range. Longer target sequence regions increase the total number of cleavage products and thereby cause more cleavage products to overlap in mass. This reduces the number of possible observations based on either additional or missing signals. Thus, the ability to locate a sequence change exactly is a more likely limitation than sensitivity and mass accuracy for long fragments,

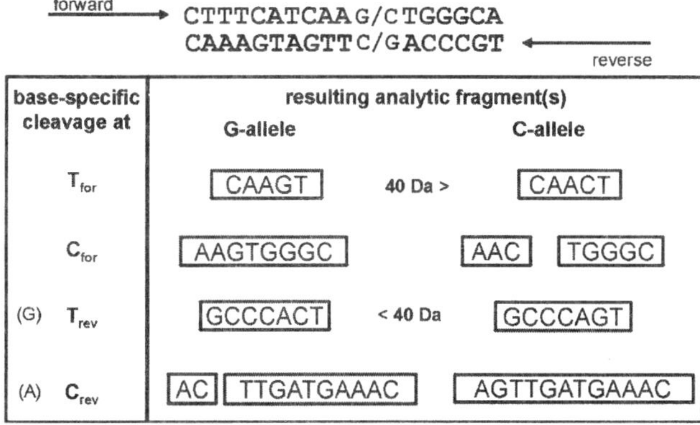

FIGURE 5.1 Allele-specific mass signal patterns generated by base-specific cleavage. Displayed is a hypothetical sequence carrying a G-C polymorphism. Assuming two transcription reactions of the forward and two from the reverse strand, both followed by either T- or C-specific cleavage, the resulting cleavage pattern shows distinct changes. The T-specific cleavage of the forward transcript yields a new fragment with 40-Da mass difference to the reference fragment when a G-C substitution is present. The substitution does not change the cleavage sites, but leads to fragments with the same length, but different composition; hence, a mass difference can be observed. In the C-specific cleavage, the same substitution introduces a new cleavage site and thus yields two new fragments of 3- and 5-nucleotide length, respectively, as opposed to the single 8-mer fragment generated by the reference. Supplementary information is generated in the reverse transcription and cleavage reactions. Again, the T-specific cleavage leads to a new fragment of same length but different composition. The compositional difference between C and G yields a 40-Da mass difference. The substitution removes a cleavage site for C-specific cleavage and generates a new fragment that is two nucleotides longer compared to the reference cleavage pattern. Combining the information contained in the four cleavage reactions, five additional mass signals can be observed in a sample heterozygous for the respective sequence change.

as encountered in the mass spectrometry of sequence ladders. An initial simulation of arbitrarily chosen 500 bp amplicons in the human genome revealed that about 90% of all theoretically possible sequence changes could be detected, characterized, and localized. An additional 10% can still be detected and characterized, leaving a fraction of nondetectable sequence changes below 1% (P. Stanssens, Methexis Genomics, personal communication, 2002). A way to further improve these numbers, especially for even longer amplicons, would be the additional evaluation of the (properly normalized) signal intensities or peak areas. This approach would enlarge the observations based on additional and missing signals by supporting observations based on significant changes in single peak intensities. The success of such an approach will largely depend on the reproducibility of overall signal-to-noise ratios between spectra and between samples, a challenge that is closely related to sample preparation and analyte homogeneity.

Several approaches have been developed that allow for the generation of such base-specific cleavage patterns.[25–27] Among these, transcription of the PCR amplicons into RNA prior to base-specific cleavage appears most promising, because the transcription process further amplifies the number of molecules available for mass spectrometric analysis and generates single-stranded templates, thereby eliminating complicated means of generating single-stranded DNA molecules from the PCR product. Moreover, RNA is more stable under MALDI-TOF-MS conditions, as discussed above.

The technique for a transcription-based approach is illustrated in Figure 5.2. PCR is performed with a primer pair carrying two different promoters at their 5′ positions. Following PCR and deactivation of the deoxy-NTPs by dephosphorylation, RNA polymerase and ribonucleotides are added to the reaction mixture. The RNA transcript is then cleaved by a base-specific RNase, and the resulting cleavage products are purified and conditioned for subsequent analysis by MALDI-TOF-MS. Typical mass spectra obtained from forward and reverse transcriptions of a 400-bp amplicon followed by G-specific cleavage with RNase T1 are depicted in Figure 5.3.

The list of potential applications for base-specific cleavage followed by mass spectrometric analysis is widespread: it ranges from pathogen-typing and SNP discovery to the analysis of methylation patterns or mutation screening. Even the

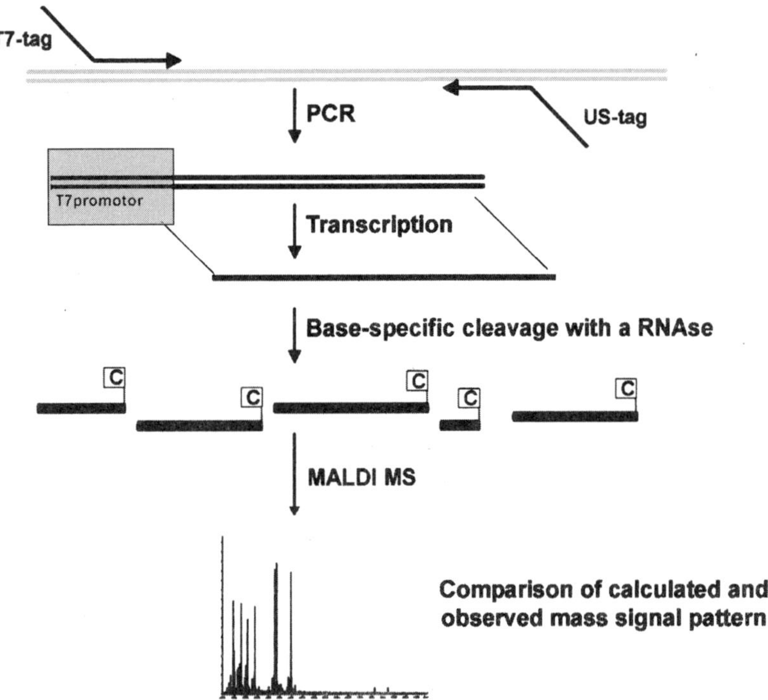

FIGURE 5.2 Process flow for base-specific cleavage by RNA transcription. The target region is amplified with PCR primer carrying promoter tags. The PCR product is subsequently transcribed into a single-stranded RNA molecule, which then is cleaved base specifically by an RNase. The resulting cleavage pattern is analyzed by MALDI-TOF-MS. Comparison of an *in silico* reference sequence-derived cleavage pattern with the experimental data allows for sequence validation and detection of sequence changes.

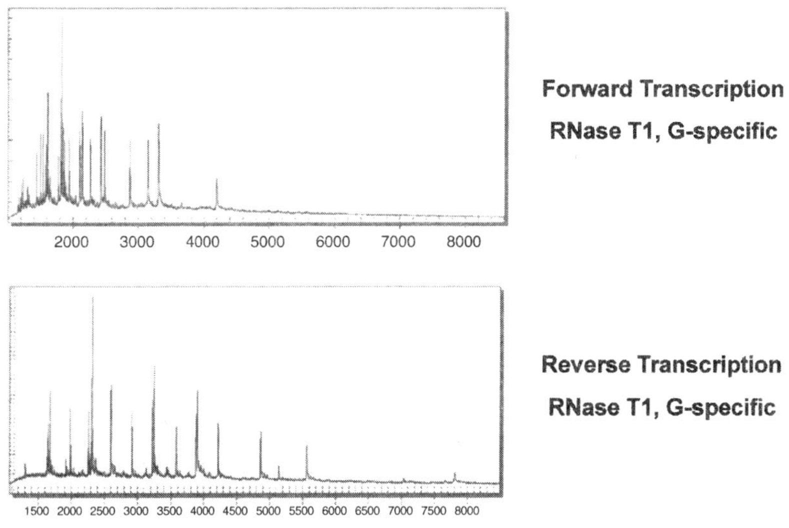

FIGURE 5.3 Representative MALDI-TOF-MS spectra of G-specific cleavage patterns generated by RNase T1 treatment of a 500-mer transcript. Both forward and reverse transcription was performed. The analyzed mass window usually ranges from 1000 to 9000 Da. Signals below 1000 Da are suppressed to minimize interfering matrix signals. Mass signals over 9000 Da usually have insufficient signal-to-noise ratios to contribute faithfully to the analysis.

large-scale characterization of cDNAs and their alternative splice variants might be approached by base-specific cleavage.

All these applications comprise large-scale comparative sequence analysis. They are a next cornerstone in the attempt to further elucidate the genetic code and its individual variations.

As shown again later, the advantage of MALDI-TOF-MS in comparative sequence analysis rests on two important features: speed of signal acquisition and accuracy. The real-time spectra acquisition and analysis of a single sample is today routinely achieved in about 1 s when chip arrays are used as launching pads. This holds for base-specific cleavage patterns as well. Even if more than just one sum spectrum is acquired for a given sample, the acquisition can be completed in about 5 s at a laser pulse repetition rate of 20 Hz. Considering four base-specific cleavage reactions and an average amplicon length of 500 bp (4 × 5 s = 20 s/500 bp), a single MALDI-TOF mass spectrometer can easily scan 1 million to 2 million base pairs per day. This compares favorably with state-of-the-art sequencing equipment.

Automated analysis of this amount of data is a challenging task, especially when single base changes have to be detected with high reliability. The combination of MS with base-specific cleavage offers some significant advantages in that respect: redundancy of information and, thereby, the reliability of the result is strongly enhanced, because the identification of a polymorphism or mutation is based on one or multiple observations, which occur at separate and distinct positions of one or more mass spectra. This is in strong contrast to standard fluorescence-based Sanger sequencing, where sequence changes have to be identified by multiple bands/colors occurring at the exact same position of the electropherogram. Moreover, the measured fragment mass is an inherent property of the analyte, in contrast to the fluorescence of tagged nucleotides.

MALDI-TOF-MS–Based SNP Scoring and Mutation Analysis

The continuing progress of genome projects[28,29] has provided the basis for identification of a very large and still increasing set of DNA markers. DNA markers are stretches of inheritable polymorphic nucleotide sequence. They proved useful in assessing inter- and intraspecies specific variations and help to understand the genetic contributions to phenotypic expression of an organism as well as its differences. DNA markers are widely used in a diverse set of applications, which include criminal suspect identification, linkage analysis, pharmacogenomics, or routine clinical diagnostics of mutations, suspected or known to cause a given disease, to name just a few.

Genetic marker analysis evolved through different stages. It first made use of restriction fragment length polymorphisms (RFLPs), then microsatellites (short tandem repeats, or STRs) and, most recently, single nucleotide polymorphisms (SNPs). Prevalence for the use of one or the other type of marker in genetic mapping and other applications was, and still is, largely a question of their availability, information content, and the availability of a suitable technology.

SNPs, the "youngest" member of the family of genetic markers, occur with a high frequency in most genomes and thus have a tremendous impact on the generation of high-resolution genetic maps. Studies performed on human genomic samples lead to an estimate of a frequency of about 1 SNP for every 1000 bp, with even higher frequencies suggested for certain gene regions.[30] Diversity in other organisms, like plants, is expected to be five to seven times larger.[31] Because of the simplicity of their bi-allelic nature, SNPs offer a high potential for process automation and fully automated data analysis, a key aspect when entering high-throughput genetic analysis projects.

The efforts to explore the genetic basis of complex inherited diseases or disease predisposition have generated an increasing demand for high-throughput marker analysis.[32] Many further research areas, like plant and animal genomics, require the large-scale identification of quantitative trait loci for molecular breeding.[33] These efforts have in common that candidate or target genes/gene regions need be identified. To accomplish this task, whole genome association or linkage disequilibria studies in large populations have to be performed. The approaches use either large STR or SNP marker sets or a combination of both.[34–36] Upon identification of candidate gene regions, further validation using a particular subset of even more densely spaced markers in affected and unaffected individual samples may be required, especially because functional allelic variants need to be explored to understand genotype-to-phenotype correlations.

To illustrate the experimental workload of such approaches, consider the scoring of about 300,000 genome-wide distributed SNPs to deduce the association of particular genomic regions in a polygenic disease. For a statistically significant association, about 1000 individuals need to be analyzed for the complete set of

300,000 SNPs, totaling to 3×10^8 data points to be acquired. Projects of this size can only be performed with reasonable time and cost using technologies with a high enough rate of data point acquisition at low cost. High reliability and accuracy of the technology are of prime importance. Both are required to avoid any manual interference for data interpretation, a crucial factor in high-throughput applications. Furthermore, both influence the required sample size, often a decisive limitation, where the number of diseased individuals is limited.

Many different platforms have been developed promising to fulfill the needs of current and future SNP scoring. Among them, MALDI-TOF-MS–based systems have gained significant attention due to their accuracy, speed, degree of automation, and their low cost per assay. Considering for example the estimated 3 to 5 M SNPs in the human genome, assay design, flexibility, and ease of assay implementation become further important aspects for a sustainable typing platform.

Initially, MALDI-TOF-MS–based DNA analysis focused on the detection of PCR products associated with disease identification.[37] The analysis of mutations in the cystic fibrosis gene is one such example, where the molecular mass of the detected PCR product indicates the presence or deletion of codon 508. The identification of mutations by MALDI-TOF-MS was also exemplified using restriction endonuclease digests,[38] ligase chain reaction, and hybridization with PNA probes.[39,40]

All these experiments proved the general feasibility of MS in DNA analysis, but they required skillful scientists to acquire the spectra in a nonroutine environment; time-consuming hunting for so-called sweet spots of MALDI matrix preparations was necessary to obtain data with good signal-to-noise ratio and high resolution. Most assays and assay formats required optimization to obtain a minimum signal-to-noise ratio and mass resolution. Additionally, sample preparation techniques were not applied in medium- to high-throughput mode.

The development of generic methods for the analysis of polymorphisms and mutations was another cornerstone in the development of MALDI-TOF-MS to a high-throughput genotyping platform. Primer extension methods are such a generic scheme, which produce diagnostic DNA fragments of suitable length for analysis by MALDI-TOF-MS. The general principle is depicted in Figure 5.4. The target region carrying the polymorphism or mutation is first amplified by PCR. In a subsequent reaction, a primer is annealed adjacent to the polymorphic site and is terminated allele-specifically upon addition of a DNA polymerase and a suitable nucleotide mix. Termination products are analyzed by MALDI-TOF-MS and the obtained molecular masses are used to assign the corresponding genotype information.

Several modifications of this basic scheme have been described in the literature. It had been introduced originally as the primer oligo base extension (PROBE) assay.[41] PROBE was a solid-phase assay. Amplification was performed with one biotinylated primer, which allowed for an immobilization of the PCR product on streptavidin-coated solid support. Following denaturation, the detection primer was annealed directly adjacent to the polymorphic region. Extension is initiated by addition of a DNA polymerase and a nucleotide mix, with at least one natural nucleotide (dNTP) replaced by its corresponding terminator (ddNTP). As depicted in Figure 5.4, this leads to an allele-specific termination and the termination products differ in mass by at least one nucleotide.

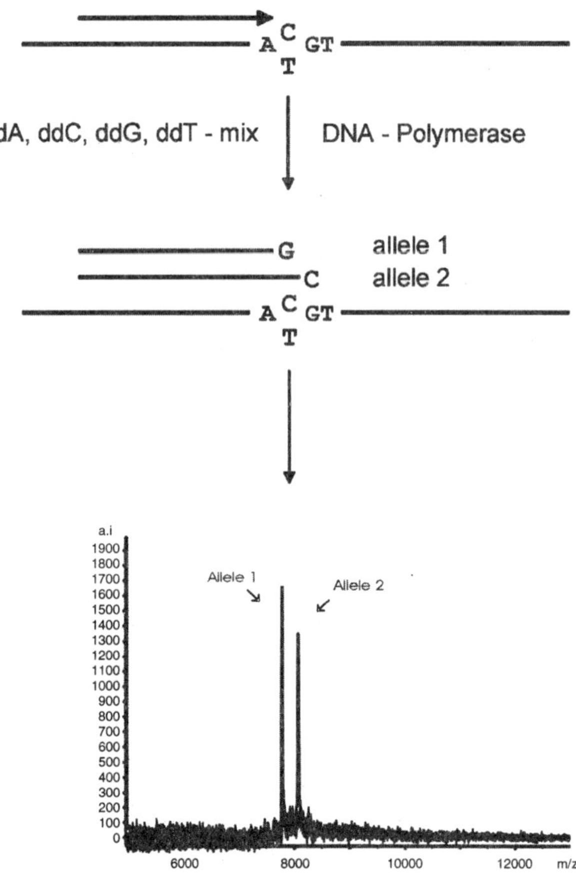

FIGURE 5.4 Principle of primer extension–based analysis of SNPs. Following amplification, a primer is annealed adjacent to the polymorphism. The extension is terminated allele-specific. The products are analyzed by MALDI-TOF-MS and unambiguous assignment of the alleles is performed based on the molecular weight information.

With a careful choice of the termination mix, this reaction allows identification of virtually any kind of single mutations or polymorphisms. In addition, polymerase pausing artifacts, often occurring in complicated sequence contexts, can be discriminated from real alleles. Pausing events lead to primer extension products not terminated by incorporation of a dideoxynucleotide. Thus, the corresponding products show mass signals of at least 16 Da difference to the allele specific products. Again, careful design of the assay, using, for example, three terminating nucleotides (ddNTPs), increases the mass difference between allele-specific termination products and unwanted polymerase artifacts and avoids misinterpretation of mass signals even in routine high-throughput environments.

The use of paramagnetic particles in high-throughput processes soon became a significant bottleneck. It required complicated pipetting and handling steps, hindered miniaturization, and added significant costs. Thus, alternative approaches were introduced.

The PinPoint™ assay, for example, used reversed-phase columns in the form of ZipTips (Millipore) for the purification of extension products.[42] ZipTips, however, required frequent replacement and showed a tendency to become clogged, a reason this purification approach did not comply with high-throughput processing.

PinPoint also employed a nucleotide mix devoid of any elongators (dNTPs). The primer is only extended by one nucleotide and alleles are discriminated by the value of the mass difference between the four ddNTPs. This assay type increases the flexibility for the design of multiplexed assays, but also imposes the risk of poor discrimination between A and T alleles (only 9 Da mass difference) and between adduct signals (for example, sodium and potassium: 23 and 39 Da, respectively) and "real" alleles (mass difference T − G is 25 Da and C − G is 40 Da, respectively).

A different approach was introduced with the GOOD assay.[43] In contrast to the other two methods, this approach did not require any purification steps prior to mass spectrometric analysis. The extension reaction is performed with an oligonucleotide primer carrying α-S-dNMPs at the 3′ end and in the presence of α-S-ddNTPs. After the extension reaction, the unmodified 5′ end of the primer is removed by degradation with phosphodiesterase II. The remaining extended oligonucleotide is backbone-alkylated with methyliodide, to suppress the otherwise abundant formation of multiple salts of monovalent and divalent cations and limit ion fragmentation, known to be initiated by zwitterion formation of the bases with their neighboring 5′-phosphate group. Chemical charge-tagging of this molecule with either a single excess positive or negative charge allows the analysis of positive or negative ions from the crude solution with high sensitivity.

Although this approach avoids purification steps, it increases the complexity of the genotyping process by adding several more reaction layers. A major drawback is also the use of the toxic methyliodide. To address these issues, a modification of the original protocol has been introduced recently, which makes use of methyl-phosphonate primers.[44] The methyl groups are already introduced at the stage of primer synthesis by the use of respective phosphoamidites so that no further post-extension modification except the phosphodiesterase treatment is required. Commonly used DNA polymerases did not extend methylphosphonate primer efficiently, but the group also introduced a new DNA polymerase, which seems to eliminate this issue.

The accuracy and efficiency of mass spectrometric separation and detection of primer extension products have recently been commercialized as an integrated system for SNP analysis.[45] Original solid-phase formats have been replaced by homogeneous single-tube assays, which are combined with the simple addition of ion-exchange resin beads for conditioning of the extension products prior to mass spectrometric analysis. To fulfill the demands of high-throughput genotyping, parallel processing in 384-well formats has been combined with miniaturized sample preparation on chip arrays as launching pads for the mass spectrometry.

Advanced nanoliquid handling based on piezoelectric pipettes or pin tools allows the preparation of as little as 15 nl matrix or sample on surface-modified silicon chips. MALDI-TOF-MS benefits from this miniaturization in that crystal formation is much more homogeneous, allowing fully automated scanning with only a few laser shots per sample. Currently, as many as 3840 samples can be analyzed in unattended mode, with an average data acquisition time of only 1.5 s per sample. This allows a single mass spectrometer to acquire up to 50,000 mass spectra per day.

Further increases in throughput can be achieved by multiplexing of the SNP assay. The analysis of a single primer extension reaction requires only a fraction of the accessible mass window. Figure 5.5 depicts a mass spectrum of a tetraplexed primer extension assay, which shows that intercalation of the extension products allows a more efficient use of the available mass window. Through intelligent assay design, even higher multitudes of primer extension reactions can be combined in a single reaction or detection. Multiplexing of as many as 12 SNPs assays has already been described,[46] although the current level of generically developed multiplexes is in the range of 6- to 8-plexes. The main factor influencing this number is the ability to perform multiplexed amplification of the target loci in the initial PCR without major deviations in PCR yield between the respective products.

The use of a genotyping platform in high-throughput environments also depends strongly on the bioinformatics capabilities. Large numbers of data sets must be automatically translated into genotype information without requirement for manual interference. Also, the assay design process and any required assay validation procedures need to be very efficient. Mass spectrometry meets both these requirements extremely well in that this technology determines directly an intrinsic molecule property, the molecular mass. Each of the primer extension products can only consist of the four natural dNTPs (dATP, dCTP, dGTP, and dTTP) as well as one or more of the terminators of the primer extension reactions (ddATP, ddCTP, ddGTP, and ddTTP).

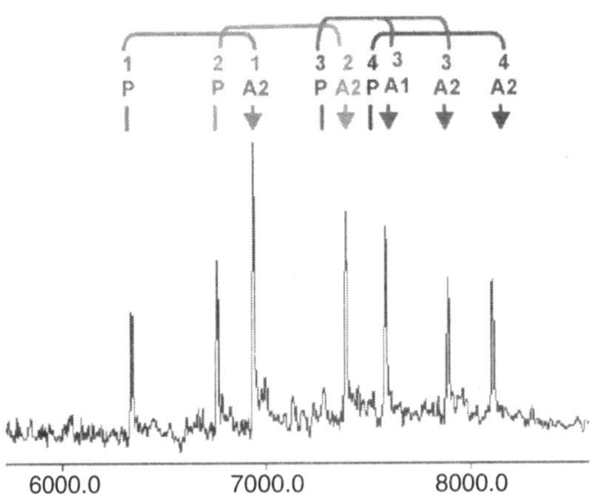

FIGURE 5.5 Multiplexed analysis of SNPs. Depicted is a MALDI-TOF mass spectrum of a 4-plex MassEXTEND reaction. The four sets of primer and primer extension products are marked with a number. 1P to 4P represent the positions of unextended primer. Primer extension efficiencies vary with the sequence context and length. In some cases no unextended primer is detectable (4P). Corresponding alleles are marked with an A. The respective DNA is homozygous for assays 1, 2, and 4 and heterozygous for assay 3.

Thus, any possible product can be precalculated and there is no need for complex algorithms to analyze the molecular mass information. Given the use of a suitable nucleotide mix, polymerase artifacts like pausing can be discriminated from "real" termination products by their molecular mass.

These features have been combined in a computer-aided assay design tool, which allows the generation of large numbers of primer extension reactions for detection by MS in high-throughput mode.

An interesting new application for MALDI-TOF-MS is the determination of allele frequencies by means of pooled DNA samples. Rather than analyzing the genotype of individual samples to determine the allele frequency in a given population, this approach uses DNA pools, which comprise equimolar amounts of up to several hundred individual DNAs. The DNA pool is subjected to amplification. Then the described primer extension reaction is performed. A careful quantitative analysis of the relative peak areas of the two alleles in the mass spectra allows us to estimate the respective allele frequency in the sample pool.

Allele frequency data generated this way can be an important feature for current aims of high-throughput SNP scoring, i.e., large-scale SNP validation. Before use of any *in silico* identified SNP (isSNP) in genome screens or candidate gene approaches, there has to be some sort of validation for this potential marker with respect to the phenotype of interest: Is the isSNP a relevant polymorphism, or is it a sequencing artifact? In this respect, validation, for example, can be performed in ethnic pools. In addition, the use of pools could provide a way to filter large sets of SNPs for those candidates worthy of follow-up, without carrying the burden of individual genotyping for all these markers.[47–49]

Combining the described features of MS in automated assay design and high-throughput SNP scoring with the concept of sample pooling has recently been shown to allow the large-scale implementation of several thousand SNP assays within a period of only 1 month.[50]

CONCLUSIONS AND FUTURE PROSPECTS

MALDI-TOF-MS has become a vital technology in current and emerging bioanalytical approaches. It plays a pivotal role in protein and carbohydrate analysis. Application of MALDI-TOF-MS to analysis of nucleic acids has also increased vastly in recent years. MALDI-TOF-MS combines core features, such as the speed of signal acquisition, determination of an analyte-specific inherent physical property, the molecular mass, a high accuracy, and a high degree of automation. These features enabled MALDI-TOF-MS to capture a leading position in the high-throughput analysis of genetic variations. As shown, the predominant application for MALDI-TOF-MS of nucleic acids is the analysis of SNPs and mutations in large-scale genetic studies. The determination of allele frequencies in DNA pools has been introduced recently and has found widespread use. Although the quantitative nature of MALDI has been a topic of debate in protein analysis, it was found that the area-under-the-curve ratios of allele-specific primer extension products provide an accurate estimate of the allele prevalence in a mixture of several individual genomic DNAs. This is mainly because nucleic acids are composed of only four building blocks (dAMP, dCMP, dGMP, and dTMP) with very similar desorption characteristics compared to amino acids/peptides. Furthermore, miniaturization of the sample preparation in chip array formats decreased the variance caused by inhomogeneous crystallization.

Primer extension–based MALDI-TOF-MS methods for genetic analysis not only have found widespread use in research laboratories; but they have also been perfected for industrial-scale processes when combined with suitable automation robotics and are now routinely used in high-throughput environments.

Expansion of the portfolio of available molecular biological methods, such as the principle of base-specific cleavage, is a significant milestone, which allows a more generic use of MALDI-TOF-MS in the field of genomics. This approach is comparatively young and is not established to the same degree as the SNP scoring approaches presented. However, it exemplifies the prospects to keep MALDI-TOF-MS on board as a competitive technology in nucleic acid analysis.

Currently, the envisioned future developments can be separated into three main categories: instrument developments, application developments, and sample generation/preparation. Multiplexed SNP analysis, as well as resequencing using base-specific cleavage, is starting to challenge the capabilities of current axial MALDI-TOF mass spectrometers. Mass accuracy, mass resolution, and sensitivity can become a limiting factor, when we attempt to fully exploit these applications on a biochemical basis. The dynamic range might be another concern relevant for those applications, where a minority of genetic information has to be detected in the presence of excess of wild-type information. Such applications include the detection of rare genetic variants in DNA pools or sample mixtures such as tumor biopsies.

The combination of a MALDI ion source with a quadrupole-TOF (orthogonal [O]-TOF mass spectrometer) has a very high potential to meet the increased need in more complex applications of nucleic acid analysis. Extremely high-level multiplexing in SNP analysis, as well as *de novo* sequencing, requires the analysis of a very high density of nucleic acid fragment species within a defined mass window and thus will depend on higher-resolution advanced instrumentation.

Further expansion of the application suite will be driven through improvements and redevelopment of biological/biochemical processes. In addition to SNP/mutation analysis by primer extension and SNP discovery by base-specific cleavage, new schemes can be envisioned for analysis of mRNA expression levels, the large-scale assessment of methylation patterns, screening of cDNA libraries, and analysis of microsatellites. This growing number of applications will help to establish MALDI-TOF-MS as one of the most versatile generic platforms in nucleic acid analysis.

Last, further technology development must synchronize with current trends in miniaturization of sample handling and sample processing. Part of this trend is already implemented through the use of miniaturized chip arrays for MALDI-TOF-MS analysis as detailed above. As little as 10 nl of analyte is currently dispensed on these chip arrays in high-throughput settings. Considering that as much as 25 µl of analyte per well is currently produced in 384 microtiter plate formats and that only a 1/2500 fraction is actually used for analysis, the potential for process miniaturization without loss in performance becomes evident.

REFERENCES

1. K Strupat. Molecular weight determination of peptides and proteins by ESI- and MALDI-MS, in Methods in Enzymology, AL Burlingame, Ed., San Diego: Academic Press, in press.
2. J Gross, F Hillenkamp. Mass Spectrometry of Nucleic Acids. Encyclopedia of Analytical Chemistry: Applications, Theory and Instrumentation, Vol. 7, RA Meyers, Ed., New York: Wiley, 2000, 5022–5051.
3. U Pieles, W Zürcher, M Schär, HW Moser, Nucleic Acids Res 21(4):3191–3196, 1993.
4. E Nordhoff, Trends Anal Chem 15(6): 67–138, 1996.
5. AP Null, LT George, DC Muddiman, J Am Soc Mass Spectrom 13(4):338–344, 2002.
6. DP Little, TJ Cornish, MJ O'Donnell, A Braun, RJ Cotter, H. Koester, Anal Chem 69(229): 4540–4546, 1997.
7. DP Little, A Braun, MJ O'Donnell, H Koster. Mass spectrometry from miniaturized arrays for full comparative DNA analysis. Nat Med 3:1413–1416, 1997.
8. M Schürenberg, C Luebbert, H Eickhoff, M Kalkum, H Lehrach, E Nordhoff, Anal Chem 72(15):3436–3442, 2000.
9. P Lecchi, LK Pannell, J Am Soc Mass Spectrom 6(10):972–975, 1995.
10. J Gross, S Hahner, M Karas, A Leisner, F Lützenkirchen, E Nordhoff, J Schäfer, F Hillenkamp, J Am Soc Mass Spectrom 9:866–878, 1998.
11. J Gross, F Hillenkamp, KX Wan, ML Gross, J Am Soc Mass Spectrom 12:180–192, 2001.
12. S Sauer, D Lechner, K Berlin, C Plancon, A Heuerman, H. Lehrach, IG Gut, Nucleic Acids Res 28(23):e100/1–e100/6, 2001.
13. CP Rodi, B Darnhofer-Patel, P Stanssens, M Zabeau, D van den Boom. A strategy for rapid discovery of disease markers using the MassARRAY system. BioTechniques Suppl: 62–69, 2002.
14. S Berkenkamp, F Kirpekar, F Hillenkamp, Science 281:260–262, 1998.
15. S Berkenkamp, M Bromirski, W Ens, KG Standing, F Hillenkamp and A Loboda, Proceedings of the 49th ASMS Conference on Mass Spectrometry and Allied Topics, Chicago, IL, May 27–31, 2001.
16. PB O'Connor, E Mirgorodskaya, CE Costello, J Am Soc Mass Spectrom 13:402–407, 2002.
17. AP Null, JC Hannis, DC Muddiman, Anal Chem 73(18):4514–4521, 2001.
18. LM Smith. The future of DNA sequencing. Science 262:530–532, 1993.
19. H Koster, K Tang, DJ Fu, A Braun, D van den Boom, CL Smith, RJ Cotter, CR Cantor. A strategy for rapid and efficient DNA sequencing by mass spectrometry. Nat Biotechnol 14:1123–1128, 1996.
20. F Kirpekar, E Nordhoff, LK Larsen, K Krisitansen, P Roepstorff, F Hillenkamp. DNA sequence analysis by MALDI mass spectrometry. Nucleic Acids Res 26:2554–2559, 1998.
21. E Nordhoff, C Luebbert, G Thiele, V Heiser, H Lehrach. Rapid determination of short DNA sequence by the use of MALDI-MS. Nucleic Acids Res 28:E86, 2000.
22. NI Taranenko, SL Allman, VV Golovlev, NV Taranenko, NR Isola, CH Chen. Sequencing DNA using mass spectrometry for ladder detection. Nucleic Acids Res 26:2488–2490, 1998.
23. Hahner S, HC Ludemann, F Kirpekar, E Nordhoff, P Roepstorff, HJ Galla, F Hillenkamp. Matrix-assisted laser desorption/ionization mass spectrometry (MALDI) of endonuclease digests of RNA. Nucleic Acids Res 25:1957–1964, 1997.

24. AM Maxam, W Gilbert. A new method for sequencing DNA. Proc Natl' Acad Sci USA 74:560–564, 1977.

25. MS Shchepinov, MF Denissenko, KJ Smylie, RJ Worl, AL Leppin, CR Cantor, CP Rodi. Matrix-induced fragmentation of P3′-N5′ phophoramidate-containing DNA: high-throughput MALDI-TOF analysis of genomic sequence polymorphisms. Nucleic Acids Res 29:3864–3872, 2001.

26. F von Wintzingerode, S Bocker, C Schlotelburg, NH Chiu, N Storm, C Jurinke, CR Cantor, UB Gobel, D van den Boom. Base-specific fragmentation of amplified 16S rRNA genes analyzed by mass spectrometry: a tool for rapid bacterial identification. Proc Natl Acad Sci USA 99:7039–7044, 2002.

27. R Hartmer, N Storm, S Boecker, CP Rodi, F Hillenkamp, C Jurinke, D van den Boom. RNase T1 mediated base-specific cleavage and MALDI-TOF MS analysis for high-throughput comparative sequence analysis. Nucleic Acids Res 31:e47, 2003.

28. FS Collins, A Patrinos, E Jordan, A Chakravarti, R Gesteland, L Walters, and the members of DOE and NIH planning groups. New goals for the US human genome project. Science 282:682–689, 1998.

29. S Broder, JC Venter. Whole genomes: The foundation of new biology and medicine. Curr Opin Biotechnol 11:581–585, 2000.

30. DA Nickerson, SL Taylor, KM Weiss, AG Clark, RG Hutchinson, J Stengard, V Salomaa, E Vartiainen, E Boerwinkle, CF Sing. DNA sequence diversity in a 9.7-kb region of the human lipoprotein lipase gene. Nat Genet 19:233–240, 1998.

31. GL Sun, O Diaz, B Salomon, R von Bothmer. Genetic diversity in *Elymus caninus* as revealed by isozyme, RAPD, and microsatellite markers. Genome 42:420–431, 1999.

32. N Rich, K Merikangas. The future of genetic studies of complex human diseases. Science 273:1516–1517, 1996.

33. R Mott, CJ Talbot, MG Turii, AC Collins, J Flint. From the cover: a method for fine mapping quantitative trait loci in outbred animal stocks. Proc Natl Acad Sci USA 97:12649–12654, 2000.

34. GC Johnson, JA Todd. Strategies on complex disease mapping. Curr Opin Genet Dev 10:330–334, 2000.

35. L Kruglyak. Prospects for whole-genome linkage disequilibrium mapping of common disease genes. Nat Genet 22:139–144, 1999.

36. MK Halushka, JB Fan, K Bentley, L Hsie, N Shen, A Weder, R Cooper, R Lipshutz, A Chakravarti. Patterns of single-nucleotide polymorphisms in candidate genes for blood-pressure homeostasis. Nat Genet 22:239–247, 1999.

37. LY Chang, K Tang, M Schell, C Ringelberg, KJ Matteson, SL Allman, CH Chen. Detection of delta F508 mutation of the cystic fibrosis gene by matrix-assisted laser desorption/ionization mass spectrometry. Rapid Commun Mass Spectrom 9:772–774, 1995.

38. D Little, A Jacob, T Becker, A Braun, B Darnhofer-Demar, C Jurinke, D van den Boom, H Koster. Direct detection of synthetic and biologically generated double-stranded DNA by MALDI-TOF MS. Int J Mass Spectrom Ion Processes 169/170:133–140, 1997.

39. C Jurinke, D van den Boom, A Jacob, K Tang, R Wörl, H Köster. Analysis of ligase chain reaction products via matrix-assisted laser desorption/ionization time-of-flight-mass spectrometry. Anal Biochem 237:174–181, 1996.

40. PL Ross, K Lee, P Belgrader. Discrimination of single-nucleotide polymorphisms in human DNA using peptide nucleic acid probes detected by MALDI-TOF mass spectrometry. Anal Chem 69:4197–4202, 1997.

41. A Braun, D Little, H Köster. Detecting CFTR gene mutations by using primer oligo base extension and mass spectrometry. Clin Chem 43:1151–1158, 1997.

42. LA Haff, IP Smirnov. Single-nucleotide polymorphism identification assays using a thermostable DNA polymerase and delayed extraction MALDI-TOF mass spectrometry. Genome Res 7:378–388, 1997.

43. S Sauer, D Lechner, K Berlin, H Lehrach, JL Escary, N Fox, IG Gut. A novel procedure for efficient genotyping of single nucleotide polymorphisms. Nucleic Acids Res 28:E13, 2000.

44. S Sauer, DH Gelfand, F Boussicault, K Bauer, F Reichert, IG Gut. Facile method for automated genotyping of single nucleotide polymorphisms by mass spectrometry. Nucleic Acids Res 30:e22, 2002.

45. C Jurinke, D van den Boom, CR Cantor, H Koster. High-throughput genotyping based on MassARRAY. In Advances in Biochemical Engineering Biotechnology, J Hoheisel, Ed., Berlin: Springer-Verlag, 2002, 57–74.

46. P Ross, L Hall, I Smirnov, L Haff. High level multiplex genotyping by MALDI-TOF mass spectrometry. Nat Biotechnol 16:1347–1351, 1998.

47. M Werner, M Sych, N Herborn, T Illig, IR Konig, M Wjst. Large-scale determination of SNP allele frequencies in DNA pools using MALDI-TOF mass spectrometry. Hum Mutat 20:57–64, 2002.

48. A Bansal, D van den Boom, S Kammerer, C Honisch, G Adam, CR Cantor, P Kleyn, A Braun. Association testing by DNA pooling: An effective initial screen. Proc Natl Acad Sci USA 99:16871–16874, 2002.

49. KL Mohlke, MR Erdos, LJ Scott, TE Fingerlin, AU Jackson, K Silander, P Hollstein, M Boehnke, FS Collins. High-throughput screening for evidence of association by using mass spectrometry genotyping on DNA pools. Proc Natl Acad Sci USA 99:16928–16933, 2002.

50. KH Buetow, M Edmonson, R MacDonald, R Clifford, P Yip, J Kelley, DP Little, R Strausberg, H Koester, CR Cantor, A Braun. High-throughput development and characterization of a genomewide collection of gene-based single nucleotide polymorphism markers by chip-based matrix-assisted laser desorption/ionization time-of-flight mass spectrometry. Proc Natl Acad Sci USA 98:581–584, 2001.

 Sequencing the Single
DNA Molecule

Kenneth D. Weston and Markus Sauer

CONTENTS

THE IDEA OF SINGLE-MOLECULE DNA SEQUENCING

The U.S. Human Genome Project was initiated in 1990 and was originally planned to last 15 years. Effective resources and technological advances have substantially accelerated the determination of the sequence of all 3×10^9 base pairs (bp) that make up human DNA and the identification of the approximately 30,000 genes in human DNA. Several types of genome maps have already been completed, and the first analysis of the working draft of the entire human genome sequence was published in February 2001.[1,2] Although Sanger's enzymatic chain termination method proved to be very reliable, the limited read length of <1000 bases per run requires the determination of overlapping subsequences to construct a "consensus" sequence of a larger DNA segment. Despite dramatic increases in speed over the past decade, existing procedures for sequencing remain labor-intensive and time-consuming. Improved sequencing methods are still needed to understand the function of each gene and genetic variations among cell types, individuals, and organisms. Furthermore, there is a growing interest in understanding the molecular basis of complex diseases and the variety of responses to drugs. The development of more effective and better-tolerated drugs, i.e., the idea of personalized medicine, requires alternative DNA sequencing techniques that are even faster, more efficient, more accurate, and more cost-effective.

A number of techniques such as tunneling microscopy,[3] hybridization of oligonucleotides,[4–6] matrix-assisted laser desorption/ionization mass spectrometry (MALDI),[7,8] and single nanopore current monitoring[9,10] have been proposed and pursued with the aim of developing novel sequencing methods. In this chapter, we focus on fluorescence-based single-molecule DNA sequencing, a very promising concept. In contrast to current DNA sequencing schemes it would allow us to sequence a single fragment of DNA, several tens of kilobases or more in length, at a theoretical rate of more than several hundred bases per second. The concept is to "watch" the stepwise incorporation of DNA bases by polymerase enzymes to a single DNA strand. If each base can be identified as it is incorporated into the DNA strand, the sequence can be obtained. Analogous to this, the reverse process could also be monitored; i.e., a exonulease could be used to degrade a single DNA strand base by base from one end while identifying each base after it is cleaved. Multiple DNA strands cannot be used because the distribution of enzymatic rates on the different DNA strands would result in rapid dephasing. To achieve this goal, single-molecule detection is required; i.e., only one nucleotide at a time should be monitored in the detection region. This elegant alternative to common sequencing techniques would require the minimum conceivable starting material to obtain a sequence.

As a result of the low fluorescence quantum yield and photostability of the four native nucleotides at room temperature, the detection of single native nucleotides has not been reported to date. To circumvent these problems, native nucleotides might be deposited after cleavage onto a moving substrate.[11] Because the fluorescence quantum yield increases at lower temperatures, the DNA sequence might be retrieved by cooling the substrate to low temperatures and detecting and identifying the immobilized bases. The use of fluorescent nucleotide analogues such as 2-aminopurine and other derivatives, fluorescence enhancement by metallic particles, or postlabeling strategies of the immobilized native nucleotides with a fluorescent dye are reasonable alternatives to consider.[12–14] However, the use of nucleotides labeled covalently with a fluorescent dye is probably the method within closest reach of currently available technology.

The idea of sequencing an individual DNA molecule by laser-induced fluorescence detection of single fluorescently labeled mononucleotide molecules was first proposed in 1989 in Los Alamos by Keller's group.[15,16] Interestingly, the method was proposed 1 year prior to the first report of single-molecule fluorescence detection of diffusing dye molecules in solution in 1990.[17] The suggested procedure for sequencing a single DNA molecule contains three major steps (Figure 6.1): (1) the DNA to be sequenced should be copied using a biotinylated primer, a DNA polymerase, and the four nucleotide triphosphates (dNTPs), each containing a different fluorescent label that exhibits a characteristic laser-induced fluorescence; (2) a single DNA fragment is bound to a microsphere or other solid support coated with avidin or streptavidin via the biotinylated primer and that microsphere is transferred into a flowing sample stream by mechanical micromanipulation or optical trapping,

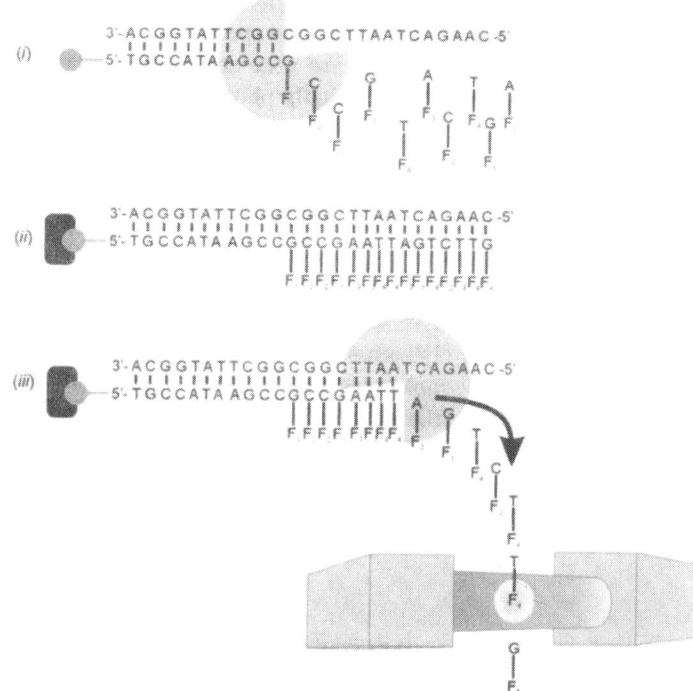

FIGURE 6.1 Illustration of the proposed procedure for sequencing a single DNA molecule. (*i*) The DNA to be sequenced should be copied using a biotinylated primer, a DNA polymerase, and the four nucleotide triphosphates (dNTPs), each base type containing a different fluorescent label with distinct laser-induced fluorescence, (*ii*) as a single DNA fragment is bound to a microsphere or other solid support coated with avidin or streptavidin by the biotinylated primer and transferred into a flowing sample stream by mechanical micromanipulation or optical trapping, and (*iii*) upon addition of a 3′→5′ exonuclease fluorescent nucleotide monophosphate molecules (dNMPs) will be cleaved and transported to the detection area down stream where they are identified based on the characteristic fluorescence properties. Finally, the DNA sequence is directly retrieved from the detected and signal sequence.

and (3) upon addition of a 3′ → 5′ exonuclease fluorescent nucleotide monophosphate molecules (dNMPs) will be cleaved and transported to the detection area downstream, where they are identified based on their characteristic fluorescence properties. The DNA sequence is obtained directly and immediately.

As simple as it sounds, the realization of single-molecule DNA sequencing constitutes one of the greatest challenges biologists, chemists, and physicists have confronted. From the biological point of view, two major problems arise due to the use of fluorescently labeled nucleotides. The bulky fluorescent dye attached to the nucleotide could hinder the polymerase- and exonuclease-mediated incorporation and degradation of nucleotides, respectively. Besides problems associated with the complete enzymatic substitution of native nucleotides by dye-labeled nucleotides, the well-defined selection of a single DNA strand and the detection and identification of each nucleotide due to the spectroscopic characteristics of the fluorescent label with high accuracy are very demanding tasks. To identify each incorporated or cleaved nucleotide, highly photostable fluorescent dyes that exhibit distinguishable fluorescence characteristics are needed. Finally, since the DNA sequence is determined by the order of detected nucleotides, misordering due to different mobilities of, for example, cleaved nucleotides on their way to the detection volume, has to be prevented.[18] In other words, the enzymatic turnover rates, flow velocities, and the distance to the detection volume represent crucial parameters that have to be carefully optimized to make the proposed DNA sequencing method work.

In this chapter we discuss (1) techniques used to detect and identify single fluorescently labeled nucleotide molecules and (2) the biological prerequisites for single-molecule sequencing, i.e., the polymerase- and exonuclease-mediated incorporation or degradation, respectively, of fluorescently modified nucleotides and DNA. Finally, we give an overview on some of the ongoing efforts worldwide in reaching the final goal: to sequence a single DNA molecule using laser-induced fluorescence.

DETECTION AND IDENTIFICATION OF SINGLE MOLECULES BY LASER-INDUCED FLUORESCENCE

Recent advances in optical spectroscopy and microscopy have made it possible not only to detect and identify freely diffusing or immobilized molecules, but also to measure spectroscopic properties and dynamic processes. Although very young, the field of single-molecule spectroscopy at ambient temperature has already been elaborated in excellent reviews.[19–25] Historically, the first attempts to detect low amounts of biologically relevant molecules under physiological conditions by optical methods can be traced to Hirschfeld, who demonstrated in 1976 the detection of a single antibody molecule labeled with 80 to 100 fluorescein molecules.[26] The quest for detection of single fluorescent dye molecules in solution began with the work of Dovichi in 1983 who tried to develop methods for the hydrodynamic focusing of sample streams in sheath flow cuvettes with the final goal of detecting single fluorescent molecules as they passed an excitation beam.[27–29] A few years later these efforts led to the first successful detection of single fluorophores in aqueous solvent.[17] Simultaneously, but independently, the groups of Moerner and Orrit developed a method to detect single dopant molecules in host crystals at cryogenic temperatures using the narrow linewidth of the zero-phonon line and the corresponding enormous absorption cross section of the relatively rigid molecule pentacene.[30] The first demonstrations of room temperature microscopy of single immobilized fluorophores by near-field[31–34] and far-field[35] scanning optical techniques stimulated and influenced researchers worldwide. Detection methods for diffusing single molecules were also rapidly improved. The detection of single fluorescent molecules in solution using a small volume of ~10^{-15} l defined by a confocal microscope was pioneered by Rigler and coworkers.[36,37] The use of an extremely small open volume element improved the signal-to-background (S/B) ratio by orders of magnitude without measurable photodestruction of the dye molecules.[38,39] As a consequence of these improvements, laser-induced fluorescence detection of single molecules in liquids has become a standard laboratory technique with applications in a number of research fields. Individual molecules are now routinely detected in liquids with S/B ratios of 100 to 200 (Figure 6.2).

It is generally accepted that the ability to measure single molecules is a technological breakthrough. However, it is still debated whether or not single-molecule techniques can teach us more than we can currently learn from ensemble measurements. Although most people think about and model molecular systems individually, our basic knowledge has been obtained from experiments on huge numbers of molecules, so-called ensembles, which we observe over long periods of time. In contrast to ensemble measurements that yield information only on a sample's average properties, single-molecule experiments provide information on individuals, distributions, and time trajectories of properties that would otherwise be hidden. Furthermore, single-molecule methods are well suited to study reaction pathways of individual members in a heterogeneous mixture and measure intermediates or follow turnover rates of single enzymes on their substrates, e.g., DNA, that are impossible to synchronize at the ensemble level. Hence, to sequence a single DNA molecule based on an enzymatic incorporation or degradation process, the detection and identification of each individual fluorescently labeled nucleotide is absolutely essential to gain the desired sequence information of a single DNA strand.

As a dye molecule in the focus of a laser beam tuned to an optical transition of the molecule it is excited from the ground state, S_0, into high-lying vibrational levels

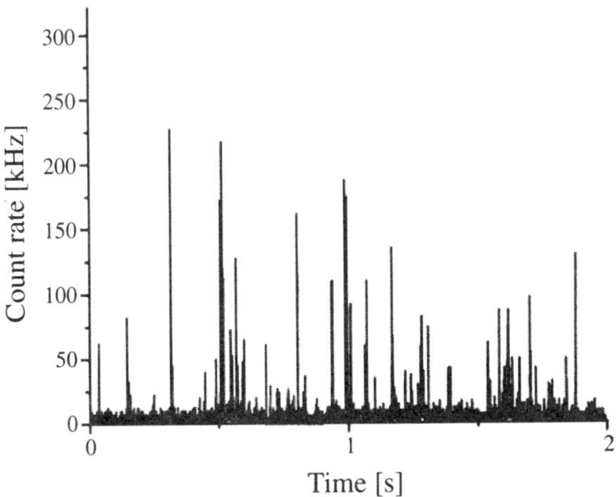

FIGURE 6.2 Fluorescence signals observed from a 10^{-10} M solution of Rhodamine 6 G in water using confocal fluorescence microscopy with a probe volume of ~1 fl. The data are binned into 1 ms time intervals.

of the first excited state, S_1. It then undergoes rapid nonradiative internal conversion to low-lying S_1 levels, and finally emits a photon during its transition back to S_0. The optical saturation limit is the maximum rate that a dye molecule can be cycled between S_0 and S_1 and is dependent on the fluorescence lifetime of the dye, τ_f. Besides irreversible photodestruction, several depopulation pathways such as intersystem crossing into the triplet state compete with fluorescence emission, thus reducing the number of emitted photons. Hence, the ability to detect a single molecule is not as much an issue of sensitive detection but of efficient background rejection. There are three major sources of background signals: (1) Rayleigh scattering, which can be efficiently suppressed with suitable bandpass filters; (2) Raman scattering, which is proportional to the probe volume applied; and (3) autofluorescence from impurities, which strongly depends on the excitation and detection wavelength. Because the background signal is proportional to the number of illuminated solvent molecules and the fluorescence signal of a molecule is independent of the probe volume, all demonstrations of single-molecule detection are based on the use of probe volumes of a few picoliters or less.

The suitability of a fluorescent dye for use in the proposed sequencing method is determined by the absorption cross section at the excitation wavelength, the fluorescence lifetime and quantum yield, the triplet lifetime and quantum yield, and the photobleaching quantum yield. The absorption cross section typically has maximum values of $\sim 10^{16}$ cm^2 in organic dyes and the fluorescence quantum yield often approaches values close to 1.00. Rhodamine dyes, which have fluorescence lifetimes in the range 2 to 4 ns are frequently used. The fluorescence brightness or detected count rate from a single molecule depends on the rate of cycling between the ground and excited states. This is why molecules with longer fluorescence lifetimes (>100 ns) are inappropriate for single-molecule detection. The brightness, or detected fluorescence count rate, is particularly important because residence time of a diffusing molecule in the laser focus is very short (<1 ms). In air-saturated ensemble solutions, the triplet state lifetimes, τ_T, of rhodamine, oxazine, and carbocyanine dyes vary from ~ 0.1 μs up to several microseconds (with intersystem crossing rates, k_{ISC}, ranging from 4.2×10^5 to 2.8×10^7 s^{-1} [40,41] Because no fluorescence photons are emitted during the lifetime of the triplet state, it is important either that the intersystem crossing yield is very low or that the triplet lifetimes are very short. The photostability of the dye is also of fundamental importance. Typical photobleaching yields are in the range of 10^{-3} to 10^{-4} for coumarin dyes and $\sim 10^{-5}$ to 10^{-6} for rhodamine dyes.[42] Irreversible photobleaching significantly limits the statistical accuracy of detection. The percentage of molecules that bleach as they pass the detection volume can be as high as 100%.[42–44] To detect as many photon counts from a single molecule as possible within a given time, relatively high excitation intensities must be applied. Because of the high excitation intensity, the molecule spends a large proportion of time in the S_1 and T_1 states. This means that the probability of exciting the chromophore to higher electronic states, S_N and T_N, by absorption of a second photon is substantial. The population of higher electronic states opens additional bleaching channels such as the formation of dye radical ions and solvated electrons.[42] This so-called two-step photolysis becomes particularly important at higher excitation intensities in the UV region.

The importance of using a small probe volume to reduce background emission and scatter from the solvent was mentioned previously. There are several techniques for minimizing the background that have been used for single-molecule DNA sequencing: (1) hydrodynamic focusing of sample streams crossed with a tightly focused excitation beam and imaging detection optics (probe volume of ~1 pl), (2) confocal excitation and detection with one- and two-photon excitation in microcapillaries or microchannels to (probe volume of ~1 fl), and (3) wide-field total-internal-reflection illumination of molecules on surfaces. In this chapter, we describe these approaches to the detection of single fluorescent deoxyribonucleotides released from or incorporated into a DNA strand.

IDENTIFICATION OF SINGLE MOLECULES

In single-molecule DNA sequencing experiments it is necessary to determine, for each detected molecule, which one of the four types of fluorescently labeled nucleotides (A, T, G, or C) was measured. To distinguish and identify individual dye molecules, several methods have been proposed and developed: (1) fluorescence intensity, (2) fluorescence decay time,[45–52] (3) emission spectrum,[53,54] (4) diffusion coefficient,[55] and (5) fluorescence anisotropy.[56,57] In general, the number of detected fluorescence photons per molecule has a large statistical distribution. Although the entire distribution can be a well-defined characteristic for a given chromophore,[58] a single event is not. In addition, photobleaching of fluorophores in the laser focus makes the identification of molecules due to their fluorescence burst sizes very difficult, if not impossible.[59] The limited number of photons that can be detected before photobleaching occurs will also prevent accurate identification based on molecular diffusion coefficients. Moreover, the small differences in molecular weights of the various types of dye-labeled nucleotides means that the diffusion coefficients will be difficult to distinguish. Recently, Yan and Myrick investigated a solution-phase steady-state polarization-based method for discriminating among the four DNA nucleotides, each labeled with tetramethylrhodamine.[57] They demonstrated that for ensemble measurements, classifying the four types of labeled nucleotides solely on the basis of fluorescence polarization in the presence of a surfactant (4.5 mM Triton X-100) is possible. Although this measurement works very well in bulk solutions, comparison with the signal levels that are expected in single-molecule sequencing efforts, i.e., a few tens to hundreds of photon counts per nucleotide, implies that fluorescence polarization is not a promising method for single-molecule DNA sequencing. Therefore, only the emission spectrum and fluorescence decay time are well-suited quantities for identification of individual labeled nucleotides.

Soper et al.[53] demonstrated a two-laser, two-detector technique to distinguish single Rhodamine 6G (R6G) and Texas Red molecules with absorption spectra centered at ~530 and ~580 nm. A 5×10^{-14} M solution containing a 1:1 mixture of the two dyes was introduced into a flow cell. R6G was excited by a frequency-doubled, mode-locked Nd:YAG laser (532 nm), and Texas Red by a mode-locked dye laser tuned to 585 nm. The emitted fluorescence was split by a dichroic beam splitter and focused onto two detectors. Each detected fluorescence burst could be identified with a high degree of confidence based on the ratio of the intensity contribution at the two detectors. Similar results were reported by Dörre et al.,[54] who distinguished single tetramethylrhodamine and Cy5 molecules with two-color irradiation and two-color detection.

Using pulsed excitation and time-correlated detection, the arrival time of a detected photon with respect to the excitation pulse can be measured with high accuracy. A histogram of arrival times is a fluorescence decay curve and provides a measure of the fluorescence lifetime. This approach was first demonstrated for single molecules in 1993 when the fluorescence lifetime of single rhodamine 110 molecules was measured (Figure 6.3).[45] The identification of single molecules by their characteristic fluorescence lifetime is an attractive approach because it requires only a single excitation laser and a single detection channel.

To use fluorescence lifetime as a distinguishing characteristic in single-molecule DNA sequencing, each of the four DNA nucleotides must be labeled with a different dye showing distinct differences in fluorescence lifetime or labeled with a so-called "intelligent" dye.[60] An intelligent dye is one that has a fluorescence lifetime that depends on the DNA base to which it is bound. The shift in lifetime is caused by excited state interactions between the fluorescent dye and the DNA base. The base-specific fluorescence quenching efficiency results in different fluorescence lifetimes that can be used for identification of the base type. A dye that is appropriate for this purpose is Coumarin-120 (C-120). Phosphothioate modified nucleotides labeled with C-120 influence the fluorescence lifetime and quantum yield of C-120 in a peculiar manner. The four C-120 conjugates have fluorescence lifetimes of 5.3 and 1.9 ns

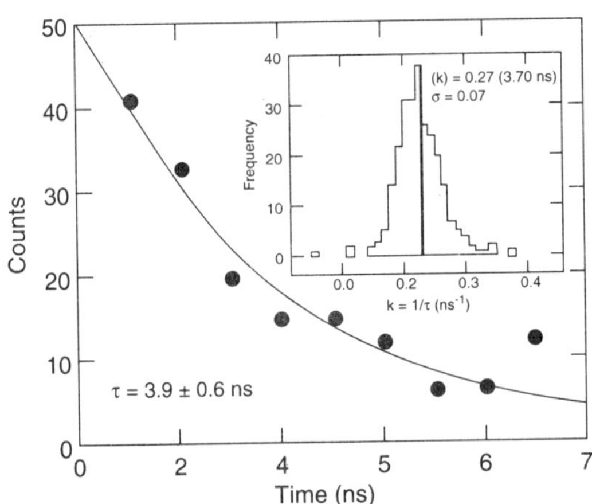

FIGURE 6.3 Fluorescence decay curve of a single-molecule Rhodamine 110. The solid line is an exponential decay derived from a maximum likelihood estimator. The inset shows the lifetime of several hundred Rhodamine 110 molecules determined at the single-molecule level. The center of the distribution corresponds to a lifetime of 3.7 ± 0.1 ns, in excellent agreement with the ensemble lifetime of 3.8 ± 0.1 ns. (From CW Wilkerson et al., Appl Phys Lett 62:2030–2032, 1993. © 1993 American Institute of Physics. With permission.)

for the C-120 adenosine and guanosine conjugate, respectively. The observed DNA base-specific quenching can be explained as a photoinduced electron transfer process.[60] Depending on the redox properties of the DNA base, the dye is reduced or oxidized in its excited state. The measured fluorescence quantum yield and lifetime strongly depend on the DNA base, as well as the length and type of linker connecting the base and chromophore.[61]

Unfortunately, most coumarins have a very low photochemical stability. The quantum yield of photobleaching under moderate one-photon excitation (OPE) conditions is on the order of 10^{-3} to 10^{-4}, which is two orders of magnitude larger than the photobleaching yield of rhodamine dyes.[42] Brand et al.[62] studied fluorescence bursts from single C-120 molecules using OPE at 350 nm and two-photon excitation (TPE) at 700 nm (Figure 6.4). They concluded that the single-molecule detection sensitivity of C-120 molecules is enhanced substantially by using TPE, primarily due to the higher background with OPE at UV wavelengths. The principles of two-photon microscopy were first elucidated by Webb and coworkers.[63] TPE cross sections are extremely small, typically on the order of 10^{-50} cm^4 s for most fluorophores,[64,65] so that high-intensity, short laser pulses (pulse widths ~100 fs),

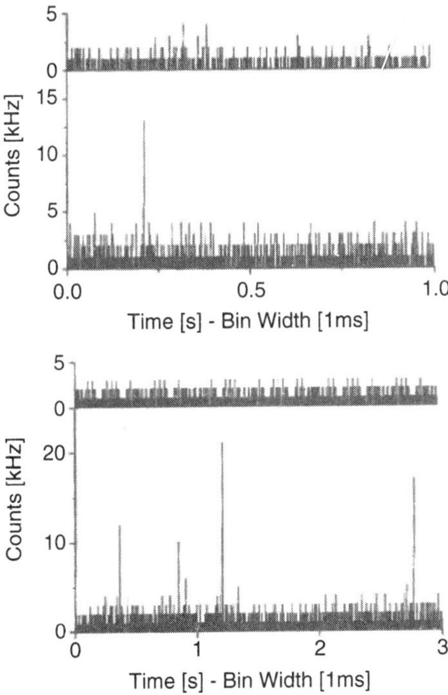

FIGURE 6.4 Time-resolved fluorescence signals observed from a 10^{-11} M solution of coumarin-120 in water using (a) OPE at 350 nm and an average irradiance of 7×10^{22} photons/(cm^2 s), and (b) TPE at 700 nm with an average irradiance of 7×10^{26} photons/(cm^2 s). The upper traces are from pure water without dye. (From L Brand et al., J Phys Chem 101:4313–4321, 1997. © 1997 American Chemical Society. With permission.)

e.g., from a mode-locked Ti:sapphire laser, are required to achieve efficient TPE. As a second-order, nonlinear process, the molecular excitation rate depends quadratically on the laser intensity. Because efficient excitation occurs only at the laser beam focus, photobleaching of out-of-focus molecules is reduced. The quadratic dependence on intensity also means that the effective probe volume for two-photon microscopy is somewhat smaller than for OPE confocal microscopy, resulting in improved spatial resolution. Furthermore, because of the large spectral separation between excitation and detection wavelengths, elastic and inelastic scatter can be efficiently suppressed. Thus, it is sometimes possible to achieve a higher detection sensitivity using TPE as compared to OPE.[66] As in OPE experiments, a rhodamine derivative, Rhodamine B, was the first single molecule to be detected using TPE. Although the burst sizes for single molecules were much lower than those observed for OPE, the background count rate was reduced by more than one order of magnitude using TPE.[67]

In some cases, the high excitation intensities required to generate fluorescence signals by TPE may lead to other nonlinear processes, e.g., continuum generation in the solvent, which can increase the background and deteriorate fluorescence sensitivity. Although the background count rate is generally lower using TPE, achievable single-molecule TPE fluorescence count rates are also generally lower compared to that obtained with OPE. This is due primarily to the lower two-photon absorption cross sections for most fluorescent dyes. Because high fluorescence count rates are essential for the single-molecule DNA sequencing scheme proposed, OPE is still the most promising method for identification of single nucleotides. This may change if appropriate DNA base labels with competitive two-photon absorption cross sections are developed.

After excitation by a short laser pulse, a dye molecule will emit a fluorescence photon after some time delay. That time delay can be measured with high precision (<100 ps) using high-speed electronic circuitry. By repeating the pulsed excitation/photon detection measurement many times, the distribution of delay times gives the fluorescence lifetime decay. This technique is called time-correlated single photon counting (TCSPC). In the simplest case, the delay time distribution is a monoexponential curve with a decay time characteristic for each type of fluorescing molecule. The TCSPC technique was successfully employed for identifying individual molecules in solution[45–52] and at interface.[35,68,69] In practice, for pulsed excitation, power-consuming and relatively expensive laser systems like frequency-doubled Ti:sapphire or Nd:YAG lasers are required. It is advantageous to utilize diode lasers that have emission wavelengths in the red spectral region. Semiconductor lasers as consumer electronic devices offer the advantage of low cost, small size, low power consumption, and long life. In addition, pulsing of semiconductor lasers is conveniently obtained by current modulation at repetition rates of up to some hundreds of megahertz (MHz). In combination with suitable dye molecules, another advantage is evident: as a result of the limited number of compounds that absorb and emit light at wavelengths >600 nm, a drastic decrease in background fluorescence results, even in biological samples.[70]

In practice, several thousand photon counts have to be acquired for the exact determination of an unknown fluorescence lifetime using the TCSPC technique.

However, in aqueous solutions typical transition times of single dye molecules through the detection volume of a few femtoliters are on the order of hundreds of microseconds to a few milleseconds. Therefore, an average of fewer than ~200 photon counts are collected from a single dye molecule during its Brownian diffusion through the laser focus. The fluorescence bursts shown in Figure 6.5 were obtained from single fluorescently labeled nucleotide molecules in water.[52] The nucleotides were labeled with four different red-absorbing fluorescent dyes: a carbocyanine dye (Cy5-dCTP), an oxazine derivative (MR121-dUTP), a rhodamine derivative (JA53-dUTP), and a bora-diaza-indacene dye (Bodipy630/650-dUTP). Because of the similar absorption and emission characteristics of the four dyes, a single excitation laser, a pulsed laser diode emitting at 635 nm with a repetition rate of 56 MHz, and a single detector (a single-photon sensitive avalanche photodiode) can be used. From ensemble measurements, the fluorescence lifetimes were determined to be 1.04 ns (Cy5-dCTP), 2.10 ns (MR121-dUTP), 2.21 ns (JA53-dUTP), and 3.87 ns (Bodipy630/650-dUTP). Several groups have developed maximum likelihood methods to estimate the fluorescence lifetime from noisy data in a finite time window.[46,48,71,72] The accuracy of the calculated fluorescence lifetimes and the confidence of identification of the different molecules are functions of the number

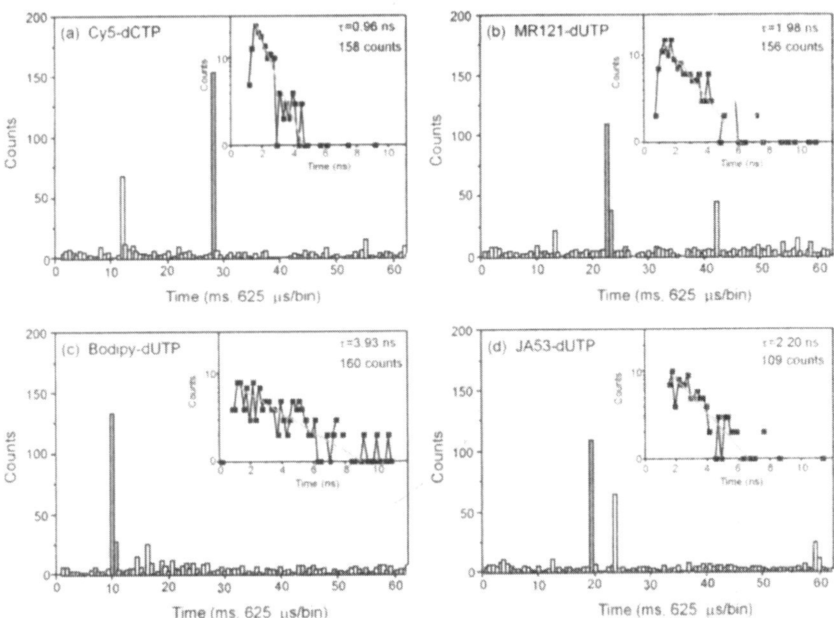

FIGURE 6.5 Fluorescence signals observed from an aqueous 10^{-11} M solution of (a) Cy5-dCTP, (b) MR121-dUTP, (c) Bodipy-dUTP, and (d) JA53-dUTP recorded with integration times of 625 μs and excitation energy of 0.63 mW at 635 nm. The insets show the time-resolved fluorescence decays (0.195 ns/channel) recorded during single-molecule bursts (marked bins) and the corresponding fluorescence lifetime calculated with using MLE. (From M Sauer et al., Bioimaging 6:145–24, 1998. With permission.)

of photons detected for each molecule and the differences in their fluorescence lifetimes. More than 250 fluorescence bursts (Figure 6.5) that contain at least 30 photon counts for each of the four types of labeled nucleotides were analyzed using a maximum likelihood estimator (MLE) algorithm. Comparable fluorescence burst size distributions demonstrate that the four labeled nucleotides are detected with nearly equal efficiency under the applied experimental conditions. The resulting fluorescence lifetime distributions (Figure 6.6) were fit with Gaussian functions revealing fluorescence lifetimes of 1.05 ± 0.33 ns (Cy5-dCTP), 2.07 ± 0.59 ns (MR121-dUTP), 2.24 ± 0.63 ns (JA53-dUTP), and 3.88 ± 1.71 ns (Bodipy-dUTP). In agreement with predictions, the relative error rate in the estimated fluorescence

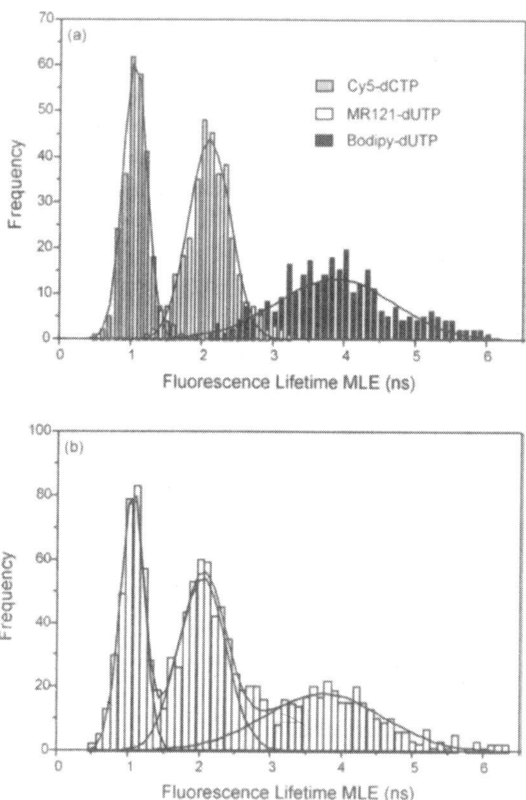

FIGURE 6.6 Distributions of the measured lifetimes of single Cy5-dCTP, MR121-dUTP, and Bodipy-dUTP molecules in water obtained from separate experiments containing only one class of labeled mononucleotides and the corresponding Gaussian fits. The fluorescence lifetimes measured for single molecules, $1.05 \pm 0:33$ ns (Cy5-dCTP), $2.07 \pm 0:59$ ns (MR121-dUTP), and $3.88 \pm 1:71$ ns (Bodipy-dUTP) are in good agreement with the lifetimes measured in bulk solutions. (b) Histogram of 1108 measured fluorescence lifetimes of a 1:1:1 mixture of Cy5-dCTP, MR121-dUTP, and Bodipy-dUTP molecules 10^{-11} M in water with a minimum of 30 collected photons per single-molecule transit and the corresponding Gaussian fits. (From M Sauer et al., Bioimaging 6:14–24, 1998. With permission.)

lifetimes exceeds the ideal $N^{-1/2}$ limit that holds for N photon counts, free of background, recorded over an infinite time window. Nevertheless, forming the convolution of the normalized Gaussian functions reveals a classification probability of 91% for the correct identification of Cy5-dCTP, MR121-dUTP, and Bodipy-dUTP at the single-molecule level.[52] By using only the two mononucleotide molecules with the shortest and longest lifetimes, i.e., Cy5-dCTP, and Bodipy-dUTP; the classification probability is higher than 99%. Hence, two fluorescence lifetime single-molecule DNA sequencing is possible with an error rate of less than 1 in 100 detect single nucleotide molecules with burst sizes >30 photon counts.

Although the MLE algorithm is an extremely efficient and accurate method for determining lifetime values for fluorescence burst data, it is disadvantageous in several respects: First, the decay-time fitting procedure is time-consuming, which can be especially problematic when applying the algorithm to online data evaluation, where one would like to identify 100 or more molecules per second. Second, the fitting procedure and estimation of error rates become significantly more complicated for molecules that have a multiexponential fluorescence decay. Finally, the error of the fitting procedure itself adds to the overall error of correctly identifying the molecules. Recently, an advanced pattern-matching algorithm was presented for single-molecule identification based on fluorescence decay characteristics.[73] This algorithm is applicable to arbitrary fluorescence decays and does not assume any knowledge of its underlying nature. Thus, no lifetime fitting or similar methods are involved. Moreover, the algorithm can be mathematically proved to be the best option for distinguishing molecules by their fluorescence decay behavior. The algorithm was applied to measured single-molecule data in solution for three different red-absorbing rhodamine derivatives with similar absorption and emission characteristics but slightly different fluorescence decay times of 3.85, 2.78, and 2.13 ns.[73] A comparison with the error rates reported in References 35 and 46 through 52, which used the more conventional method of lifetime fitting with subsequent identification, showed that the advanced pattern-matching algorithm yields two to three times smaller error rates for similar differences in fluorescence lifetime and number of photons detected per molecule.

A technique that will increase the identification accuracy of single molecules is a combined analysis of the fluorescence emission maxima and lifetimes of the dyes used. Recently, four different dyes embedded in a polymer film were identified with TCSPC spectrally resolved in two channels.[74] In the following discussion, the emission maximum is obtained by splitting the emitted fluorescence with a dichroic beam splitter and detecting it at two independent detectors. The emission maximum is estimated using the ratio of the detected counts at the long-λ channel to the total counts and a careful analysis of the optical filters used in the detection path. Monte Carlo simulations showed that fewer than 500 photon counts are needed to assign an observed single molecule to one of four species with a confidence level higher than 99.9%. More recently, Sauer's group[75] undertook an experimental feasibility study with the goal of identifying freely diffusing single dye-labeled mononucleotide molecules in solution using both fluorescence emission maxima and lifetimes (Table 6.1). As shown in Table 6.1, the fluorescence emission maxima of the four conjugates differ between 644 nm (JA133-dUTP) and 692 nm (JA242-dUTP) with lifetimes differences between 1.32 ns (Cy5-dCTP) and 3.96 ns (JA133-dUTP).

TABLE 6.1
Ensemble and Single-Molecule (SM) Spectroscopic Characteristics
(λ_{abs}, λ_{em}, τ, F_2) of Fluorescently Labeled Nucleotides JA133-dUTP,
JA169-dUTP, Cy5-dCTP, and JA242-dUTP (10^{-6} M) at 25°C in the Solvent
Mixture (3% PVP, 20 mM Tris-borate buffer pH 8.4, 0.1% Tween 20,
30% glycerol)

	λ_{abs} (nm)	λ_{em} (nm)	τ (ns)	F_2	$\tau_{(SM)}$ (ns)	$\sigma(\tau)_{(exp)}$ (ns)	$F_{2\,(SM)}$	$\sigma((F_2)_{exp}$
JA133-dUTP	622	644	3.96	0.44	4.02	0.88	0.40	0.06
JA169-dUTP	635	659	3.04	0.50	3.09	0.51	0.46	0.05
Cy5-dCTP	651	670	1.32	0.58	1.36	0.18	0.62	0.05
JA242-dUTP	673	692	2.17	0.85	2.15	0.36	0.83	0.05

Note: The fractional intensity, F_2, is the ratio of the signal measured at the long-λ (detector 2) to the total signal. The values in the table were predicted based on the emission spectra of the conjugates and the transmission curves of the beam splitters and bandpass filters in the measurement system.

A dichroic beam splitter was used to separate the fluorescence emission at ~670 nm into the short-wavelength and long-wavelength fraction. As calculated from the ensemble spectra and the transmission of the filter set used, JA133-dUTP fluorescence bursts occur predominantly at the short-λ detector 1, whereas JA242-dUTP signals are detected almost only at the long-λ detector 2. Fluorescence bursts from JA169-dUTP molecules are detected with nearly equal efficiency at the two detectors, while Cy5-dCTP bursts appear higher at the long-λ detector 2 (Figure 6.7). For statistical analysis, ~4000 fluorescence bursts were measured for solutions of JA133-dUTP, JA169-dUTP, Cy5-dCTP, or JA242-dUTP. In Figure 6.8, the distribution of the fractional intensities, F_2 ($F_2 = I_1/[I_1 + I_2]$), and fluorescence decay times calculated from the fluorescence photon counts registered at both detectors is shown. The resulting histograms were fit with Gaussian functions. Shown in Table 6.1, the mean single-molecule fluorescence lifetimes ($\tau_{(SM)}$) and spectral characteristics (F_2) were in excellent agreement with the ensemble data. The symmetric shape of the decay rate and fractional intensity distributions signify spectroscopic homogeneity of the labeled mononucleotides in solution. Although the spectral distributions are relatively narrow for all four conjugates investigated, independent of the emission maxima, the experimental standard deviations of the calculated fluorescence decay times increases with decay time (Table 6.1).

Figure 6.8 also shows the correlated fluorescence decay rate ($k_f = 1/\tau_{(SM)}$) and fractional intensity (F_2) data as two-dimensional histograms (scatter plots) with darker shades of gray indicating increasing number of events. Four distinct populations with symmetric shape are evident, corresponding to each of the species. Assuming independent distributions with respect to the fluorescence decay rate, k_f, and the fractional intensity, F_2, the data was sampled in a scatter plot ranging from 0 to 1.2 for k_f on the x-axis and from 0 to 1 for F_2 on the y-axis with a sampling value of 0.01. These scatter plots were fit using the superposition of two independent Gaussian distributions.

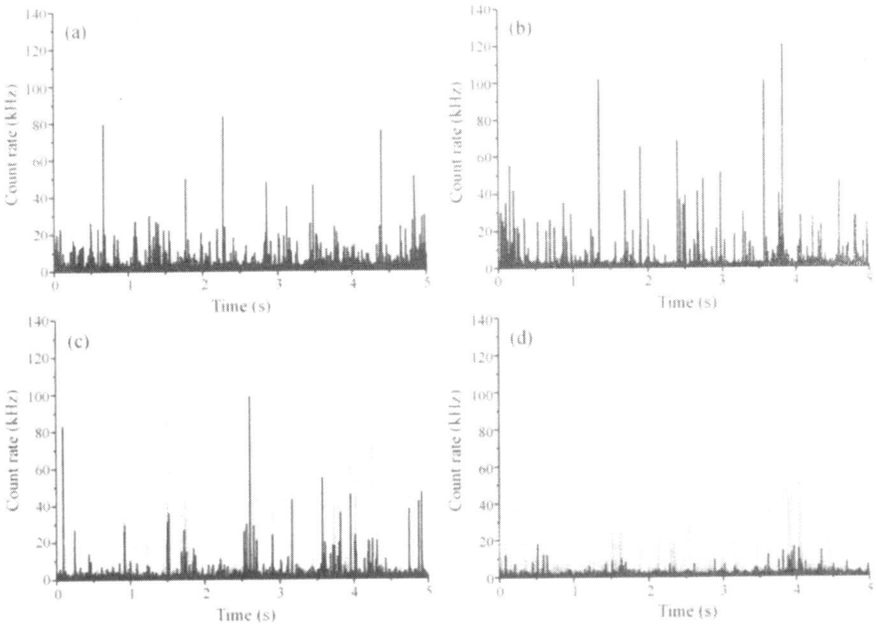

FIGURE 6.7 Time-dependent fluorescence signals recorded at the short-λ detector 1 (black) and long-λ detector 2 (gray, dotted line) from 10^{-11} M solutions of (a) JA133-dUTP, (b) JA169-dUTP, (c) Cy5-dCTP, and (d) JA242-dUTP (solvent: 3% PVP, 20 mM tris-borate buffer pH 8.4, 0.1% (v/v) Tween 20, 30% glycerin). The data were binned into 1-ms time intervals. The average laser power at the sample was 300 µW. (From DP Herten et al., Appl Phys B 71:765–771, 2001. With permission.)

This analysis yielded significantly higher classification probabilities for the four nucleotide conjugates. Applying the two-dimensional superposition of two independent Gaussians, three mononucleotide molecules were identified with a classification probability of ~99%. This demonstrates that the use of spectrally resolved, time-correlated single-photon counting improves the accuracy of species identification and provides information about the heterogeneity or homogeneity of analyte solutions. The technique is a simple and sensitive tool and gives additional confidence that DNA sequencing at the single-molecule level will eventually become a reality.

INCORPORATION AND DEGRADATION OF FLUORESCENTLY LABELED NUCLEOTIDES BY DNA POLYMERASES AND EXONUCLEASES

For the proposed method of single-molecule sequencing, labeling the bases of one strand of duplex DNA with distinctly coding fluorescent dyes is a necessary prerequisite. High-density labeling of DNA relies on the acceptance and proper incorporation of fluorescently labeled deoxynucleoside triphosphates (dNTPs) by

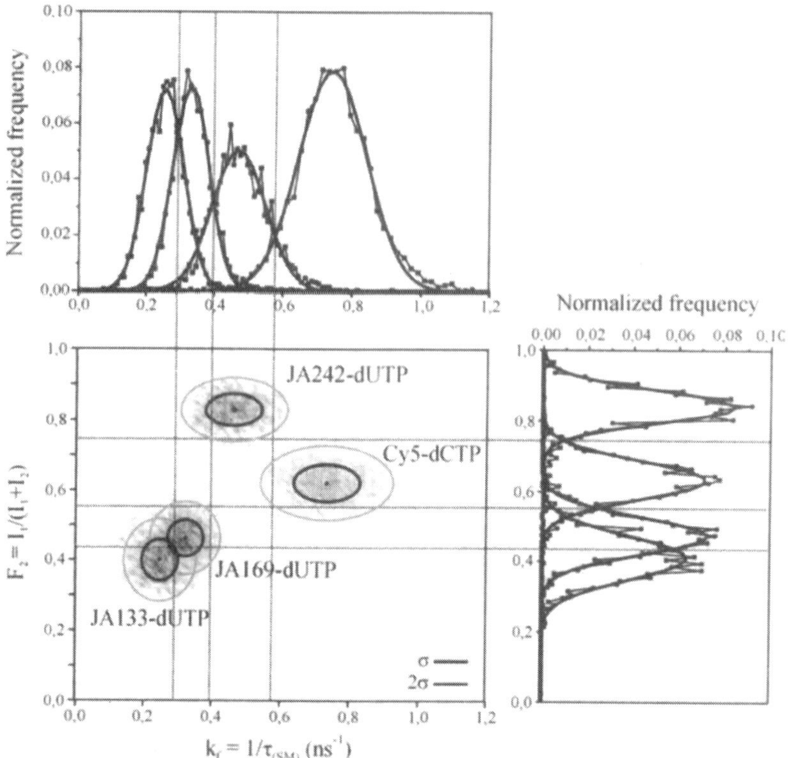

FIGURE 6.8 Scatter plot showing the correlated spectrally resolved (fractional intensity at the long-λ detector 2, F_2) and time-resolved (fluorescence decay rate, $k_f = 1/\tau(\mathrm{SM})$) data and corresponding mean values and standard deviations of the superposition of two Gaussian fits. The data are presented as a two-dimensional histogram with darker shades of gray indicating increasing number of events. Histograms were constructed from separate experiments containing only one class of labeled mononucleotides. In each experiment, about 4000 fluorescence bursts with a burst size >50 photon counts were used. Average excitation power at the sample was 300 μW. (From DP Herten et al., Appl Phys B 71:765–771, 2001. With permission.)

the $5' \rightarrow 3'$ polymerase activity. The incorporation of fluorescently labeled dNTPs by DNA polymerases into newly synthesized DNA, e.g., via polymerase chain reaction (PCR), is a standard technology of molecular biology. Most labeling protocols use the dNTPs attached to a fluorescent dye (via a spacer compound) only in mixtures with the respective natural dNTPs.[76–79] In practice, most natural DNA polymerases have been found to discriminate against dye-labeled nucleotides; this may be because of steric hindrance at the active site of the polymerase due to the bulkiness of the label, and/or because the fluorescently labeled dNTPs typically have a net charge that differs from that of the natural substrates.[76,77]

Waggoner's group[80] studied the synthesis of labeled DNA probes by PCR using 60 mM dUTP attached to the fluorescent dyes Cy3 and Cy5 as a function of the linker length (Figure 6.9). Although the efficiency of incorporation of labeled

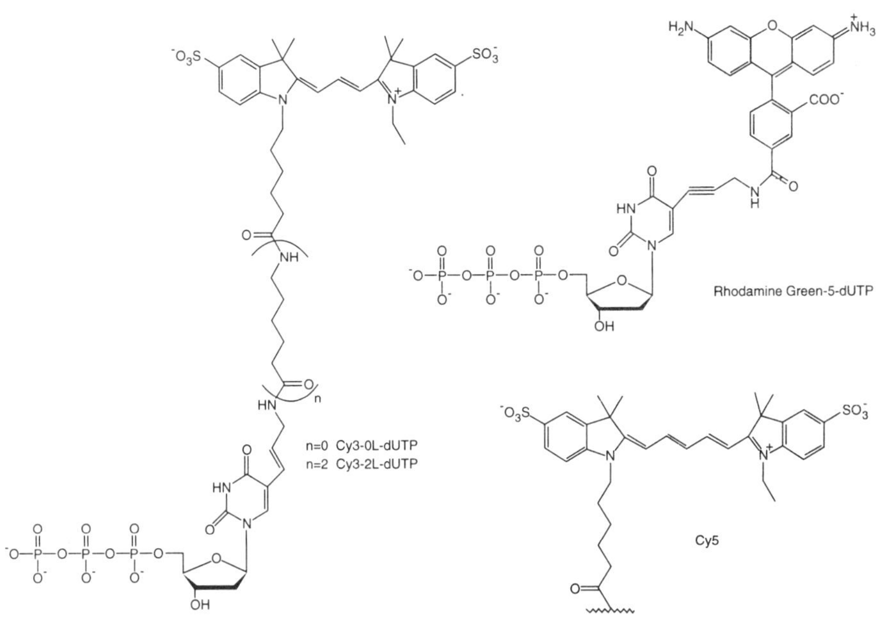

FIGURE 6.9 Molecular structures of fluorescently labeled nucleotides.

nucleotides increased with the length of the linker, the yield of labeled probe generated by PCR dropped significantly. In this study, as many as seven successive Cy3-2L-dUTPs could be incorporated. However, the fraction of complete chains in the product decreased below 20% after incorporation of only two neighboring labeled nucleotides (Cy3-0L-dUTP) with a short linker. Other groups reported a similar decrease in PCR yield.[81,82] The higher incorporation efficiency of modified nucleotides with longer linkers can be explained by the reduced steric and/or electrostatic interactions between the polymerase and the fluorescent dye. These results imply that the interactions decrease with the distance between the active center of the polymerase and the incorporated fluorophores. In addition, it is assumed that dye-modified bases destabilize the double helix.[80] During chain extension, the duplex structure may be stabilized with the help of the polymerase.[83] If the duplex structure is destabilized only after the polymerase releases the DNA, it may be that chain extension cannot be resumed by another polymerase, even with longer extension times. Other studies have also found that the dye structure itself strongly influences the incorporation efficiency.[84] Using Taq DNA polymerase for low-density labeling of DNA with various fluorescently labeled nucleotides, it has been shown that the incorporation efficiency of dye-dNTPs decreased in the following order: rhodamine-green-5-dUTP > tetramethylrhodamine-4-dUTP > Cy5-dCTP (Figure 6.9).[84] Because of the various problems, a complete labeling of all four bases in long DNA strands, i.e., 100% substitution with fluorescent dNTPs, has yet not be achieved. An alternative approach is to synthesize the DNA to be sequenced with amino-modified nucleotides, e.g., aminoallyl-dUTP.

This DNA could be then be labeled with fluorescent dyes in a chemical postlabeling step. Unfortunately, quantitative chemical postlabeling of all aminoallyl groups with fluorescent dyes is complicated and difficult to control.

Very recently, the enzymatic incorporation of modified dNTPs into growing DNA strands has been intensely studied using various reporter groups such as digoxigenin, biotin, fluorophores, or aliphatic side chains covalently attached to dUTP. Incorporation efficiencies were determined using various DNA polymerases.[85] The linear primer-extension reactions were followed by polyacrylamide gel electrophoresis (PAGE) for high-resolution detection. It was possible to incorporate up to 40 successive bases with complete substitution of all four natural dNTPs using various modified nucleotides.

The incorporation efficiency of dye-labeled nucleotides might be improved by using mutant polymerases. Some success in the complete labeling of DNA with fluorescently labeled nucleotides using a mesophilic, $3' \rightarrow 5'$ exonuclease-deficient mutant of the T4 DNA polymerase has been reported.[86,87] However, only a few mutant DNA polymerases have been identified that exhibit an increased capacity to incorporate modified nucleotides for the synthesis of long chains of complementary fluorophore-labeled DNA. Recently, the enzymatic incorporation of dNTPs by a new, thermostable, $3' \rightarrow 5'$ exonuclease-deficient mutant of the Tgo DNA polymerase was studied.[88] PCR based copying of 217-bp "natural" DNA in which fluorescent-dUTP was substituted completely for the normal dTTP was demonstrated using a relatively low nucleotide concentration of 50 μM.[88] In contrast to other exonuclease-deficient B-type enzymes, this polymerase has a high thermostabilty, and offers the possibility of thermal cycling. The full-length replication and sequence integrity was demonstrated by preparative mobility-shift electrophoresis, reamplification in a subsequent PCR with normal dNTPs and resequencing. Fluorescently labeled nucleotide with 12 linker atoms between the base and the dye (rhodamine-green-5-dUTP, Figure 6.9) was base specifically incorporated in 82 to 88 positions out of 92 possible positions in a 217-bp DNA. The labeling efficiency of 0.89 to 0.96 indicates that nearly all of the possible incorporation sites of a thymine contained a fluorescently modified uracil, but that a fraction of the substrate analogue was not bearing fluorophores. The labeling efficiency of 0.89 to 0.96 indicates that nearly all the possible incorporation sites of a thymine contained a fluorescently modified uracil.

Brakmann and Nieckchen[89] used the well-documented power of direct evolution to identify mutant DNA polymerases that incorporate labeled nucleotides with high efficiency and retain a sufficient incorporation fidelity. A functional screening system allowed for the assessment of individual clones that show an increased acceptance of fluorescently labeled dNTPs. They discovered that a cloned natural polymerase, the unmodified exonuclease-deficient Klenow fragment of *Escherichia coli* DNA polymerase I, polymerized 55 template-instructed tetramethylrhodamine-4-dUTPs using an artificial (dA)55 template at a dUTP concentration of 25 mM. This result is quite surprising because 55 bp are expected to build up more than five turns of the DNA double helix, which should cause immense steric and electronic constraints due to the bulky rhodamine dyes. In a next step they applied the same enzyme to the analogous reaction with a natural primer-template with a length of 2700 bp (2.7 kb), substituting two of the natural substrates, dCTP and dTTP, by their

rhodamine labeled analogues, and they obtained a full-length product.[90] Even more surprising was the finding that the exonuclease-deficient Klenow polymerase retained its replication fidelity (error rate < 1/10,000).

From the biological point of view, it is unexpected that a growing primer-template containing bulky rhodamine dyes can be elongated with additional labeled nucleotides by a DNA polymerase that has evolved with the purpose of incorporating native unmodified nucleotides. The recently solved structure of bacteriophage T7 DNA polymerase shows that the enzyme's active site forms contacts to the bases through numerous van der Waals interactions, and therefore, can precisely recognize the geometry of each base pair.[91] Some experimental evidence exists that suggest that duplex DNA with one completely labeled strand undergoes a transition from a right-handed helix (B-DNA) to its left-handed form (Z-DNA).[90] Obviously, some polymerases exhibit an extremely flexible catalytic cleft to surround a "swollen" DNA structure, and do not discriminate against the modified forms of duplex DNA. Furthermore, it has been shown that highly labeled DNA strands exhibit dramatically altered physical and chemical properties in comparison to native DNA.[84,88,92]

The impressive results obtained during the last few years strongly indicate that a complete error-free labeling of a DNA strand is within reach using new mutant polymerases. However, the sequence of a single DNA molecule can also be retrieved in several sequencing steps if the DNA is only partially labeled, e.g., if two kinds of fluorescently labeled nucleotides are substituted completely in varying combinations. A processive exonuclease is an exocnuclease that remains attached to and successively cleaves nucleotides from the same DNA strand. By using a processive exocnuclease that has the same cleavage rate on labeled and native nucleotides, it might be possible to obtain the complete sequence in only two sequencing runs if the dark gaps between detection of subsequent fluorescent signals can be interpreted as time allotted to the cleavage of unlabeled bases. If the cleavage rate of native and fluorescently labeled nucleotides differs, at least six sequencing reactions including all possible nucleotide combinations would need to be performed.

To test and optimize the many steps (incorporation, optical detection, etc.) required for single-molecule DNA sequencing, it is useful to have model DNA with a known and planned sequence. Figure 6.10 shows such a model 218-mer DNA sequence developed by Rigler's group.[84] It contains 6 adenosine and 6 guanosine residues at well-defined positions with 15 filling nucleotides (thymidine and cytosine) between in a thermodynamically optimized and highly degenerate sequence. This is necessary to prevent, for example, possible inter- or intramolecular priming, because PCR-driven amplification of this model DNA is desired to produce large amounts of this sequence.

```
5'- TCT TCA CCT CTC TCT CTT TCT GTC TCT CTC TTT CTT TGT CCT CTT CCT CTT
    CCA TCT TCT TTC TTC TTT ATT CCT CTC TTC CTC CGC CCT TTT CCT TCT CTG
    TCT CTT TCC TTT CCT ATC TTC TTC CTC TCT CAC CTC CTC TTC TT CTG CTC
    CTT CCT TTC TCT GCT CCT CCT TTC TTC TAT CCT TTC TTC TTC CTG GCG ATC
    TCC GCT TAG AAT TG -3'
```

FIGURE 6.10 Sequence of the synthetic model 218-mer DNA developed by Rigler and co-workers.[84] The adenosine (black) and guanosine (gray) residues indicate the incorporation positions for complementary labeled nucleotides dCTP and dUTP.

Thus, by using fluorescently labeled dUTP or dCTP derivatives alone or in combination, a total of up to 12 fluorescent labels (6 dUTP and 6 dCTP derivatives) can be incorporated enzymatically into one DNA strand in a defined sequence. For example, in sequencing a single DNA strand, the following sequence should be retrieved: UCCUUCCUUCCU. Due to the large distance between the label positions, the polymerase-mediated reaction occurs with various fluorescently labeled nucleotides without difficulties.

The availability of such model DNA will allow for a well-defined calibration and optimization of all other steps involved in the development of a new single-molecule DNA sequencing strategy, including (1) selection of single DNA strands, (2) determination of cleavage rates of different exonucleases on single- and double-stranded DNA, (3) determination of the influence of the dye structure (charge, hydrophobicity and linker arms) on the incorporation fidelity and cleavage rate, (4) determination of the frequency of misordering due to differences in the rate of transport of cleaved labeled dNTPs through the detection area, and (5) determination of the error rate of identification of the nucleotides by the spectroscopic properties of the fluorescent labels. An advantage of this system for optimization is that, even if more than a single DNA molecule is selected, a 1:1 ratio of the two fluorescently labeled dUMPs and dCMPs must be found. In addition, the number of detected and identified events should always be 12 (6 labeled dUMP and 6 labeled dCMP molecules) or a multiple thereof if more than a single DNA-strand was selected.

To test and compare the cleavage rates of exonuclease enzymes on double- or single-stranded labeled DNA under various conditions, simple fluorescence intensity measurements in a conventional fluorescence spectrometer can be used. Due to intermolecular energy transfer between closely spaced chromophores via the Förster[92] or Dexter[93] mechanisms, highly labeled DNA exhibits relatively low fluorescence intensity. Hence, the efficiency of exonucleolytic cleavage can easily be monitored by the increase in fluorescence intensity with time using, for example, simple doubly labeled oligonucleotides (Figure 6.11).

As pointed out previously, highly labeled DNA strands are expected to have dramatically different physical and chemical properties in comparison to native DNA.[84,88,94] This might seriously deteriorate the successive cleavage of fluorescently labeled nucleotides by a DNA exonuclease. Although relatively little is known about the cleavage rate of exonuclease enzymes on highly labeled DNA, several reports have investigated this topic. For example, the turnover rate of exonuclease III on rhodamine-labeled double-stranded DNA was measured with three to seven nucleotides per DNA fragment per second at 36°C using enzyme excess.[87] The cutting rate of T7-DNA polymerase on double-stranded rhodamine and Cy5-labeled DNA of 1 to 2 Hz at 16°C have been reported.[95] However, the reported values do not represent the final cleavage rate of the exonuclease attached at the DNA strand. Besides the rate of cleavage itself, the overall measured cleavage rate is controlled by the rate that the exonuclease attaches to and detaches from the DNA strand, as well as the processivity. The attachment of the exonuclease involves the recognition of the DNA strand, and this is expected to be altered when the DNA is fluorescently labeled. Once attached to the DNA strand, all measurements performed thus far have indicated that highly processive exonucleases have substantially higher cleavage

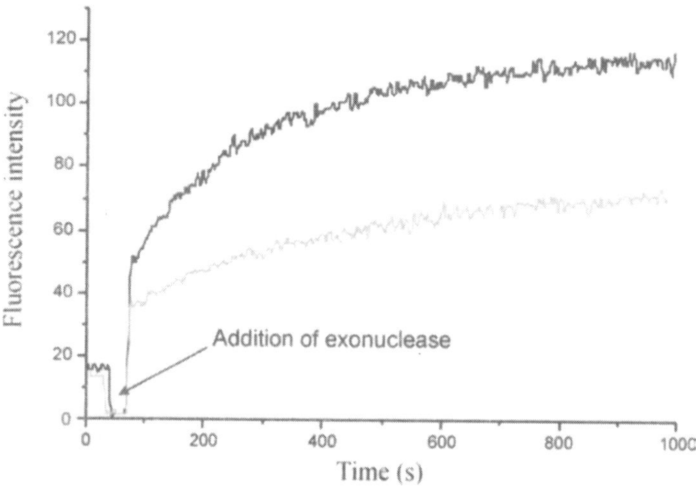

FIGURE 6.11 Test of exonuclease cleavage rates on doubly labeled single-stranded 28-mer oligonucleotide 5′-T*GTAAAAT*GACGAGTTTACTAGTGAACT-3′ (T*: tetramethylrhodamine labeled thymidine). Fluorescence intensity of the oligonucleotide monitored at the emission maximum of tetramethylrhodamine upon addition of 30 units exonuclease I (black) and T7 DNA polymerase (gray) to 600 µl of a 10^{-6} M solution of single-stranded oligonucleotide. Conditions: 25°C, 20 mM Tris-borate, pH 8.4, 1 mM MgCl$_2$.

rates (the measured average rates include the attachment). Furthermore, in single-molecule sequencing experiments based on the exonucleolytic cleavage of single nucleotides from an immobile DNA strand, the exonuclease may be accelerated by the flow gradient applied for the purpose of moving the cleaved nucleotides to the detection area. By using the sheath flow technique, exonuclease molecules are accelerated by the relatively high sheath flow velocities of up to 1 cm/s.[96,97] Application of electrical fields induces comparable accelerations of the exonuclease molecules.[92,98,99] Independent of the method used, exonuclease recognition of the highly labeled DNA as substrate might be problematic. Furthermore, once attached, it might be detached from the DNA strand due to forces associated with the required flow.

To avoid hairpin structures that might disturb the recognition and attachment of an exonuclease to the DNA, digestion of double-stranded DNA is generally preferred over single-stranded digestion. On the other hand, fluorescently labeled dNTPS exhibit a higher tendency to bind nonspecifically to double-stranded DNA as well as to proteins, e.g., streptavidin. Therefore, without careful purification steps, erroneous sequences might be obtained when adsorbed and released fluorescent dNTPs are detected along with enzymatically cleaved fluorescent dNMPs. To simplify purification, it would be much easier if the enzymatic degradation could be performed using single-stranded DNA. An exonuclease that shows high single-strand activity is exonuclease I (Exo I).[94] Exo I digests single-stranded DNA in a 3′ → 5′ direction and is active under a wide variety of buffer conditions. In addition, Exo I is a highly processive exonuclease on single-stranded native DNA and has a high cleavage rate of 275 bases per second.[100] Sauer and coworkers[94] used Exo I to degrade

single-stranded labeled DNA in a micropipette and found cleavage rates of 3 to 24 Hz at room temperature using an aqueous buffer containing 30% glycerol and 0.1% (v/v) Tween 20.

SINGLE-MOLECULE DNA SEQUENCING IN HYDRODYNAMICALLY FOCUSED SAMPLE STREAMS

An important prerequisite for single-molecule sequencing is the efficient detection of each fluorescently labeled nucleotide molecule with high S/B ratio. The importance of using small probe volumes to distinguish fluorescence from a single molecule from background emission from the solvent was already discussed. Hydrodynamic focusing of the sample stream in a sheath flow cuvette,[29] an excitation laser beam focused to ~10 μm, and a spatial filter in the detection path can be used to attain a detection volume of approximately 1 pl. Hydrodynamic focusing occurs when a sample stream is introduced into a rapidly flowing sheath stream from a small orifice. During focusing, the sample stream accelerates to catch up with the sheath stream keeping the volume flow rate of the sample constant. The focused sample stream is typically in the center of the flow cell where the flow profile is relatively constant. A diagram of an apparatus for single-molecule DNA sequencing used by Keller and coworkers[96] is shown in Figure 6.12. The sheath flow cell is mounted on a three-axis translation stage to allow precise alignment of the sample stream to the focused excitation laser. Fluorescence is collected at 90° to the flow and excitation laser axes using a 40×, 0.85 numerical aperture (NA) microscope objective. Photomultiplier tubes (PMTs) or avalanche photodiodes (APDs) are used as detectors. The use of photon counting APDs for single-molecule detection is discussed in detail by Li and Davis.[101] The main advantage of APDs is their high quantum efficiency in the visible region of up to 70%. Typical overall detection efficiencies with APDs can be as high as 5% when combined with high numerical optics and appropriate filters.

The 1 pl volume contains approximately 3×10^{13} water molecules. Despite the small Raman scattering cross-section of a single water molecule of $~10^{-28}$ cm^2 at 488 nm,[19] the large number of molecules in this volume contribute to a total Raman scattering area that is larger than the absorption cross-section of a typical dye ($~10^{-16}$ cm^2). For background rejection, it is helpful to use a pulsed laser system and time-gated detection. Fluorescence decay times of highly fluorescent molecules are typically in the range of a few nanoseconds while elastic and inelastic scattering occurs only during the laser pulse. A time-to-amplitude converter (TAC) can be used to process only photons arriving with a delay longer than, for example, 1 ns. This time-gating can result in a reduction of the detected Raman and Rayleigh scattered light of more than two orders of magnitude (Figure 6.13).[19]

Even with small probe volumes, there is often considerable background emission associated with fluorescent impurities in the solvent. This is an important limitation, especially in single-molecule sequencing experiments, where biological buffers and enzymes have to be used. For a 1 pl volume, fluorescent impurities present at a concentration of $~10^{-12}$ *M* will give an average of one impurity molecule in the detection volume at any time. If the impurity is strongly fluorescent in the same spectral

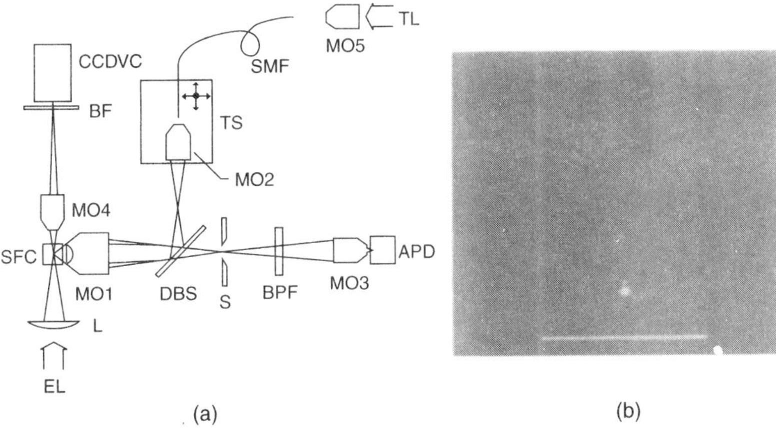

FIGURE 6.12 (a) Diagram of the experimental setup used to detect single molecules eluting from a microsphere. Legend: APD, single-photon counting avalanche photodiode; BF, blocking filters; BPF, bandpass filter; CCDVC, charge-coupled device video camera; DBS, dichroic beam splitter; EL, excitation laser; L, focusing lens; MO1, optical trapping/fluorescence collection objective; MO2, trapping laser fiber output coupling objective; MO3, fluorescence imaging objective; MO4, imaging objective; MO5, trapping laser fiber input coupling objective; S, slit; SFC, sheath flow cuvette; SMF, single mode fiber; TL, trapping laser; TS, three-axis translation stage. (b) The sheath flow channel viewed along the excitation laser axis. The width of the square-bore flow channel is indicated by the white 250 μm scale bar near the bottom of the picture. Sheath fluid flows from top to bottom. An optically trapped 1 μm microsphere, illuminated with a HeNe laser (633 nm), is visible less than 20 μm upstream of the focused (16 μm e^{-2} diameter) excitation laser beam. The end of the sample delivery capillary (90 μm o.d., 20 μm i.d.) used for delivery of microspheres to the optical trap is visible ~200 μm upstream of the excitation laser. (From NP Machara et al., Bioimaging 6:33–42, 1998. With permission.)

region as the analyte of interest, fluorescence bursts from impurities can be mistaken for analyte detection. If the impurity is only weakly fluorescent, it will contribute to a quasi-continuous background. It has been shown that the background from impurities can be reduced by one order of magnitude or more by photobleaching the sheath fluid before introducing the analyte.[102] As a result of these improvements, single fluorescent molecules can be easily detected in hydrodynamically focused sample streams with high S/B ratios.[19,24,45,46,55,59,96]

Detection of each fluorescent molecule present in the sample stream is an important prerequisite for single-molecule DNA sequencing. Efficient sample delivery to 1 pl volumes requires sample stream diameters less than 10 μm. To reduce broadening of the sample stream due to radial diffusion, flow velocities of ~1 cm/s are needed. This means that the transition times of single molecules through the laser focus are in the range of 1 ms, comparable to typical transition times of freely diffusion molecules through confocal probe volumes of ~1 fl.[37–39] For rapidly diffusing analyte molecules, such as fluorescently labeled nucleotides, one positions a tapered sample injection capillary with a inner diameter of ~1 μm as close to the detection volume

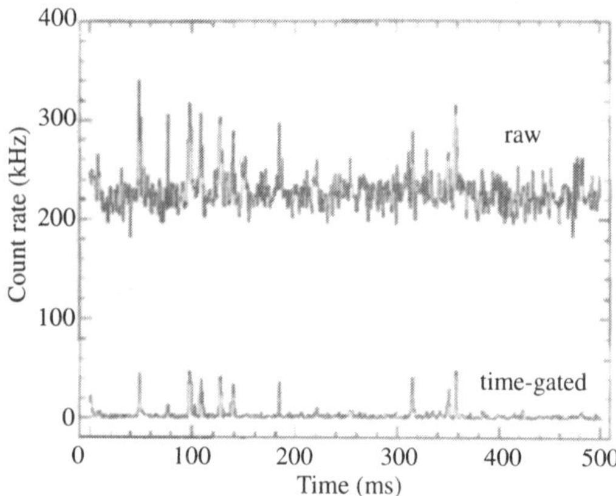

FIGURE 6.13 Detection of single R6G molecules using pulsed excitation and TCSPC to discriminate against Raman scattering background. The top curve shows 500 ms of raw data binned into 1-ms intervals. The bottom curve (time-gated) is the same data processed to remove photons detected within ~1 ns of the excitation laser pulse. (From WP Ambrose et al., Chem Rev 99:2929–2956, 1999. With permission.)

as possible.[103,104] Because of increase in background associated with scattered light and fluorescence from the capillary tip, it is normally preferable if the capillary output and detection volume do not overlap. In a typical experiment, the analyte is dissolved in a salt buffer and is delivered electrokinetically into the sheath flow from a pulled microcapillary located approximately 50 μm upstream from the focused excitation laser beam.[19] Burst size distributions detected from single rhodamine molecules under these conditions showed a peak at ~40 photon counts. When the sample stream diameter is larger than the detection volume, the burst size distribution decreases monotonically from zero photon counts. For a detection threshold set to 20 photon counts, simulation predicts that >90% of the molecules leaving the capillary are detected.[105] The good agreement found between experiment and simulation corroborates the assertion that single molecules are indeed detected.[19] According to the simulation, approximately 10% of the molecules photobleach while crossing the probe volume; about half of these are detected before they photobleach.

In addition to scattered light and fluorescence from the capillary tip, there are other problems associated with capillary sample introduction. Perturbations of the sheath fluid flow by the capillary can broaden the sample stream. Also, adsorbed fluorescent impurities released from capillary surfaces contribute to the background. Therefore, new sample delivery methods have been investigated that avoid scattering from the capillary tip and simultaneously ensure a short diffusion time of released analyte molecules to the detection volume.[96] In Reference 96, efficient sample delivery was achieved from a 1-μm-diameter microsphere optically trapped ~20 μm upstream of the detection volume. Because of the short transport time to the detection volume, diffusional broadening of the sample stream is significantly reduced, thereby enabling single-molecule detection efficiencies greater than 90%. Microspheres were introduced into the sheath flow by a 20-µm i.d. capillary positioned approximately 200 µm upstream of the detection volume (Figure 6.12). A water immersion microscope objective was used for both optical trapping and fluorescence collection.[106,107] A laser power of approximately 500 mW provided by a ND$^+$:YAG laser (1.06 µm) was used for trapping. Fluorescence of single rhodamine 6G (R6G) and tetramethylrhodamine isothiocyanate (TRITC) molecules was excited using a mode-locked Ar$^+$ laser operated at 514.5 nm at a repetition rate of 82 MHz. This laser system provided pulses with a length of <200 ps (FWHM). The fluorescence signal was isolated by a bandpass filter and focused with a long working distance objective (32×, 0.6 NA) onto the active area of an APD. The detected photons were processed using time-correlated single-photon counting electronics.

To demonstrate the potential of the method for single-molecule DNA sequencing, 1-µm streptavidin-coated polystyrene microspheres were stained with R6G and TRITC by nonspecific adsorption in aqueous dye solutions (10^{-4} M) for several dyes. Stained microspheres were separated from the concentrated dye solution by centrifugation and transferred into the injection capillary. With the excitation laser blocked, microspheres were injected into the sheath stream. After a microsphere was trapped, the sample stream was shut off, the sample capillary moved laterally from above the probe volume, and the excitation laser unblocked. After some seconds, the rate of dye elution from the trapped microsphere fell to a level where single-molecule bursts are visible in the data stream. After a few minutes the burst rate approached the background level. Figure 6.14 shows unprocessed fluorescence bursts

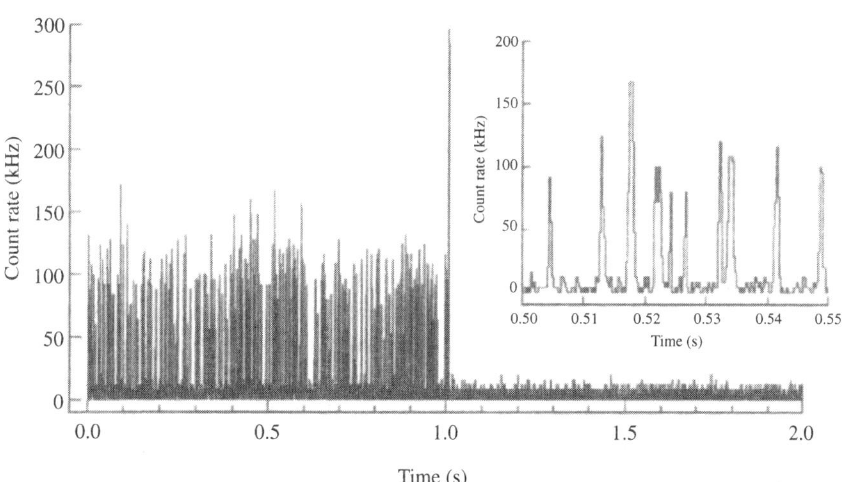

FIGURE 6.14 Unprocessed data showing photon bursts of single R6G molecules eluting from a dye-stained microsphere optically trapped ~20 µm upstream of the laser beam. The data are binned into 250 µs intervals. At ~1 s the microsphere is released from the optical trap and falls through the detection laser, causing the large photon burst. The inset shows an expanded view of the time axis near 0.5 s. (From NP Machara et al., Bioimaging 6:33–42, 1998. With permission.)

of single R6G molecules released from a microsphere with count rates of up to 100 kHz. At ~1 s, the trapped microsphere is released and flowed through the probe laser, confirmed by the large fluorescence burst. To increase the S/B ratio, time-gated photon detection is used; i.e., only those photons are recorded that arrive with a delay greater than 1 ns with respect to the laser pulse. A photon burst is identified by a series of successive gated photons recorded at a high rate (~100 kHz) compared to the background count rate of ~4 kHz. To search for fluorescence bursts, the authors used a burst search threshold time of 0.1 ms, that is, successive gated photons recorded at time intervals of less than 0.1 ms are considered as a photon burst. For each photon burst, two parameters are saved: the number of fluorescence photons, i.e., the burst size, and the duration of the burst. In addition, bursts were time filtered; that is, those with durations significantly shorter or longer than the mean molecular transit time across the detection volume were discarded. Figure 6.15 shows burst duration and burst size distributions obtained from a microsphere with and without R6G staining. Fluorescence bursts from single R6G molecules eluting from the microsphere and crossing the detection volume gave a peak in the burst duration distribution centered at ~2.5 ms. It was concluded that burst durations in the range from 1.0 to 3.8 ms are due mainly to single R6G molecules. Shorter bursts are due to background and longer bursts arise from accidental R6G molecule coincidences. The peak in the burst duration distribution at ~2.5 ms corresponds to a peak in the burst size distribution of single R6G molecules centered at ~100 photon counts. For a burst size threshold of 45 photon counts, it was estimated that 92% of the R6G molecules eluting off the microsphere are detected. Furthermore, the authors presented experiments to distinguish between single TRITC and R6G molecules eluting from a microsphere solely due to different burst sizes.[96] The results demonstrate the potential of the method for single-molecule DNA sequencing: (1) replacement of the micropipette with an optical trap greatly simplifies sample loading; (2) the background fluorescence burst rate is decreased considerably, probably because that most background was from fluorescent impurities adsorbed to the surface of the micropipette; and (3) elution from a microsphere positioned 20 μm upstream results in smaller stream diameters and improved single-molecule detection efficiencies.

The detection of individual fluorescently labeled nucleotides cleaved from DNA attached to an optically trapped polystyrene microsphere suspended in the flow apparatus is shown in Figure 6.16.[24] In these experiments, approximately 100 double-stranded DNA fragments were attached to a microsphere and transferred into the sheath flow stream. Each DNA fragment contained 40 bp with 8 of the last 22 bases being R6G-dUMP. Exonucelase III was added to the flow buffer for digestion of the double-stranded DNA. To control the cleavage of the DNA, Mg^{2+} ions present in the buffer were complexed by addition of EDTA. Because Exo III requires Mg^{2+} ions for the degradation of nucleotides, Exo III can controllably activated by the addition of Mg^{2+} ions. At ~100 s (Figure 6.16) the Mg^{2+} ions reached the microsphere and activated Exo III to initiate DNA degradation. The time-gated count rate increased abruptly and then dropped to the background level as the DNA was digested. From the experimental signal and the estimated ~100 DNA fragments bound to the microsphere, the authors calculated Exo III cleavage rates of ~1 nucleotide/s at 20°C and ~5 nucleotides/s at 36°C. At about 250 s the microsphere was released, which resulted

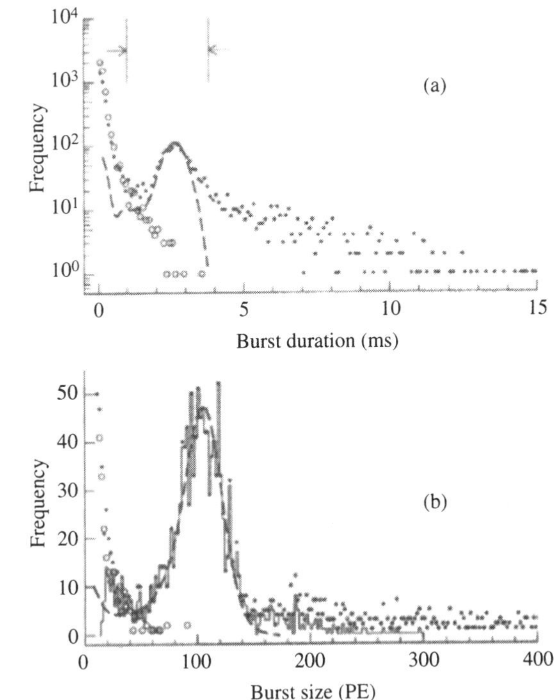

FIGURE 6.15 (a) A semilog plot of burst duration distributions (BDDs) compiled from photon burst data collected with a R6G stained microsphere upstream of the detection laser beam (•) and from data collected after release of the microsphere (○). Both distributions were compiled from 55 s of data. Vertical lines and arrows denote the range of burst durations due primarily to single R6G fluorescence bursts (1.0 to 3.8 ms). The dashed curve shows the BDD generated by the simulation for single R6G molecules without accidental coincidences. (b) Burst size distributions (BSDs) compiled from the same data used for (a). BSD compiled from data recorded with R6G stained microsphere upstream of the detection volume (•). The peak at 100 photon counts is due to single R6G molecules; bursts greater than 140 photon counts are accidental coincidences. BSD compiled from data without microsphere (○). The BSD shown with the solid line is compiled from the subset of bursts with durations between 1.0 and 3.8 ms to discriminate background and accidental coincidences. The dashed curve shows the BSD generated by the simulation for single R6G molecules without accidental coincidences. (From NP Machara et al., Bioimaging 6:33–42, 1998. With permission.)

in a decrease in the count rate. In Figure 6.16b through d the timescale is expanded to show fluorescence bursts of individual molecules. The data in Figure 6.16 show that even before addition of Mg^{2+} ions, fluorescence bursts from impurities are detected. Figure 6.16c shows fluorescent bursts associated with individual nucleotide molecules cleaved from the DNA on the microsphere, and Figure 6.16d shows bursts after the microsphere was released. In the absence of exonucleolytic degradation, there were approximately two background fluorescence bursts per second. This small, but nonzero background burst rate is similar to the burst rate expected from

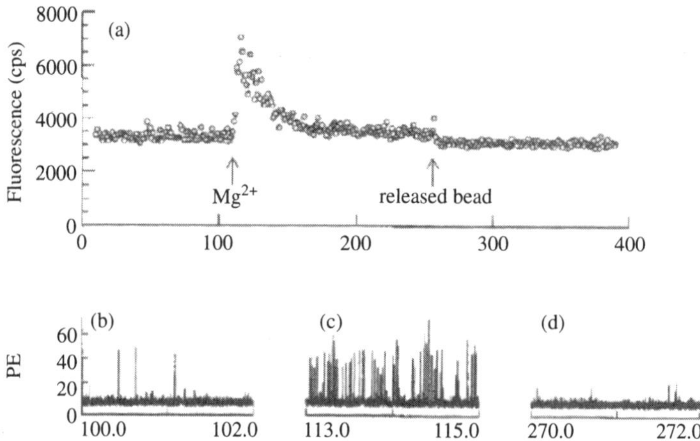

FIGURE 6.16 Detection of single, R6G labeled nucleotides cleaved from fluorescently labeled DNA in flow. (From WP Ambrose et al., Chem Rev 99:2929–2956, 1999. With permission.)

exonucleolytic degradation of a single DNA strand (~1/s).[19,97] In addition, the fluorescence bursts detected before activation of the exonuclease, i.e., before addition of Mg^{2+} ions, would be indistinguishable from enzymatically cleaved nucleotides from a single DNA strand. These fluorescence bursts are most likely due to fluorescently labeled nucleotides or free dyes noncovalently bound to the DNA. This problem, which is of minor importance for high cleavage rates of the exonuclease on fluorescently labeled DNA (~100 nucleotides/s), will seriously distort the sequence information obtainable from a single DNA strand.

It should be pointed out that the flow cytometric single-molecule detection technique is ideally suited for sizing of DNA fragments labeled with intercalating dyes that react stoichiometrically with the DNA.[108–110] The sizing of DNA fragments created by a restriction digest is an important analytical tool in medical diagnostics and forensics. By using specific restriction enzymes, the length of the DNA fragments is characteristic for a particular individual and is useful for forensic identification.

SINGLE-MOLECULE DNA SEQUENCING IN MICROCHANNELS AND SUBMICROMETER CAPILLARIES

To attain smaller probe volumes in the femtoliter range, confocal excitation and detection can be used. The use of confocal techniques to detect single molecules in solution was pioneered by Rigler and coworkers.[36–38] The background emission from the solvent scales with the volume, so that single fluorescent molecules can be detected with high S/B ratios. As in confocal microscopy, the excitation laser is reflected from a dichroic beam splitter and focused by microscope objective with high numerical aperture to a submicron diameter (waist) in the sample. Fluorescence is collected by the same objective, spectrally filtered, focused onto a pinhole (generally 50 to 100 μm), and imaged onto an APD. Emission from analyte molecules outside of the image of the pinhole (is out of focus) does not pass through the pinhole. The diffusion time of a typical fluorescent dye through a femtoliter probe volume is ≈100 μs to 1 ms. Unfortunately, there is a drawback associated with the use of such small volumes in applications requiring efficient detection of all analyte molecules such as single-molecule DNA sequencing. To detect every molecule, analyte solution must be confined to flow through a channel with a dimension of <1 μm (the detection volume). Although the refractive index differences at the outer walls of such channels can be matched by the use of the appropriate index-matching oil, the refractive index differences at the inner wall and deviations of the beam profile generally result in higher background rates and smaller photon bursts. In addition, the use of channels with such small volume-to-surface ratios can result in problematic adsorption of analyte molecules to the channel walls.

The use of capillaries to confine the sample stream to a small probe volume suitable for single-molecule detection (picoliter volume) was first reported by Winefordner and coworkers.[111,112] An excitation laser emitting at 780 nm was used excite single IR140 molecules dissolved in methanol (to prevent adsorption on the capillary walls) passing through a 11-μm i.d. capillary. The excitation laser was focused to a $1/e^2$ diameter of 11 μm to irradiate the entire cross section of the capillary. However, as a consequence of the Gaussian spatial distribution of the laser beam, the Poiseuille flow velocity profile, and photobleaching of dye molecules, large variations in the burst sizes detected from individual molecules were observed.

Because of dynamic adsorption, burst durations of up to 60 ms have been measured for single rhodamine 6G molecules in aqueous buffer in submicrometer channels.[113] Although longer burst durations can be beneficial for increasing the number of photons detected from a single molecule, the fluorescence properties of a chromophore change upon surface adsorption and identification becomes more difficult. Adsorption also adds an unpredictable delay to the transport time of the labeled nucleotide from the DNA strand to the detection volume. Surface adsorption can be reduced in polymethylmethacrylate (PMMA) microchannels with diameters of about 10 μm in the detection area.[54,114,115] To excite all molecules passing the channel efficiently, the laser beam was shaped by a cylindrical lens and focused by a microscope objective to irradiate the entire channel. To reduce Raman scattering, the volume element was imaged onto a glass fiber bundle where seven fibers were aligned. Each fiber was connected to its own separate detector, thus producing seven overlapping femtoliter volume elements. Eigen and Rigler[54,116] proposed the use of small channels and multiple confocal probe volumes for DNA sequencing based on exonuclease-mediated cleavage of fluorescently labeled nucleotide molecules from a single DNA fragment (Figure 6.17). A single 5′-biotinylated, fluorescently labeled DNA strand is immobilized on a streptavidin-coated microsphere. This microsphere was selected due to its fluorescence brightness, and manipulated in an optical trap.[117] Alternatively, the increase in viscous drag caused by the attachment of a DNA strand to a bead can be used to monitor the selection of a single strand.[11] This method allows one to distinguish between one and two DNA fragments bound to the bead.

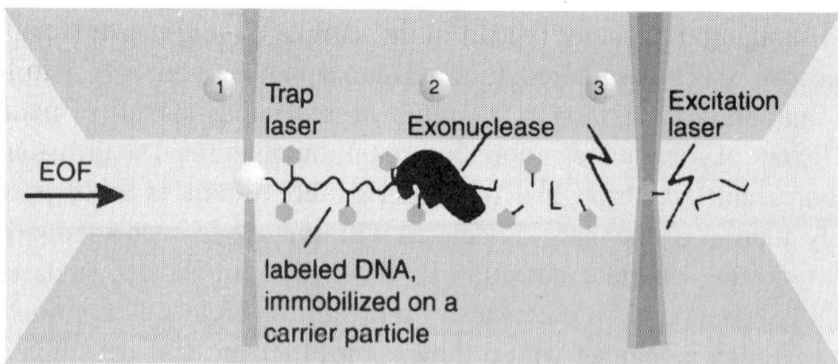

FIGURE 6.17 The principle of single-molecule sequencing. (1) A bead loaded with a labeled DNA molecule is held by a trap laser ($\lambda = 1064$ nm) inside a transparent microstructure. (2) The DNA is degraded sequentially by an exonuclease. The liberated monomers are transported to the detection focus via EOF. (3) Passing the focus, the labeled monomers are excited by a laser ($\lambda = 532$ nm) and emit photon bursts. The fluorescence characteristics for each burst can be used to identify the label. (From K Dörre et al., Bioimaging 6:139–152, 1997. With permission.)

Most importantly, this technique does not require fluorescence detection, so there is no bleaching of fluorescent dyes during the selection process.

As described in Reference 115, microstructures were manufactured from transparent, chemically inert material (PMMA) with microchannels generated by laser ablation at $\lambda = 157$ nm. This provides steep and extremely smooth walls, which do not adversely affect the detection efficiency. The direction and velocity of the beads and cleaved nucleotides in the channel are controlled by an electrical field, which induces an electroosmotic flow (EOF) toward the cathode. The narrow part of the microstructure serves as the detection volume with a cross section of 5×5 µm. A 218-mer model DNA sequence containing either six tetramethylrhodamine (TMR) or Rhodamine Green labeled dUTPs was used as a template for enzymatic digestion.[84] Photobleaching of the exonuclease solution (T7 DNA polymerase) was reduced the concentration of fluorescent impurities from approximately 2 nM before to about 0.8 pM after bleaching. The buffer contained 10 mM MgCl$_2$, 5 mM NaCl, and 0.5% Triton X-100. This buffer minimizes the adhesion of beads to the walls of the microstructure while maintaining a high activity of the T7 DNA polymerase. Before a typical DNA sequencing experiment was started, a microstructure is filled with purified reaction buffer by capillary forces. For the initial tests, the ratio of DNA (218-mer, single-stranded, and labeled with six TMR molecules) to beads during preparation was 100:1. The DNA-labeled bead solution (10 µl) was placed in the microstructure at the anode. Upon application of an electrical field of 60 V/cm, the resulting EOF (~100 µm/s) moves the beads toward the cathode. After the beads travel some distance into the channel, the electrical field is switched off for a few seconds and one of the beads is trapped by an IR laser (100 mW, 1064 nm). At this point, the microstructure can be moved as needed, and the trapped bead remains stationary in the channel. Inside the narrow detection channel, the bead is attached

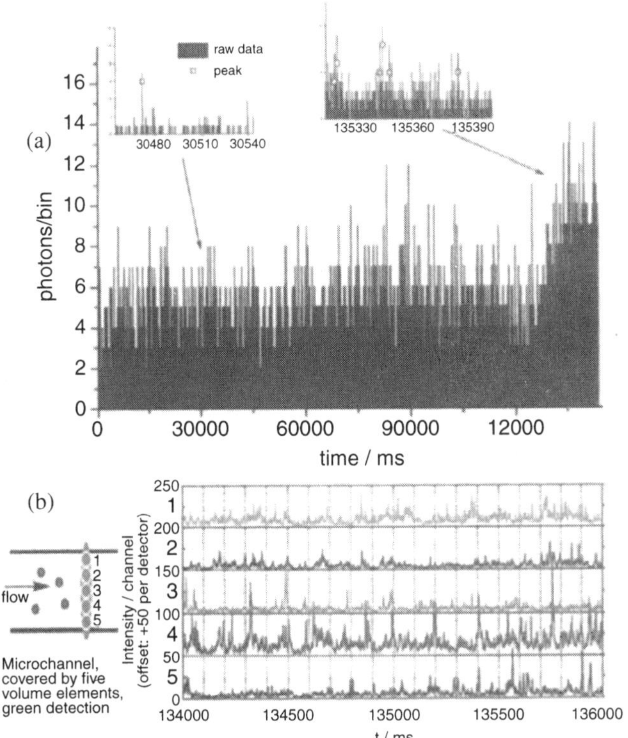

FIGURE 6.18 (a) Recorded raw data of one detector trace. The experimental trace was started 120 s after injection of photobleached enzyme solution. The two insets show a part of the fluorescence trajectory before (left) and after the sequencing process (right). Here, the timescale is enlarged in to show the single-molecule events. (b) Lee-filtered and squared data of all five detected traces of the experiment. For better visualization, the traces of neighboring detection elements are plotted on top of each other with an offset of 50 intensity units per detector. (From J Stephan et al., J Biotechnol 86:255–267, 2001. With permission.)

to the wall by adhesion, approximately 15 μm in front of the detection volume. In this case, the excitation beam was elliptically shaped and covered several overlapping detection volumes.

Digestion of the DNA strands is induced by injection of a large excess of T7 DNA polymerase (if there are no mononucleotides available T7 and other polymerases act as exonucleases). To allow the enzyme to reach the DNA strands, data recoding is started some time after (~120 s) application of an electrical field of 400 V/cm. In these experiments, background fluorescence bursts are registered with a frequency of ~14 bursts/s. When the enzyme reaches the DNA, degradation began and the fluorescence burst frequency increased to 90 bursts/s (Figure 6.18). After less than 1 min, the burst frequency dropped back down to 14 bursts/s, indicating that the 218-mer was completely degraded. The authors concluded that in this experiment approximately 300 DNA strands, each containing six TMR molecules, were degraded in less than 1 min by an excess of exonuclease enzymes. This amounts to an average cleavage rate of approximately three nucleotides per second.[116] These results, with the background burst rate of ~14 bursts/s and exonucleolytic cleavage rates of only a few nucleotides per second, indicate that much work is still needed before a single DNA fragment can be observed and sequenced.

To circumvent the problems associated with background fluorescence from impurities, excitation and detection in the red spectral range is a valuable alternative. Shifting to longer-wavelength excitation dramatically reduces the efficiency of both Rayleigh and Raman scattering (these processes scale with the $1/\lambda^4$). Likewise, the number of fluorescent impurities is significantly reduced with longer excitation and detection wavelengths.[102,118] In addition to reduced background, a further advantage is that low-cost, energy efficient, rugged diode lasers can be used in place of the more expensive and shorter-lived gas lasers. The advantages of red-absorbing fluorophores has prompted current efforts to develop new fluorescent dyes that absorb and emit above 620 nm but still exhibit a sufficient fluorescence quantum yield, especially in aqueous surrounding.[119–121] Among these new red-absorbing dyes are rhodamine,[120,122] bora-diaza-indacene,[52] oxazine,[120,123] squaraine,[124,125] and indocarbocyanine dyes.[126–129]

Soper and coworkers[130] first demonstrated the detection of single near-infrared fluorescent molecules dissolved in methanol. The relatively low fluorescence quantum yield of most known near-infrared dyes makes detection at the single-molecule level in aqueous solutions difficult. Recently Sauer and coworkers[52,131] showed efficient detection and time-resolved identification of single red-absorbing dyes in aqueous solvents. Using a combination of new rhodamine and oxazine dyes together with pulsed diode laser excitation at 635 nm enables the detection of single molecules with high S/B ratio. It was also shown that by using a suitable combination of excitation wavelength and fluorescent dye, even single fluorescently labeled antibody molecules can be detected in undiluted human serum samples.[132] This method has that important advantage that the number of fluorescence bursts from impurities during DNA sequencing can be significantly reduced.

Zander and Drexhage[133] demonstrated that single R6G molecules dissolved in pure ethylene glycol in a 1-μm-diameter capillary can be sequentially counted. Under the conditions used, adsorption of dye to the capillary glass walls did not pose a problem. If dynamic adsorption of fluorescent dyes to channel walls can be suppressed efficiently, we should be able to precisely control the movement of single molecules. To restrict solution flow so that all mononucleotide molecules travel through the detection area in a microchannel with an inner diameter <1 μm, techniques known from conventional capillary electrophoresis (CE) can be applied. While in conventional CE the analyte molecules are separated based on charge and/or mass of the analytes, in single-molecule DNA sequencing it is preferable that each labeled mononucleotide be transported to the detection area with the same velocity. This is because the interpreted sequence will be incorrect if there are large differences in the rate of transport such consecutively cleaved bases are not detected in the order they were cleaved. The misorder probability depends strongly on the distance between the position of release and the detection area as well as the cleavage rate of the

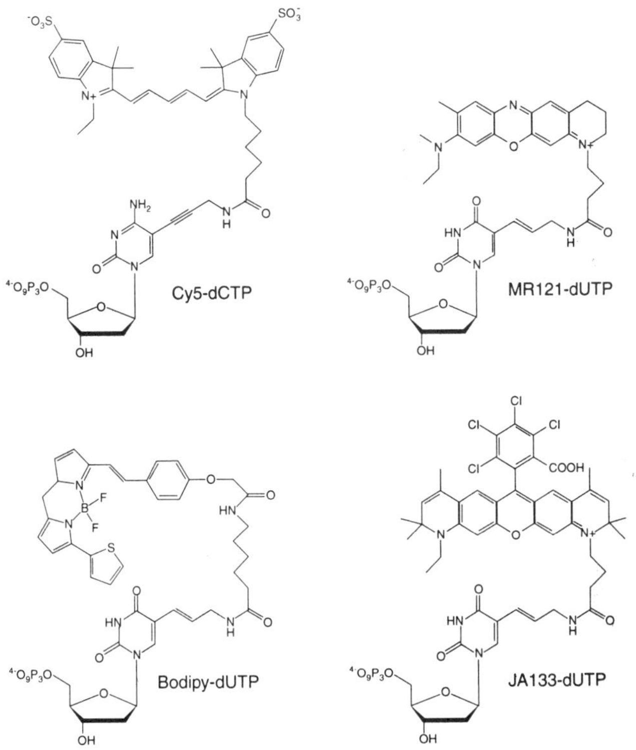

Cy5-dCTP

MR121-dUTP

Bodipy-dUTP

JA133-dUTP

FIGURE 6.19 Molecular structures of DNA nucleotides labeled with four different red-absorbing fluorescent dyes. (From M Sauer et al., J Biotech 86:181–201, 2001. With permission.)

exonuclease. With a short cleavage–detection separation distance (and time) in combination with a slow exonuclease cleavage rate (a few nucleotides per second), the misorder probability can be rendered insignificant. At higher cleavage rates, the misorder probability will increase if the velocities of the differently labeled nucleotide molecules are not efficiently matched. We emphasize again here the importance of preventing adsorption of nucleotides to the glass walls. In general, a dynamic coating that prevents surface adsorption is critical.

In considering the nucleotide transport velocity, there are two effects to consider: (1) the electrophoretic mobility, μ_{ep}, of charged mononucleotide molecules, and (2) the electroosmotic mobility, μ_{eo}. The total charge of the various dNTPs is strongly affected by the choice of dye. Although the four fluorescently labeled mononucleotide molecules in Figure 6.19 have similar absorption and emission characteristics, they exhibit different total charges at neutral pH. After exonucleolytic cleavage from a DNA strand, the dNMP molecules have a total charge of –3 (Cy5-dCMP), –1 (MR121-dUMP), –1 (Bodipy-dUTP), and –2 (JA133-dUMP), respectively. Therefore, they exhibit different electrophoretic mobilities.

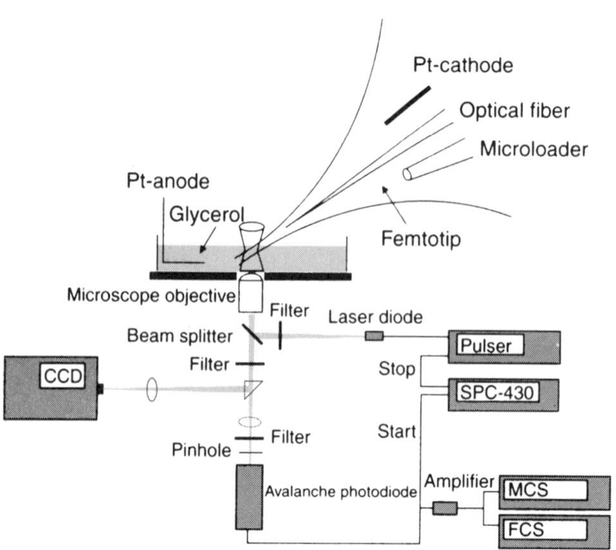

FIGURE 6.20 Schematic diagram of the optical and electronic setup. A pulsed diode laser emitting at 635 nm served as excitation source. The laser system provided light pulses with duration of less than 400 ps full width at half maximum (FWHM) at a repetition rate of 57 MHz. The laser light is coupled into a high numerical aperture oil-immersion objective by a dichroic beam splitter. Fluorescence is collected by the same objective, filtered by bandpass filters and imaged onto a 100 μm pinhole oriented directly in front of an APD. The detector signal is split between two PC plug-in cards, one for online monitoring of the fluorescence intensity and the other for data collection and TCSPC. The instrument response function of the entire system was 420 ps. From TCSPC-data, multichannel-scalar (MCS) traces are generated. All photons of a decay curve are summed for each bin of the MCS-trace. The diameter of the detection volume, approximately 1 μm, is defined by the pinhole size used. To ensure detection of all fluorescent molecules, a microcapillary with an inner diameter of about 500 ± 200 nm at the tapered end is used. The capillary dips into a tissue culture dish containing pure glycerol, which is well index matched to the capillary glass and minimizes vibrations of the capillary. The capillary is adjusted so that the focus of the excitation laser beam is about 1 μm in front of the tip orifice. A three-axis electrostrictive actuator and a CCD camera are used for this alignment. The capillary is filled with a solution of 20 m*M* Tris-borate pH 8.4, 3% (w/v) polyvinylpyrolidone (PVP), containing 69.9% water, 30% glycerol, and 0.1% (v/v) Tween 20. The flow of the negatively charged, labeled mono-nucleotide molecules through the capillary is established by electrokinetic forces using two platinum electrodes; the cathode (67 μm diameter) is inserted into the capillary and the anode dips into the glycerol outside of the capillary. (From M Sauer et al., J Biotech 86:181–201, 2001. With permission.)

Recently, the time-resolved identification of individual fluorescent dyes as they flow through a tapered glass capillary with an inner diameter of 500 ± 200 nm at the sharp end was been demonstrated (Figure 6.20).[134,135] Because the inner diameter of the channel is smaller than the diameter of the detection volume (~1 μm), every molecule passing through the tip orifice can be the detected, with the exception of

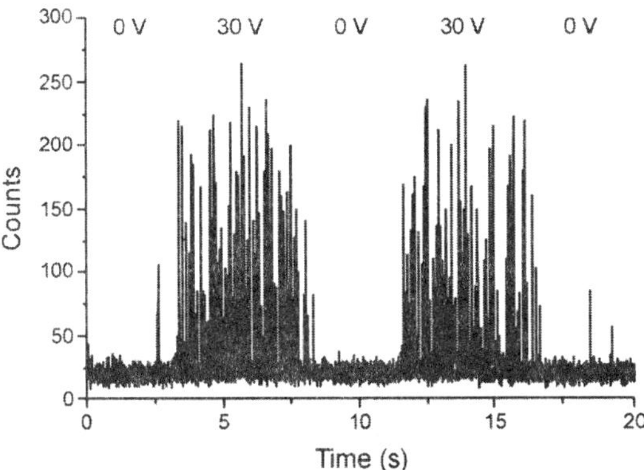

FIGURE 6.21 Fluorescence signals (3 ms/bin) observed from MR121-dUTP molecules at a range of applied voltages (anode outside of the capillary). (From M Sauer et al., J Biotech 86:181–201, 2001. With permission.)

those that photobleach very quickly. The addition of 3% polyvinyl pyrrolidone (PVP), 30% glycerin, and 0.1% (v/v) Tween 20 (a nonionic detergent) to a 20 mM Tris-borate buffer, pH 8.4, efficiently suppressed the adsorption of analyte molecules to the glass surface of the capillary and reduced the EOF. As described in Reference 136, PVP matrices have a very low viscosity at moderate concentrations (27 cP at 4.5%) and an excellent self-coating property that virtually eliminates EOF. As shown by the fluorescence intensity trace in Figure 6.21, precise control of the movement of MR121-dUTP molecules toward the anode by electrophoretic forces is possible. The velocity of MR121-dUTP molecules increases with increasing applied voltage. Cy5-dCTP and MR121-dUTP molecule have similar electrophoretic mobilities in a 3% PVP matrix even though they have different total charges.[99] The transport velocity of these two labeled bases was tested by the following procedure. A 1:1 mixture of the two labeled nucleotides was adsorbed onto a streptavidin-coated fiber, transferred into the microcapillary, and released by application of an electrical field of 15 V. The fluorescence decay times of the first events reaching the detection volume could be identified as both Cy5-dCTP and MR121-dUTP molecules. This indicates that both conjugates have similar mobilities. Hence, the use of the tapered glass capillary, when used with appropriate buffer conditions (which prevent surface adsorption) appears to be a valuable alternative to hydrodynamic focusing or planar microchannel devices for single-molecule DNA sequencing.

Figure 6.22 shows time-resolved fluorescence signals observed in a tapered micro-capillary after transfer of a 3-µm etched optical fiber loaded with Cy5-dCTP molecules. The etched optical fiber was streptavidin coated and dipped into 10^{-6} M dye solutions for 1 s. The fiber was positioned in the microcapillary approximately 50 µm in front of the detection volume at the thin end of the capillary. Upon application of an electrical field of a few volts, single nucleotide molecules are released and moved to the detection volume. With the setup described, an average background count rate was 2.5 kHz. Single-molecule fluorescence bursts were analyzed using the following burst recognition procedure.

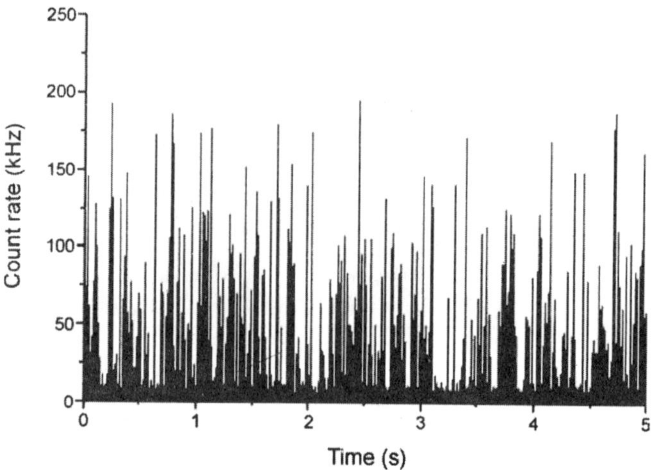

FIGURE 6.22 Raw data showing fluorescence bursts of single Cy5-dCTP molecules in the submicrometer capillary released from a streptavidin-coated, etched glass fiber at 15 V. The fiber was positioned in front of the detection area. The data were binned into 500 μs time intervals. The streptavidin-coated fiber was dipped for 1 s into a 10^{-6} M solution Cy5-dCTP. Solvent in the capillary: 3% PVP, 20 mM Tris-borate pH 8.4, containing 30% glycerin, and 0.1% (v/v) Tween 20. Average excitation power at the sample: 300 μW at 635 nm, repetition rate: 57 MHz, pulse length: ~300 ps (FWHM). (From M Sauer et al., J Biotech 86:181–201, 2001. With permission.)

First, background was suppressed by selecting only regions of the data with count rates higher than 40 kHz. The start and end point of a burst was defined by a count rate of less than 10 kHz. When two count rate maxima are detected close together in time, the burst was split at the minimum count rate between the two maxima. Each recognized burst is characterized by three parameters: (1) the number of detected photon counts per burst (burst size), (2) the duration time of the burst, and (3) the fluorescence lifetime. Statistics were accumulated for 10,000 single-molecule bursts from pure labeled nucleotide molecules Cy5-dCTP, MR121-dUTP, and JA133-dUTP. Figure 23a and b show the distributions of burst duration and burst size, respectively. Photon bursts larger than ~200 counts or longer than ~6 ms are assumed to be due to two or more dNTP molecules simultaneously passing the detection volume. Each type of labeled dUTPF had comparable burst size maxima (80 for Cy5-dCTP, 100 for MR121-dUTP, and 105 for JA133-dUTP) indicating that the detection efficiency is nearly the same for all three conjugates.[94] The burst duration maxima are located at ~2 ms for all three conjugates, indicating similar velocities in the microcapillary under an applied electrical field of 15 V.

The fluorescence lifetimes of the bursts were calculated using the MLE algorithm.[46,48,71,72] The experimental standard deviation σ_{exp} was obtained from the distribution of the calculated fluorescence lifetimes. The distributions of fluorescence lifetimes (Figure 6.23c) were fit with Gaussians. The results demonstrate that three labeled mononucleotides can be identified during their flow through the 500-nm

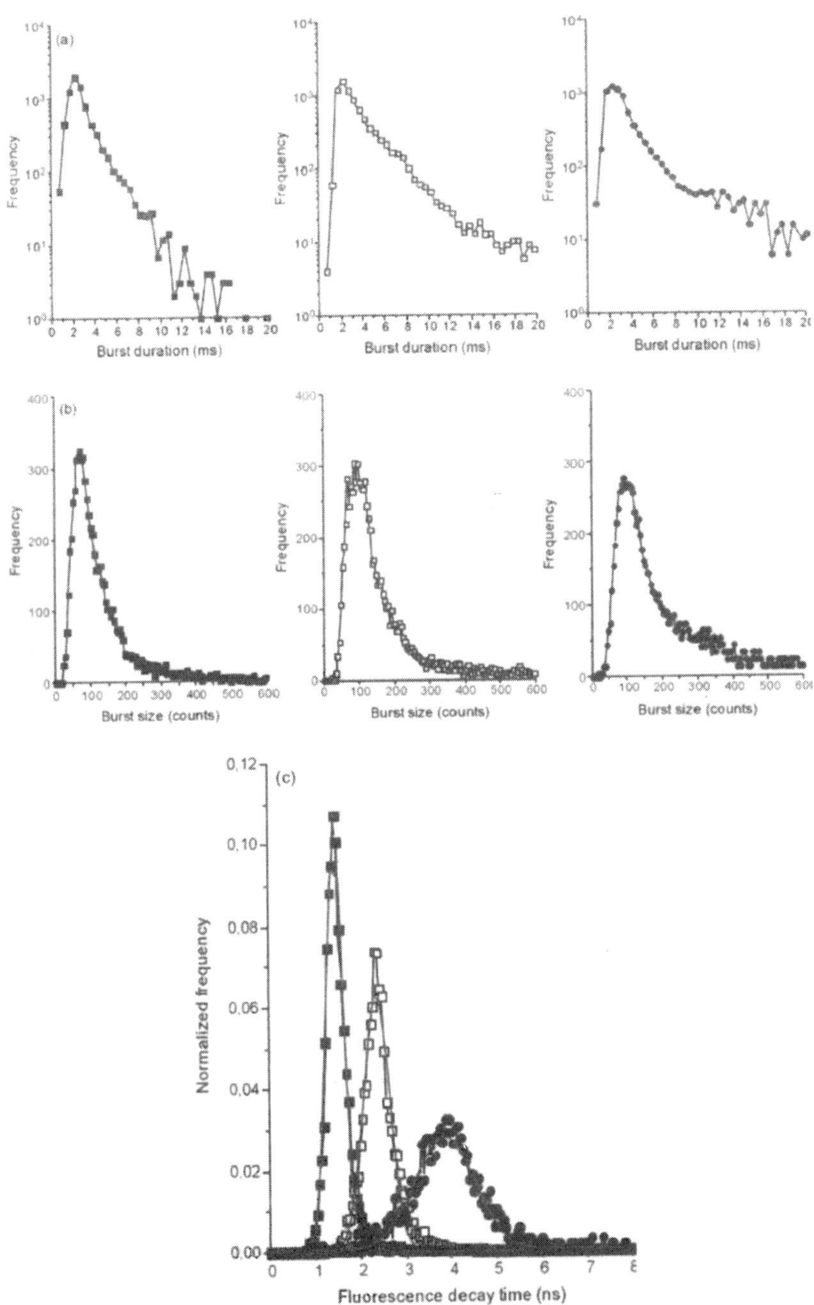

FIGURE 6.23 (a) Burst duration (semilog plot), (b) burst size, and (c) fluorescence lifetime distributions of fluorescence bursts detected from Cy5-dCTP (solid squares), MR121-dUTP (open squares), and JA133-dUTP (solid circles) released from an optical fiber in the microcapillary. (From M Sauer et al., J Biotech 86:181–201, 2001. With permission.)

TABLE 6.2
**Spectroscopic Characteristics of the Conjugates Cy5-dCTP,
MR121-dUTP, Bodipy-dUTP, and JA133-dUTP at 25°C in the
Solvent Mixture (3% PVP, 20 mM Tris-borate buffer pH 8.4,
0.1% (v/v) Tween 20, 30% glycerin)**

	$\lambda_{abs, max}$ (nm)	$\lambda_{em, max}$ (nm)	τ (ns)
Cy5-dCTP	652	671	1.32
MR121-dUTP	661	673	2.31
Bodipy-dUTP	635	652	3.92
JA133-dUTP	624	644	3.96

capillary orifice by their characteristic fluorescence decay times of 1.43 ± 0.19 ns (Cy5-dCTP), 2.35 ± 0.29 ns (MR121-dUTP), and 3.83 ± 0.67 ns (JA133-dUTP). By forming the convolution of the normalized Gaussians, the probability of correct classification is 83% (6% for misclassification of Cy5-dCTP and MR121-dUTP, 9% for misclassification of MR121-dUTP and JA133-dUTP, and 2% for misclassification of Cy5-dCTP and JA133-dUTP) (Table 6.2).

Using the microcapillary technique and time-resolved fluorescence detection in the red spectral range, Sauer and coworkers[94] demonstrated partial (two base) DNA sequencing by single-molecule detection: the order of the detected nucleotides released during exonuclease degradation of several 218-mer single-stranded model DNA fragments labeled with fluorophores with different fluorescence lifetimes at the U and C positions reflected the known two-base sequence. In these experiments, the etched optical fiber was dipped for 1 s in a 10^{-10} M single-stranded DNA solution containing 6 Cy5-dCTP and 6 MR121-dUTP as fluorescent nucleotides at well-defined positions. After transfer into the microcapillary (positioned about 50 μm upstream of the detection volume) an electric potential of 15 V was applied between the inside and side of the microcapillary. As shown by the fluorescence intensity trajectory in Figure 6.24, during the first ~300 s a fluorescence burst rate of ~0.05 Hz, i.e., 1 burst/20 s, was recorded. The calculated fluorescence decay times of these bursts indicate that even after dipping into a 10^{-10} M DNA solution, unincorporated fluorescently labeled nucleotides have been adsorbed to the fiber tip. Approximately 600 s after the start of the experiment, exonuclease I solution was injected into the microcapillary. About 150 s after addition of exonuclease solution, the first fluorescent nucleotides are detected. During the next 420 s (980 to 1400 s after start of the experiment) 86 fluorescence bursts were recognized. After 1400 s the burst count rate dropped to the background level. For further analysis each detected burst with a fluorescence decay time shorter than 1.85 ns (point of intersection of the Gaussians in Figure 6.23c) was assigned as a Cy5-dCMP molecule, whereas all bursts with decay times longer than 1.85 ns were assigned as MR121-dUMP molecules. By using these assignment criteria, 43 Cy5-dCMP and 43 MR121-dUMP molecules were identified at a fluorescence burst rate of 0.5 Hz. Considering the background burst rate of 0.05 Hz, 21 of the 86 detected fluorescence bursts originate from the background. Hence, an estimated 5 to 6 DNA-strands (65 signals) were bound

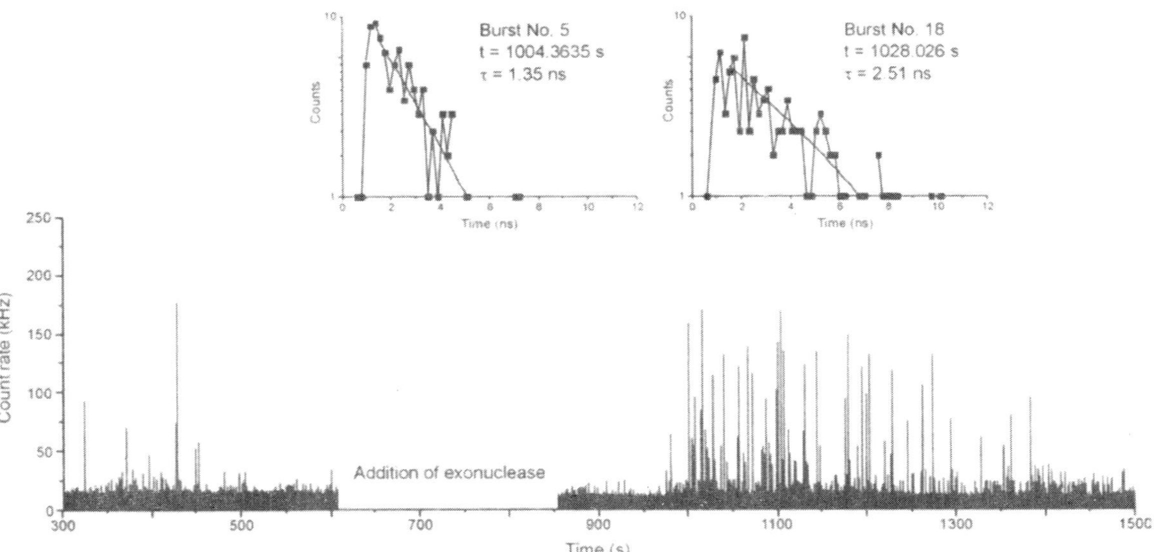

FIGURE 6.24 Fluorescence signals (0.5 ms/bin) recorded after transfer of a fiber loaded with five to six DNA strands (300 to 1500 s). The streptavidin-coated fiber was dipped for ~1 s into a 10^{-10} M solution containing synthetic model DNA (six Cy5-dCTP and six MR121-dUTP labels). The tip of the fiber was placed ~50 mm in front of the detection area and flow was induced by applying 15 V between the cathode and anode. After ~300 s, a burst count rate of ~0.05 Hz was measured. Exonucleolytic cleavage of nucleotides was initiated by adding a dilute exonuclease I solution. Between 980 and 1400 s, 86 fluorescence bursts were registered by using the burst recognition procedure. The two insets show typical fluorescence decays monitored during the experiment. (From M Sauer et al., J Biotech 86:181–201, 2001. With permission.)

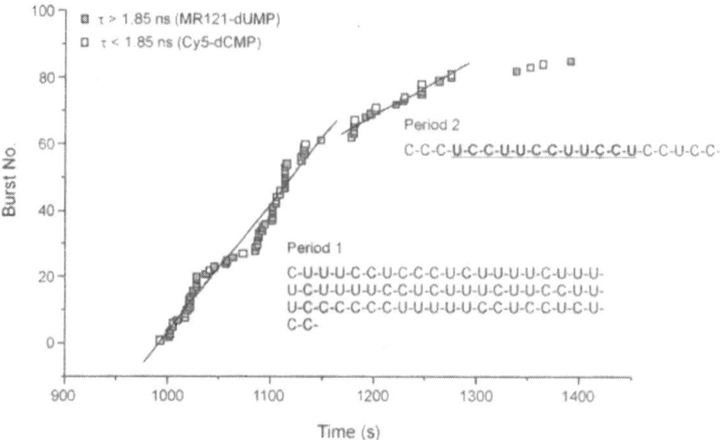

FIGURE 6.25 Detected burst number vs. time for the time interval between 980 and 1400 s. The sequence was derived from the calculated fluorescence decay times of each burst. From the detected burst count rate, the data can be roughly divided in two areas with different burst count rate (for more details, see text). (From M Sauer et al., J Biotech 86:181–201, 2001. With permission.)

to the fiber tip in the experiment. Figure 6.25 shows the sequence of detected fluorescence bursts (determined by the different fluorescence lifetimes of Cy5 and MR121) in the time period between 900 and 1450 s. Two periods with different average burst rates can be recognized. During the first period, an overall cleavage rate of 7 Hz was observed for exonuclease I on single-stranded labeled DNA under the experimental conditions (3% PVP gel, 20 mM Tris-borate buffer, pH 8.4 containing 30% glycerol, and 0.1% (v/v) Tween 20, room temperature, 15 V applied). The measured sequence during the first period did not reproduce the known DNA sequence. The expected sequence, UCCUUCCUUCCU, of a single DNA-strand was obtained during the next period (Figure 6.25 period 2). Assuming that these 12 signals originate from a single DNA-strand, a cleavage rate of 3.3 Hz for Exo I on a single labeled DNA strand was measured. Because of the relatively large time delays between subsequent bursts, this data indicate nonprocessive behavior of Exo I on single-stranded labeled DNA. To demonstrate that the detected signals did not originate from impurities from the exonuclease solution, the same experiment was carried out without DNA. In this case, the background fluorescence burst rate of 0.05 Hz did not increase.[94]

FUTURE PROSPECTS

The complete, error-free labeling of a DNA strand is within reach using new mutant polymerases. The current bottleneck of exonuclease-mediated sequencing of a fluorescently labeled DNA fragment is due to the low cleavage rates of exonucleases on modified DNA. As long as the frequency of fluorescent bursts from the background occur at the same order of magnitude as the cleavage rates, single-molecule sequencing will be not possible. Exonucleases with high processivity and turnover rates on fluorescently labeled DNA are required. By using very dilute exonuclease concentrations, it is more likely that only a single exonuclease will bind the DNA fragment to be cleaved. Despite the high identification accuracy of time-resolved single-molecule detection using two spectrally separated detectors, it should be pointed out that that errors in single-molecule identification on the order of a few percent will be very difficult to eliminate. It may be necessary to sequence the DNA fragment several times to locate and remove sequence errors. In the absence of background fluorescence bursts and assuming all bases are detected and identified with an error rate of 10%, the sequence can be obtained with an accuracy of 99.99% with ten runs.[97] To compete with current highly parallel capillary DNA sequencing, at least one order of magnitude increase in digestion rate should be accomplished.

To test and compare different enzymes and detection techniques, the availability of model DNAs, such as the 218-mer DNA used by several groups, is critical. They will provide a means to calibrate and optimize each of the steps in the development of new single-molecule DNA sequencing strategies.

A new single-molecule based DNA sequencing strategy based exclusively on DNA polymerase activity is currently being developed in the laboratories of Webb and Craighead at Cornell University.[137] Their approach is based on observing the sequential base additions during DNA polymerization of a single nucleic acid molecule in real time, i.e., polymerase-mediated single-molecule DNA sequencing. Their efforts thus far have focused on optimizing the enzymatic system and the nanostructured devices needed.[138,139]

As already mentioned, DNA polymerases are usually inhibited when native nucleotides are replaced by fluorescently labeled nucleotide analogues. In collaboration with Amersham Biosciences, researchers are searching for an answer to this problem. Their approach is to develop novel nucleotide analogues in which the fluorophore is attached to the gamma-phosphate. This is analogous to nucleotide analogues that have been developed for RNA polymerase studies.[140] The fluorophore is cleaved from the nucleotide by the enzymatic activity of the polymerase, thereby yielding native, unmodified DNA. Because polymerases synthesize DNA most efficiently at micromolar nucleotide concentrations, zero-mode waveguide nanostructures were developed, which effectively reduce the optical observation volume to tens of zeptoliters (1 zl = 10^{-21} l), over three orders of magnitude smaller than the size of a diffraction-limited laser focus.[141] Zero-mode waveguides thereby enable an inversely proportional increase in the upper limit of fluorophore concentrations amenable to single-molecule detection, extending the range of biochemical reactions that can be studied on a single-molecule level into the micromolar range. The technical challenges that remain are the development of suitable enzymatic systems and in the recognition of individual sequential base additions. The approach should lead to a very fast sequencing protocol with long read lengths. Furthermore, the sequencing can probably be run in many parallel lanes on the same device, in an integrated system with extremely high throughput.

Although DNA sequencing has been a primary goal, each development step toward that goal is productive for the generation and improvement of analytic research systems capable of following biochemical processes (e.g., enzymatic activities) at the single-molecule level. The optical tools will undoubtedly enable characterization of these processes previously unattained by conventional biochemical analysis.

REFERENCES

1. JC Ventor et al. The sequence of the human genome. Science 291:1304–1351, 2001.
2. DR Bently et al. The physical maps for sequencing human chromosomes 1, 6, 9, 10, 13, 20, and X. Nature 15:942–943, 2001.
3. SM Lindsay, M Philipp. Can the scanning tunneling microscope sequence DNA? Gen Anal Tech Appl 8:8–13, 1991.
4. W Bains, GC Smith. A novel method for nucleic acid sequence determination. J Theor Biol 135:303–307, 1988.
5. R Drmanac, I Labat, I Brukner, R Crkvenjakov. Sequencing of megabase plus DNA by hybridisation: theory of the method. Genomics 4:114–128, 1989.
6. SPA Fodor. Massively parallel genomics. Science 277:393–395, 1997.
7. F Hillenkamp, M Karas, RC Beavis, BT Chait. Matrix-assisted laser desorption/ionisation mass spectrometry of biopolymers. Anal Chem 63:1193A–1203A, 1991.
8. KK Murray. DNA sequencing by mass spectrometry. J Mass Spectr 31:1203–1215, 1996.
9. A Meller, L Nivon, E Brandin, J Golovchenko, D Branton. Rapid nanopore discrimination between single polynucleotide molecules. Proc Natl Acad Sci USA 97:1079–1084, 2000.
10. W Vercoutere, S Winters-Hilt, H Olsen, D Deamer, D Haussler, M Akeson. Rapid discrimination among individual DNA hairpin molecules at single nucleotide resolution using an ion channel. Nat Biotechnol 19:248–252, 2001.
11. J Dapprich, N Nicklaus. DNA attachment to optically trapped beads in microstructures monitored by bead displacement. Bioimaging 6:25–32, 1998.
12. S Wennmalm, H Blom, L Wallerman, R Rigler. UV-fluorescence correlation spectroscopy of 2-aminopurine. Biol Chem 382:393–397, 2001.
13. F Seela, M Zulauf, M Sauer, M Deimel. 7-Substituted 7-deaza-2′-deoxyadenosines and 8-aza-7-deaza-2′-deoxyadenosines: fluorescence of DNA-base analogues induced by the 7-alkynyl side chain. Helv Chim Acta 83:910–927, 2000.
14. JR Lakowicz, B Shen, Z Gryczynski, S D'Auria, I Gryczynski. Intrinsic fluorescence from DNA can be enhanced by metallic particles. Biochem Biophys Res Commun 286:875–879, 2001.
15. JH Jett, RA Keller, JC Martin, BL Marrone, RK Moyzis, RL Ratliff, NK Seitzinger, EB Shera, CC Stewart. High-speed DNA sequencing: an approach based upon fluorescence detection of single molecules. J Biomol Struct Dyn 7:301–309, 1989.
16. JD Harding, RA Keller. Single-molecule detection as an approach to rapid DNA sequencing. TIBTECH 10:55–57, 1992.
17. EB Shera, NK Seitzinger, LM Davis, RA Keller, SA Soper. Detection of single fluorescent molecules. Chem Phys Lett 174:553–557, 1990.
18. LR Pratt, RA Keller. Estimate of the probability of diffusional misordering in high-speed DNA sequencing. J Phys Chem 97:10254–10255, 1993.
19. PM Goodwin, WP Ambrose, RA Keller. Single-molecule detection in liquids by laser-induced fluorescence. Acc Chem Res 29:607–613, 1996.
20. T Plakhotnik, EA Donley, UP Wild. Single-molecule spectroscopy. Annu Rev Phys Chem 48:181–212, 1997.
21. XS Xie, JK Trautman. Optical studies of single molecules at room temperature. Annu Rev Phys Chem 49:441–480, 1998.
22. S Weiss. Fluorescence spectroscopy of single biomolecules. Science 283:1676–1683, 1999.

23. WE Moerner, M Orrit. Illuminating single molecules in condensed matter. Science 28:1670–1676, 1999.
24. WP Ambrose, PM Goodwin, JH Jett, A van Orden, HJ Werner, RA Keller. Single molecule fluorescence spectroscopy at ambient temperature. Chem Rev 99:2929–2956, 1999.
25. AA Deniz, TA Laurence, M Dahan, DS Chemla, PG Schultz, S Weiss. Ratiometric single-molecule studies of freely diffusing biomolecules. Annu Rev Phys Chem 52:233–253, 2001.
26. T Hischfeld. Optical microscopic observation of single small molecules. Appl Opt 15:2965–2966, 1976.
27. NJ Dovichi, JC Martin, JH Jett, RA Keller. Attogram detection limit for aqueous dye samples by laser-induced fluorescence. Science 219:845–847, 1983.
28. NJ Dovichi, JC Martin, JH Jett, M Trkula, RA Keller. Laser-induced fluorescence of flowing samples as an approach to single-molecule detection in liquids. Anal Chem 56:348–354, 1984.
29. F Zarrin, NJ Dovichi. Sub-picoliter detection with the sheath flow cuvette. Anal Chem 57:2690–2692, 1985.
30. WE Moerner, L Kador. Finding a single molecule in a haystack: Optical detection and spectroscopy of single absorbers in solids. Anal Chem 61:1217A–1223A, 1989.
31. E Betzig, RJ Chichester. Single molecules observed by near field scanning optical microscopy. Science 262:1422–1425, 1993.
32. JK Trautman, JJ Macklin, LE Brus, E Betzig. Near-field spectroscopy of single molecules at room temperature. Nature 369:40–42, 1994.
33. XS Xie, RC Dunn. Probing single molecule dynamics. Science 265:361–364, 1994.
34. WP Ambrose, PM Goodwin, JC Martin, RA Keller. Single molecule detection and photochemistry on a surface using near-field optical excitation. Phys Rev Lett 72:160–163, 1994.
35. JJ Macklin, JK Trautman, TD Harris, LE Brus. Imaging and time-resolved spectroscopy of single molecules at an interface. Science 272:255–258, 1996.
36. R Rigler J Widengren. BioScience 3:180–188, 1990.
37. R Rigler, U Mets, J Widengren, P Kask. Fluorescence correlation spectroscopy with high count rate and low background: analysis of translational diffusion. Eur Biophys J 22:169–175, 1993.
38. Ü Mets, R Rigler. Submillisecond detection of single rhodamine molecules in water. J Fluoresc 4:259–264, 1994.
39. S Nie, DT Chiu, RN Zare. Probing individual molecules with confocal fluorescence microscopy. Science 266:1018–1021, 1994.
40. J Widengren, P Schwille. Characterization of photoinduced isomerization and back-isomerization of the cyanine dye Cy5 by fluorescence correlation spectroscopy. J Phys Chem A 104:6416–6428, 2000.
41. R Menzel, R Bornemann, E Thiel. Influence of chemical substitution and electronic effects on the triplet state kinetics of xanthene dyes. Phys Chem Chem Phys 1:2435–2442, 1999.
42. C. Eggeling, L. Brand, CAM Seidel. Laser-induced fluorescence of coumarin derivatives in aqueous solution: Photochemical aspects for single molecule detection. Bioimaging 5:105–115, 1997.
43. T Hirschfeld. Quantum efficiency independence of the time integrated emission from a fluorescent molecule. Appl Opt 15:3135–3139, 1976.

44. RA Mathis, K Peck, L Stryer. Optimization of high-sensitivity fluorescence detection. Anal Chem 62:1786–1791, 1990.

45. CW Wilkerson, PM Goodwin, WP Ambrose, JC Martin, RA Keller. Detection and lifetime measurement of single molecules in flowing sample streams by laser-induced fluorescence. Appl Phys Lett 62:2030–2032, 1993.

46. J Enderlein, PM Goodwin, A Van Orden, WP Ambrose, R Erdmann, RA Keller. A maximum likelihood estimator to distinguish single molecules by their fluorescence decays. Chem Phys Lett 270:464–470, 1997.

47. JR Fries, L Brand, C Eggeling, M Köllner, CAM Seidel. Quantitative identification of different single molecules by selective time-resolved confocal fluorescence spectroscopy. J Phys Chem A 102:6601–6613, 1998.

48. C Zander, M Sauer, KH Drexhage, DS Ko, A Schulz, J Wolfrum, L Brand, C Eggeling, CAM Seidel. Detection and characterization of single molecules in aqueous solution. Appl Phys B 63:517–523, 1996.

49. R Müller, C Zander, M Sauer, M Deimel, DS Ko, S Siebert, J Arden-Jacob, G Deltau, NJ Marx, KH Drexhage, J Wolfrum. Time-resolved identification of single molecules in solution with a pulsed semiconductor diode laser. Chem Phys Lett 262:716–722, 1996.

50. M Sauer, C Zander, R Müller, B Ullrich, S Kaul, KH Drexhage, J Wolfrum. Detection and identification of individual antigen molecules in human serum with pulsed semiconductor lasers. Appl Phys B 65:427–433, 1997.

51. R Müller, KH Drexhage, DP Herten, U Lieberwirth, M Neumann, M Sauer, A Schulz, S Siebert, J Wolfrum. Efficient DNA sequencing with pulsed semiconductor lasers and a new fluorescent dye set. Chem Phys Lett 279:282–288, 1997.

52. M Sauer, J Arden-Jacob, KH Drexhage, F Göbel, U Lieberwirth, K Mühlegger, R Müller, J Wolfrum, C Zander. Time-resolved identification of individual mononucleotide molecules in aqueous solution with pulsed semiconductor lasers. Bioimaging 6:14–24, 1998.

53. SA Soper, LM Davis, EB Shera. Detection and identification of single molecules in solution. J Opt Soc Am B 9:1761–1769, 1992.

54. K Dörre, S Brakmann, M Brinkmeier, KT Han, K Riebeseel, P Schwille, J Stephan, T Wetzel, M Lapczyna, M Stuke, R Bader, M Hinz, H Seliger, J Holm, M Eigen, R Rigler. Techniques for single molecule sequencing. Bioimaging 6:139–152, 1997.

55. A Van Orden, RA Keller. Fluorescence correlation spectroscopy for rapid multicomponent analysis in a capillary electrophoresis system. Anal Chem 70:4463–4471, 1998.

56. J Schaffer, A Volkmer, C Eggeling, V Subramanian, G Striker, CAM Seidel. Identification of single molecules in aqueous solution by time-resolved fluorescence anisotropy. J Phys Chem A 103:331–336, 1999.

57. Y Yan, ML Myrick. Identification of nucleotides with identical fluorescent labels based on fluorescence polarization in surfactant solutions. Anal Chem 73:4508–4513, 2001.

58. P Kask, P Kaupo, D Ullmann, K Gall. Fluorescence-intensity distribution analysis and its application in biomolecular detection technology. Proc Natl Acad Sci USA 96:13756–13761, 1999.

59. A van Orden, NP Machara, PM Goodwin, RA Keller. Single-molecule identification in flowing sample streams by fluorescence burst size and intraburst fluorescence decay rate. Anal Chem 70, 1444–1451, 1998.

60. CAM Seidel, A Schulz, M Sauer. Nucleobase specific quenching of fluorescent dyes. 1. Nucleobase one-electron redox potentials and their correlation with static and dynamic quenching efficiencies. J Phys Chem 100:5541–5553, 1996.

61. KT Han, M Sauer, A Schulz, S Seeger, J Wolfrum. Time-resolved fluorescence studies of labeled nucleosides. Ber Bunsenges Phys Chem 97:1728–1730, 1993.

62. L Brand, C Eggeling, C Zander, KH Drexhage, CAM Seidel. Single-molecule identification of coumarin-120 by time-resolved fluorescence detection: comparison of one- and two-photon excitation in solution. J Phys Chem 101:4313–4321, 1997.

63. W Denk, JH Strickler, WW Webb. 2-photon laser scanning fluorescence microscopy. Science 248:73–76, 1990.

64. A Fischer, C Cremer, EHK Stelzer. Fluorescence of coumarines and xanthenes after two-photon absorption with a pulsed titanium-sapphire laser. Appl Opt 34:1989–2003, 1995.

65. C Xu, WW Webb. Measurement of two-photon excitation cross-sections of molecular fluorophores with data from 690 nm to 1050 nm. J Opt Soc Am B 13:481–491, 1996.

66. JB Shear, EB Brown, WW Webb. Excited fluorescence of fluorogen-labeled neurotransmitters. Anal Chem 68:1778–1783, 1996.

67. J Mertz, C Xu, WW Webb. Single molecule detection by two-photon excited fluorescence. Opt Lett 20:2532–2534, 1995.

68. P Tinnefeld, V Buschmann, DP Herten, KT Han, M Sauer. Confocal fluorescence lifetime imaging microscopy (FLIM) at the single molecule level. Single Mol 3:215–223, 2000.

69. P Tinnefeld, DP Herten, M Sauer. Photophysical dynamics of single dye molecules studied by spectrally-resolved fluorescence lifetime imaging microscopy (SFLIM). J Phys Chem A 105:7989–8003, 2001.

70. M Sauer, C Zander, R Müller, B Ullrich, S Kaul, KH Drexhage, J Wolfrum. Detection and identification of individual antigen molecules in human serum with pulsed semiconductor lasers. Appl Phys B 65:427–433, 1997.

71. J Tellinghuisen, PM Goodwin, WP Ambrose, JC Martin, RA Keller. Analysis of fluorescence lifetime data for single molecules in flowing sample streams. Anal Chem 66:64–72, 1994.

72. J Tellinghuisen, CW Wilkerson, Jr. Bias and precision in the estimation of exponential decay parameters from sparse data. Anal Chem 65:1240–1246, 1993.

73. M Sauer, J Enderlein. Optimal algorithm for single molecule identification with time-correlated single-photon counting. J Phys Chem A 105:48–53, 2001.

74. M Prummer, CG Hübner, B Sick, B Hecht, A Renn, UP Wild. Single-molecule identification by spectrally and time-resolved fluorescence detection. Anal Chem 72:443–447, 2000.

75. DP Herten, P Tinnefeld, M Sauer. Identification of single fluorescently labeled mononucleotide molecules in solution by spectrally resolved time-correlated single photon counting. Appl Phys B 71:765–771, 2001.

76. H Yu, J Chao, D Patek, R Mujumdar, S Mujumdar, AS Waggoner. Cyanine dye dUTP analogs for enzymatic labeling of DNA probes. Nucleic Acids Res 22:3226–3232, 1994.

77. Z Zhu, J Chao, H Yu, AS Waggoner. Directly labeled DNA probes using fluorescent nucleotides with different length linkers. Nucleic Acids Res 22:3418–3422, 1994.

78. H Makiko, H Shigeru. Assay of DNA denaturation by PCR-driven fluorescent label incorporation and fluorescence resonance energy transfer. Anal Biochem 221:306–311, 1994.

79. T Ried, A Baldin, TC Rand, DC Ward. Simultaneous visualization of seven different DNA probes by in situ hybridization using combinatorial fluorescence and digital imaging microscopy. Proc Natl Acad Sci USA 89:1388–1392, 1992.

80. Z Zhu, AS Waggoner. Molecular mechanism controlling the incorporation of fluorescent nucleotides into DNA by PCR. Cytometry 28:206–211, 1997.

81. U Finckh, PA Lingenfelter, D Myerson. Producing single-stranded DNA probes with the Taq DNA polymerase: A high yield protocol. BioTechniques 10:35–39, 1991.

82. P Hentosh, JC McCastlain, P Grippo, BY Bugg. Polymerase chain reaction amplification of single-stranded DNA containing a base analog, 2-chloroadenine. Anal Biochem 201:277–281, 1992.

83. R Raid, E Mar, E Huang, MD Topal. Insertion of extension of acyclic, dideoxy, and ara nucleotides by herpesviridae, human α and human β polymerases. J Biol Chem 263:3898–3904, 1988.

84. Z Földes-Papp, B Angerer, P Thyberg, M Hinz, S Wennmalm, W Ankenbauer, H Seliger, A Holmgren, R Rigler. Fluorescently labeled model DNA sequences for exonucleolytic sequencing. J Biotech 86:203–224, 2001.

85. MA Augustin, W Ankenbauer, B Angerer. Progress towards single-molecule sequencing: enzymatic synthesis of nucleotide-specifically labeled DNA. J Biotech 86:289–301, 2001.

86. M Goodman, L Reha-Krantz. Synthesis of fluorophores-labeled DNA. University of Southern California, Los Angeles, and University of Alberta, Edmonton. WO 97/39150 (PCT/US97/06493), 1997.

87. PM Goodwin, H Cai, JH Jett, SL Ishaug-Riley, NP Machara, DJ Semin, AV Orden, RA Keller. Application of single molecule detection to DNA sequencing. Nucleosides Nucleotides 16:543–550, 1997.

88. Z Földes-Papp, B Angerer, W Ankenbauer, R Rigler. Fluorescent high-density labeling of DNA: error-free substitution of a normal nucleotide. J Biotech 86:237–253, 2001.

89. S Brakmann, P Nieckchen. The large fragment of *Escherichia coli* DNA polymerase I can synthesize DNA exclusively from fluorescently labeled nucleotides. ChemBioChem 10:773–777, 2001.

90. S Brakmann, S Löbermann. High-density labeling of DNA: preparation and characterization of the target for single molecule sequencing. Angew Chem Int Ed Engl 40:1427–1429, 2001.

91. S Doublie, S Tabor, AM Long, CC Richardson, T Ellenberger. Nature 391:251–258, 1998.

92. Th Förster. Zwischenmolekulare Energiewanderung und Fluoreszenz. Ann Phys 2:55–75, 1948.

93. DL Dexter. A theory of sensitized luminescence in solids. J Chem Phys 21:836–850, 1953.

94. M Sauer, W Ankenbauer, B Angerer, Z Földes-Papp, F Göbel, KT Han, R Rigler, J Wolfrum, C Zander. Single molecule sequencing in submicrometer channels: state of the art and future prospects. J Biotech 86:181–201, 2001.

95. Z Földes-Papp, P Thyberg, S Björling, A Holmgren, R Rigler. Exonuklease degradation of DNA studied by fluorescence correlation spectroscopy. Nucleosides Nucleotides 16:781–787, 1997.

96. NP Machara, PM Goodwin, J Enderlein, DJ Semin, RA Keller. Efficient detection of single molecules eluting off an optically trapped microsphere. Bioimaging 6:33–42, 1998.

97. JH Werner, H Cai, PM Goodwin, RA Keller. Current status of DNA sequencing by single molecule detection. Proc SPIE 3602:355–366, 1999.

98. K Dörre, S Brakmann, M Brinkmeier, KT Han, K Riebeseel, P Schwille, J Stephan, T Wetzel, M Lapczyna, M Stuke, R Bader, M Hinz, H Seliger, J Holm, M Eigen, R Rigler. Techniques for single molecule sequencing. Bioimaging 6:139–152, 1997.

99. M Sauer, B Angerer, KT Han, C Zander. Detection and identification of single dye labeled mononucleotide molecules released from an optical fiber in a microcapillary: first steps towards a new single molecule sequencing technique. Phys Chem Chem Phys 1:2471–2477, 1999.

100. RS Brody, KG Doherty, PD Zimmerman. Processivity and kinetics of the reaction of exonuclease I from *Escherichia coli* with polydeoxyribonucleotides. J Biol Chem 261:7136–7143, 1986.

101. LQ Li, LM Davis. Single photon avalanche diode for single molecule detection. Rev Sci Instrum 64:1524–1529, 1993.

102. RL Affleck, WP Ambrose, JN Demas, PM Goodwin, JA Schecker, M Wu, RA Keller. Reduction of luminescent background in ultrasensitive fluorescence detection by photobleaching. Anal Chem 68:2270–2276, 1996.

103. LQ Li, LM Davis. Rapid and efficient detection of single chromophore molecules in aqueous solution. Appl Opt 34:3208–3217, 1995.

104. JA Schecker, PM Goodwin, RL Affleck, M Wu, JC Martin, JH Jett, RA Keller, JD Harding. Flow-based continuous DNA sequencing via single molecule detection of enzymatically cleaved fluorescent nucleotides. Proc Soc Photo-opt Ins 2386:4–12, 1995.

105. J Enderlein, DL Robbins, WP Ambrose, PM Goodwin, RA Keller. Statistics of single-molecule detection. J Phys Chem B 101:3626–3632, 1997.

106. W Wang, Y Liu, GJ Sonek, MW Berns, RA Keller. Optical trapping and fluorescence detection in laminar flow streams. Appl Phys Lett 67:1057–1059, 1995.

107. M Brenner. Imaging dynamic events in living tissue using water immersion objectives. Am Lab 26:14–19, 1994.

108. A Castro, FR Fairfield, EB Shera. Fluorescence detection and size measurement of single DNA molecules. Anal Chem 65:849–852, 1993.

109. PM Goodwin, ME Johnson, JC Martin, WP Ambrose, JH Jett, RA Keller. Rapid sizing of individual fluorescently stained DNA fragments by flow cytometry. Nucleic Acids Res 21:803–806, 1993.

110. Z Huang, JH Jett, RA Keller. Bacteria genome fingerprinting by flow cytometry. Cytometry 35:169–175, 1999.

111. YH Lee, RG Maus, BW Smith, JD Winefordner. Laser-induced fluorescence detection of a single molecule in a capillary. Anal Chem 66:4142–4149, 1994.

112. RD Guenard, LA King, BW Smith, JD Winefordner. Two-channel sequential single-molecule measurement. Anal Chem 69:2426–2433, 1997.

113. WA Lyon, S Nie. Confinement and detection of single molecules in submicrometer channels. Anal Chem 69:3400–3405, 1997.

114. M Brinkmeier, K Dörre, K Riebeseel, R Rigler. Confocal spectroscopy in microstructures. Biophys Chem 66:229–239, 1997.

115. K Dörre, J Stephan, M Lapczyna, M Stuke, H Dunkel, M Eigen. Highly efficient single molecule detection in microstructures. J Biotechnol 86:225–236, 2001.

116. J Stephan, K Dörre, S Brakmann, Th Winkler, T Wetzel, M Lapczyna, M Stuke, B Angerer, W Ankenbauer, Z Földes-Papp, R Rigler, M Eigen. Towards a general procedure for sequencing single DNA molecules. J Biotechnol 86:255–267, 2001.

117. C Hoyer, S Monajembashi, KO Greulich. Laser manipulation and UV-induced single molecule reactions of individual DNA molecules. J Biotech 52:65–73, 1996.

118. JE Aubin. Autofluorescence of viable cultured mammalian cells. J Histochem Cytochem 27:35–43, 1979.

119. G Patonay, MD Antoine. Near-infrared fluorogenic labels: new approach to an old problem. Anal Chem 63:321A–327A, 1991.

120. M Sauer, KT Han, V Ebert, R Müller, A Schulz, S Seeger, J Wolfrum, J Arden-Jacob, G Deltau, NJ Marx, C Zander, KH Drexhage. New fluorescent dyes in the red region for biodiagnostics. J Fluoresc 5:247–261, 1995.

121. DC William, SA Soper. Ultrasensitive near-IR fluorescence detection for capillary gel electrophoresis and DNA sequencing applications. Anal Chem 67:3427–3432, 1995.

122. J Arden-Jacob, NJ Marx, KH Drexhage. New fluorescent probes for the red spectral region. J. Fluoresc. 7:91S–93S, 1997.

123. U Lieberwirth, J Arden-Jacob, KH Drexhage, DP Herten, R Müller, M Neumann, A Schulz, S Siebert, G Sagner, S Klingel, M Sauer, J Wolfrum. Multiplex dye DNA sequencing in capillary gel electrophoresis by diode laser-based time-resolved fluorescence detection. Anal Chem 70:4771–4779, 1998.

124. E Terpetschnig, H Szmacinski, A Ozinskas, JR Lakowicz. Synthesis of squaraine-*N*-hydroxysuccinimide esters and their biological application as long-wavelength fluorescent labels. Anal Biochem 217:197–204, 1994.

125. B Oswald, L Patsenker, J Duschl, H Szmacinski, OS Wolfbeis, E Terpetschnig. Synthesis, spectral properties, and detection limits of reactive squaraine dyes, a new class of diode laser compatible fluorescent protein labels. Bioconjugate Chem 10:925–931, 1999.

126. RB Mujumdar, LA Ernst, SR Mujumdar, AS Waggoner. Cyanine dye labeling reagents containing isothiocyanate groups. Cytometry 10:11–19, 1989.

127. PL Southwick, LA Ernst, EV Tauriello, SR Parker, RB Mujumdar, SR Mujumdar, HA Clever, AS Waggoner. Cyanine dye labeling reagents: carboxymethylinindocyanine esters. Cytometry 11:418–430, 1990.

128. RB Mujumdar, LA Ernst, SR Mujumdar, CJ Lewis, AS Waggoner. Cyanine dye labeling reagents: sulfoindocyanine succinimidyl esters. Bioconjugate Chem 4:105–111, 1993.

129. JH Flanagan, Jr., SH Khan, S Menchen, SA Soper, RP Hammer. Functionalized tricarbocyanine dyes as near-infrared fluorescent probes for biomolecules. Bioconjugate Chem 8:751–756, 1997.

130. SA Soper, QL Mattingly, P Vegunta. Photon burst detection of single near infrared fluorescent dye molecules. Anal Chem 65:740–747, 1993.

131. M Sauer, C Zander, KH Drexhage, J Wolfrum. Diode laser-based detection of single molecules in solution. Chem Phys Lett 254:223–228, 1996.

132. M Sauer, C Zander, R Müller, B Ullrich, S Kaul, KH Drexhage, J Wolfrum. Detection and identification of individual antigen molecules in human serum with pulsed semiconductor lasers. Appl Phys B 65:427–433, 1997.

133. C Zander, KH Drexhage. Sequential counting of single molecules in a capillary. J Fluoresc 7:37S–39S, 1997.

134. C Zander, KH Drexhage, KT Han, J Wolfrum, M Sauer. Single-molecule counting and identification in a microcapillary. Chem Phys Lett 286:457–465, 1998.

135. WP Becker, H Hickl, C Zander, KH Drexhage, M Sauer, S Siebert, J Wolfrum. Time-resolved detection and identification of single analyte molecules in microcapillaries by time-correlated single-photon counting (TCSPC). Rev Sci Instrum 70:1835–1841, 1999.

136. QF Gao, ES Yeung. A matrix for DNA separation-genotyping and sequencing using poly(vinylpyrrolidone) solution in uncoated capillaries. Anal Chem 70:1382–1388, 1998.

137. J Korlach, WW Webb, et al. A method of sequencing nucleic acids by direct measurement of temporal order of base incorporation on a single molecule. PCT Int Appl Wo, 0070073, 64 pp, 2000.

138. J Korlach, M Levene, SW Turner, M Foquet, HG Craighead, WW Webb. A new strategy for sequencing individual molecules of DNA. Biophys J 80(1):147a, 2001.
139. J Korlach, M Levene, SW Turner, M Foquet, HG Craighead, WW Webb. Single-molecule analysis of DNA polymerase activity using zero- mode waveguides. Biophys J 82(1):507a, 2002.
140. LR Yarbrough, JG Schlageck, M Baughman, et al. Synthesis and properties of fluorescent nucleotide substrates for DNA-dependent RNA polymerises. J Biol Chem 254:12069–12073, 1979.
141. SM Turner, M Levene, WW Webb, HG Craighead. Confinement of Fluorescence Excitation for Single Molecule Detection at High Concentrations. MicroTotal Analysis System, Amsterdam: Kluwer Academic, 2001, 259–261.

7 DNA Sequencing for Genome Analysis

*Jeffrey P. Tomkins, Todd C. Wood,
and Dorrie Main*

CONTENTS

INTRODUCTION

Genome analysis has developed over time through the various fields of genetics, cytogenetics, biophysics, biochemistry, and molecular biology. Each of these disciplines has contributed to our understanding of the nature of inheritance and how genes contribute toward an organism's phenotype. We may briefly define a genome as the complete set of DNA instructions for a given organism, organized into chromosomal units and containing the genes which code for the organism's traits. As a result, historically separate fields of biological study find union within the arena of genomics. Deciphering the genetic code or precise order of nucleotides represents one of the most fundamental steps in genomic analysis. Genome sequencing in its various forms serves as a foundation for analyses of transcription, gene regulation, chromosome structure, genetic pathologies, biochemical pathways, and evolution.

There are a number of approaches to genome sequencing that may be taken depending on the size of the genome, its complexity, and the availability of funds. In addition, there are several different types of sequencing approaches that form a preliminary framework for subsequent large-scale stages of sequencing. In this chapter, we discuss the various strategies that produce usable genomic sequence for a wide variety of analyses and applications. Specifically, we discuss expressed sequence tag (EST) sequencing, development of sequence-ready genomic frameworks, whole genome sequencing approaches, and annotation of genomic sequence.

EST SEQUENCING

Technological advances associated with high-throughput sequencing facilitated the development of EST DNA sequence data in the early 1990s.[1,2] An EST is a sequence derived from an expressed gene or messenger RNA (mRNA) transcript. Typically, mRNA is harvested from various types of tissues of interest in an organism and the derived cDNAs are cloned into a high copy vector. More commonly, plasmid vectors are being used to generate cDNA libraries because of their utility in high-throughput applications. The EST data are typically generated by consecutively sequencing through a randomly picked and arrayed cDNA library (in microtiter plates) to an appropriate level. The level of redundancy for sequencing depends on the complexity of the library, the amount of repeated sequences, and the amount of funding. In the past, typical ESTs consisted of about 300 to 600 high-quality bases of sequence. With current sequencing platforms, it is possible to routinely generate reads containing more than 800 very high quality bases. In the final analysis, the ESTs are used to investigate the diversity of genes expressed by an organism, tissue, or cell. In some cases, the same types of tissue are being compared to identify differences arising from disease, environmental stress, or other factors that alter gene expression. By looking at only expressed sequences we can accomplish a variety of research goals. The data permit the analysis of differential gene expression by comparing stage or tissue specific datasets, and finally, ESTs confirm splicing and coding predictions when compared to genomic sequence. Compared to the time and expense of whole-genome sequencing, sequencing only the expressed regions of the genome is relatively trivial (no introns or intergenic DNA are sequenced) and can quickly lead to the discovery of novel protein coding genes and selectively expressed genes. The value of this approach in gene discovery is evident in the growth of public and private databases. As of June 4, 2004 the NCBI dbEST repository (http://www.ncbi.nlm.nih.gov/dbEST) contained more than 21 million ESTs that are publicly available for download and data mining. It should be noted that while ESTs are an invaluable resource for gene discovery they cannot on their own provide information on gene structure, genomic organization, or evolution.

The type of cDNA library used for EST sequencing is an important consideration. Full-length cDNA libraries are not necessarily an optimal source of sequence tags for several reasons. It is often difficult to obtain good sequences through the poly(A) stretch at the 3'-end; therefore, the best sequences are obtained in the direction 5'-3' of the transcript, thus avoiding the poly(A) region. Second, many full-length cDNAs are really only partial-length inserts that do not reach the 5'-end of the transcript. Therefore, if we sequence an insert from the 5'-end, it does not mean that we are at the 5'-end of the transcript. As a result, we can have different non-overlapping tags related to the same transcript, which makes the assembly and annotation process more difficult. Additionally, the frequency of clones in full-length cDNA libraries does not generally reflect the relative abundance of the original transcript, as the efficiency of cloning may be influenced by insert length and sequence characteristics. Alternative splicing can result in the same gene generating multiple different transcripts, which can lead erroneously to multiple gene prediction. Finally, a cDNA library restricted to a specific region of the transcript (such as the ~500 bases at the 3'-end) is less complex, thus simplifying the entire project. Sequence processing, assembly, and annotation are discussed later in the chapter.

DEVELOPMENT OF SEQUENCE-READY GENOMIC FRAMEWORKS

The development of a large-insert genomic library is a primary resource needed to construct a framework to facilitate genome sequencing and is briefly discussed. The development of large-insert DNA libraries initially began with the use of the yeast artificial chromosome (YAC) in the late 1980s.[3] With YACs, cloning of high-molecular-weight DNA in the 100 to 350 kb range became possible, and library-based exploitation of even the largest genomes was feasible. However, it soon became apparent that the YAC cloning system contained some serious problems that affected its use in a number of genomic applications.[4] For example, roughly 50% of YAC clones are chimeric or possess insert rearrangements.[3–6] Such clones are unsuitable for sequencing and mapping applications, and a great deal of effort is devoted to identifying chimeras and clones with rearranged inserts.[4,6,7] Additionally, manipulation and isolation of YAC inserts are difficult and tedious.[8,9]

In the early 1990s, "bacterial artificial chromosomes" (BACs) became a viable alternative to YACs.[10] Contrary to their name, BACs are not really artificial chromosomes per se, but modified bacterial F factors. Although BACs can carry inserts approaching 500 kb in length, insert sizes between 80 and 300 kb are more common.[10–12] Most BAC vectors possess traditional plasmid selection features such as an antibiotic resistance gene and a polycloning site within a reporter gene allowing insertional inactivation. BAC clones have several important advantages over YACs. In particular, BACs are considerably less prone to chimerism and insert rearrangements.[7,9,13] The stability of BAC inserts appears to be due, in part, to F factor genes (*parA* and *parB*) that prevent more than one BAC from simultaneously occupying a bacterium.[10,13,14] An additional advantage of BAC clones is that they are relatively easy to manipulate and propagate compared with viral- or yeast-based clones. Consequently, BACs have supplanted YACs as the dominant vector used in large-scale physical mapping and sequencing.

Although physical mapping is not the focus of this text, an understanding of the development of sequence-ready physical maps is important as these resources form the essential substrate for a number of genome sequencing applications. Therefore, the following paragraphs briefly describe how these essential frameworks are developed.

The BAC system is ideal for creating physical frameworks of both large and small genomes. These frameworks may be used for a number of applications such as the sequencing of whole genomes or selected regions of genomes. In animals, human, *Drosophila,* and mosquito were sequenced using BAC-based frameworks.[15–17] In plants, the model genomes of *Arbidopsis* and rice were both sequenced using BAC-based physical frameworks.[18,19] The BAC-based framework is generally developed from various types of data that are integrated from different high-throughput applications. The extensiveness of the framework and its nature are obviously limited by the budget at hand. Some high-throughput framework applications (e.g., BAC end-sequencing) are inherently more expensive than others.

Typically, the framework is begun by digesting BAC clones with frequent cutting enzymes such as *Hin*dIII or *Eco*RI, gel electrophoresis, digitizing the gel images, and assembling the contigs using a computer program. Although there are a number of variations on this general theme, the majority of BAC fingerprinting has been done using a high-resolution agarose system.[20] The fingerprint gel images are directly converted to a digitized format and then used for assembly into contigs with the software program FPC (Figure 7.1).[21] A new method of fingerprinting BACs is becoming more common and involves the use of capillary electrophoresis platforms such as the ABI 3730 (Applied Biosystems). This method is commonly referred to as High-Information Content Fingerprinting (HICF) and is now becoming the new standard for physical mapping applications.[22] With this system, the BAC clones are cut with five different frequent cutting restriction enzymes, the fragments are dye-labeled, and the products are electrophoresed with an internal size standard. The HICF technique uses multiple color fragment data from the capillary-based automated electrophoresis platforms generated by GeneScan fragment sizing application by Applied Biosystems.

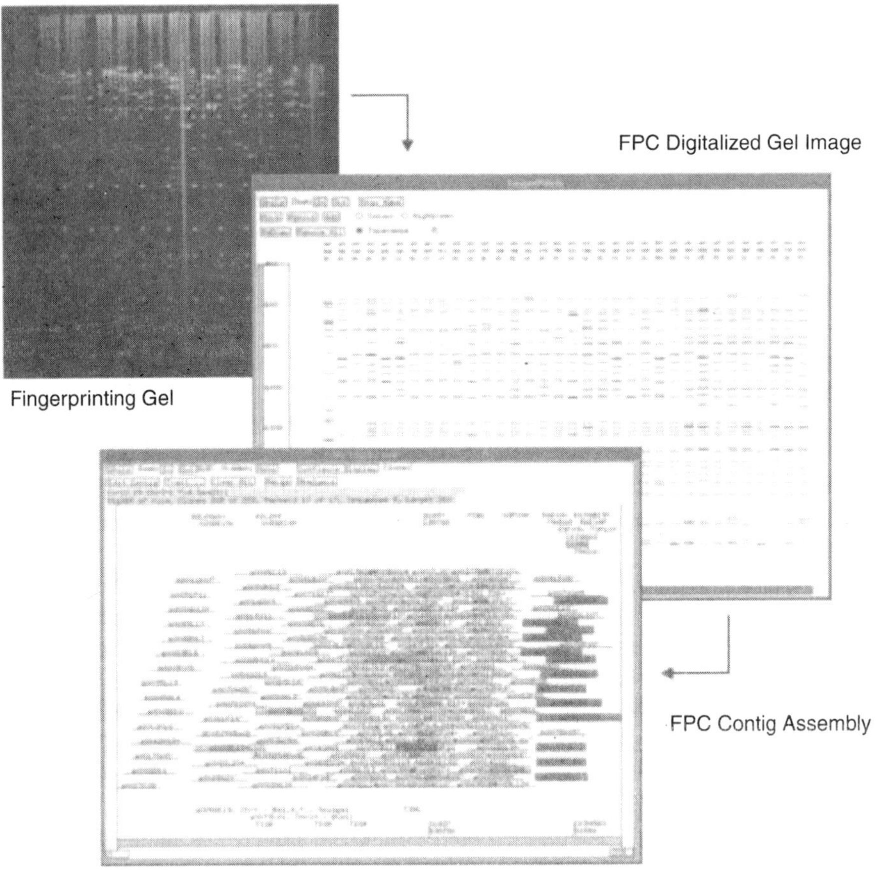

FIGURE 7.1 (Color Figure 7.1 follows page 84.) Physical mapping (high-resolution agarose system).

The GenoProfiler software package (http://wheat.pw.usda.gov/PhysicalMapping/tools/ genoprofiler /manual/) takes the output files of fragment size data created by Gene-mapper (Applied Biosystems) and removes the need for manual gel handling, providing significant automation and productivity gains. The ability to include multiple labels (colors) also increases the number of data points included per clone. Finally, measuring fragment sizes on automated instruments provides for near single-base resolution, significantly enhancing the accuracy of the band size information. Information about both fragment size and the identity of the labeled base at the end of the type IIS cut site can be extracted from the chromatograms. Together, these improvements not only make the process easier, but they also greatly increase the robustness of the statistical analysis. Consequently, the minimal overlap lengths needed between clones before they can be identified are reduced, along with the amount of work needed to construct meaningful maps.

Initially, the HICF form of fingerprinting using fluorescent tags was very expensive. However, lower reagent costs associated with bulk purchases coupled with improvements in the overall protocol and the reduction in labor required also make it financially feasible. Clearly, major costs are associated with equipment purchases, leases, and service contracts. The HICF approach is currently being used to great advantage in the development of the plant species *Mimulus guttatus* physical framework in progress at the Clemson University Genomics Institute (www.genome. clemson.edu/mimulus/).

Once the fingerprint database is established, an investigator can readily assemble the fragments into contigs using FPC. Typically, multiple assemblies are run at various stringencies until a satisfactory outcome is obtained. A previous report by the software developer describes in detail stringency-related aspects associated with contig assembly using FPC.[21] In general, the investigator must set stringency levels so that there is a minimal amount of error in producing the contigs. It only takes for one false clone combination to produce a large level of error in the assembly process. This is because one falsely assembled clone will subsequently draw in a whole host of other related clones, thus greatly compounding the original error. Hence, stringency levels must be rigorous. There is a tendency to lower stringency levels to reduce the number of contigs, but care must be taken to prevent the inclusion of errant data. In large fingerprinting and assembly operations, the FPC analysis can also be automated and the output or physical map updated on a weekly basis as more clones are added to the system.

In addition to the fingerprinting of clones, BACs may also be subjected to end sequencing.[11,12,23] Typically, BACs are end-sequenced in forward and reverse directions and will yield a high quality sequence of about 300 to 500 bases. The new sequencing platforms, such as the ABI 3730, show promise for values 50% longer than this. Sequencing is typically done consecutively in the library for either the whole library or a part of it. Because BACs are single-copy plasmids, high-throughput DNA preps are generally not as clean as typical high-copy plasmid preps. Thus, the delicate nature of early capillary sequencing systems were not conducive to running samples derived from BACs and the sequencing reactions had to be electrophoresed on slab-gel sequencers. Now that improved DNA prep protocols have been developed along with improvements in capillary sequencers, this is no longer a problem and BAC ends can be sequenced nearly as fast and efficiently as ESTs.

Data stream pipelines have been developed to process BAC end sequences in a routine and seamless system. Typically, base-calling is performed automatically using PHRED[24,25] and vector sequences removed by CROSS-MATCH (http://www. genome.washington.edu). High-quality BAC end sequences are usually defined as those having >100 nonvector bases with a PHRED quality value >20. The high-quality sequences are used as database queries and the results sorted in various formats. The BAC end sequences are ultimately submitted to GenBank and assigned accession numbers. Database queries can be very comprehensive as there are now a variety of protein and DNA sequence databases available for querying. Some databases, such as SWISSPROT,[26] offer a high level of curation and putative functional information. In addition, most major plant and animal species now have large EST databases that can be queried to electronically anchor ESTs to specific BACs through sequence similarity.

The BAC end sequences provide what has been commonly referred to in the genomics community as sequence tagged sites (STSs) or sequence tagged connectors (STCs). This strategy was originally proposed by Venter et al.[7] as a way to provide a genomic framework for sequencing the human genome. An archived collection of BAC end sequences is often referred to as a STS or an STC database. Having both an FPC and an STC database for a particular genome is very advantageous. Each BAC end sequence can now be traced back to specific contigs generated through FPC through clone addresses or IDs. The combination of the FPC database with the STC database now provides a powerful framework tool for dissecting and studying large genomes that have large amounts of repetitive DNA.

The STC database is highly useful for BAC sequencing approaches for whole chromosomes and selected genomic regions. When an original seed BAC has been shotgun sequenced, the finished assembled sequence can be queried against the STC database to pull out the adjoining BACs. Once the adjoining BACs are identified, fingerprints can be compared for verification and to determine overlap. In general, an overlap of about 5% is desired to avoid redundancy and keep sequencing costs low.

Physical frameworks can be further enhanced by anchoring genetic markers such as cDNAs, genomic fragments, and simple sequence repeats (SSRs) to the BAC libraries. The DNA probes may be anchored by hybridizing the insert DNA from the clones to high-density BAC colony filter arrays.[11,12,23] Polymerase chain reaction (PCR)-based markers can be anchored by amplifying DNA obtained from pools of rows and columns of the BAC library.[27] However, another high-throughput approach makes use of overgo technology.[28] An overgo probe is a set of ~25-mer single-copy sequences that overlap by about five bases. When labeling, the overhangs are filled in with radioactive nucleotides. The denatured overgos are then used for hybridizing to BAC colony filters. Overgos may be derived from any type of electronic sequence data such as SSRs and or ESTs. The ultimate consequence of anchoring molecular genetic markers is that the genetic and physical maps for a given organism become integrated. Furthermore, FPC contigs can now be ordered to physically reconstruct entire linkage groups.[19]

Another useful anchoring scheme is to hybridize cDNAs to the BAC library and develop transcript maps of the genome. Strategies to pool radiolabeled cDNAs in bulk based on pools and columns of a cDNA/EST library are now being implemented so that thousands of gene sequences can be anchored in single experiments.[27] To facilitate these projects, computer programs have been developed that read autoradiographs and identify addresses (Incogen Corp, Williamsburg, VA). In addition, computer programs are also being developed that deconvolute the resulting hybridization data to sort out the positive signals. It is noteworthy that overgos may also be hybridized in the same row/column pool manner.

One of the primary benefits of publicly funded physical frameworks is that the data are accessible via the World Wide Web. At the Clemson University Genomics Institute, we have sought to make all of our framework data available in searchable formats (www.genome.clemson.edu). All of our STC databases are fully searchable by clone name or sequence homology. Also, with the addition of WebFPC, users can remotely access and manipulate fingerprint databases at their leisure.

WHOLE-GENOME SEQUENCING

Shotgun sequencing is a critical technology employed in genome sequencing. The first step in shotgun sequencing is the subcloning of the DNA sequencing target, also called library construction. The target sequence can be a selected and prepped BAC clone or total genomic DNA. In this step, the target DNA is randomly sheared into smaller pieces, which are then cloned into vectors that can be used in sequencing reactions. Any collection of clones that represent complete coverage of a larger piece of DNA is called a library. To create a random sequence-ready shotgun library, the DNA of the target molecule is commonly physically sheared in specialized instrumentation or sonicated. Physical shearing generally produces a more randomized sample of the target sequence and is highly suitable for high throughput sequencing centers. After the DNA is sheared, the ends of the fragments are repaired and then ligated into a sequencing vector.

In the past, two general types of cloning vectors were used for shotgun sequencing. The first is bacteriophage M13, which naturally exists as single-stranded DNA. Because DNA sequencing templates must be single-stranded DNA, M13 naturally provides sequence-ready DNA templates. M13 clones exhibit a bias, in that repetitive sequences are poorly represented in M13 libraries. Alternatively, double-stranded plasmid clones can also serve as the vector for subcloning shotgun libraries. Plasmids do not exhibit the same cloning bias as M13 vectors, and thus represent the target DNA sequence more evenly. In the past, large genome sequencing centers have used a combination of M13 and plasmid subclones for different sequencing needs. However, plasmid sequencing is becoming the more predominant form of sequencing. This is because plasmids are easier to handle in a high-throughput automated format. In fact, most cDNA libraries are now also being cloned into plasmids for the same reason.

Having created a suitable shotgun library, the next step is typically referred to as "production sequencing," because it mimics the factory production typical of an assembly line. A selection of shotgun clones is mini-prepped, reacted, and then sequenced in a 96- or 384-well format en masse. Advances in automated DNA preps have greatly reduced the amount of cost and labor involved. In fact, some current automated platforms will perform the DNA prep and the PCR-based sequencing reactions all within the same instrument, greatly reducing the amount of hands-on labor involved.

The number of clones necessary to reconstruct the original target sequence depends on the average length of sequence obtained from a single shotgun clone, the length of the target sequence, and the desired accuracy of the completed sequence. For most production sequencing applications, genome centers typically strive for a sixfold to tenfold coverage of the target sequence; that is, each nucleotide of the target is in theory sequenced on average six to ten different times. This level of redundancy assures that most of the target sequence will be covered by the positioned shotgun clones, and whatever gaps or ambiguities remain after the shotgun sequencing can be quickly resolved.

After generating the shotgun sequence, the sequence traces are then processed computationally. First, the traces are base-called using PHRED or a comparable program. Second, the sequences are "assembled" into contiguous sequences called contigs. A contig is composed of two or more sequence reads that originate from an overlapping region of the target sequence. Contig assembly begins with an automated step, and concludes with a manual editing stage. The automated step is carried out by an assembler program such as PHRAP (http://bozeman.mbt.washington.edu/phrap.docs/phrap.html), the TIGR Assembler (http://www.tigr.org/software/assembler/), or CAP3.[29] The assembler examines the sequence reads for regions of near identity (allowing for errors in the sequence read) and attempts to reconstruct the original target sequence. If the sequence reads were a perfect and redundant random representation of the target sequence, the assembler could theoretically reconstruct the entire target sequence from the shotgun sequence. In reality, variations in the quality of sequence reads, regions of the target with low representation, and repetitive sequences can result in assembly of many different contigs (depending on the size of the target sequence and the redundancy of the shotgun clones). A schematic of the overall approach to genome sequencing is shown in Figure 7.2.

All assembly programs work in very similar ways. First, regions of overlap are identified by doing standard sequence comparisons between the sequence reads. Unlike a normal sequence comparison program, however, assemblers also consider the quality values of the sequence reads in calculating similarity scores between reads. From the overlap information, the assembler then reconstructs sequence contigs and some also calculate quality values of each nucleotide in the contig, based on the number of reads available at that position. To be certain that the contig comes only from the target sequence, sequence reads are typically filtered before assembly to remove contaminating DNA, such as vector sequence that often appears at the 5′ end of a sequence read.

After automatic assembly is completed, the sequence rarely resolves into a single contig of uniform quality. Reasons for obtaining multiple contigs include failure of the assembly program, cloning bias, regions that are difficult to sequence, and regions that were poorly represented in shotgun sequence reads by chance. To resolve these problems and to produce a single contig of uniform quality, manual inspection, and editing of the contigs is necessary, together with additional sequencing reactions where appropriate. This manual phase is called either gap closure or, more commonly, finishing.

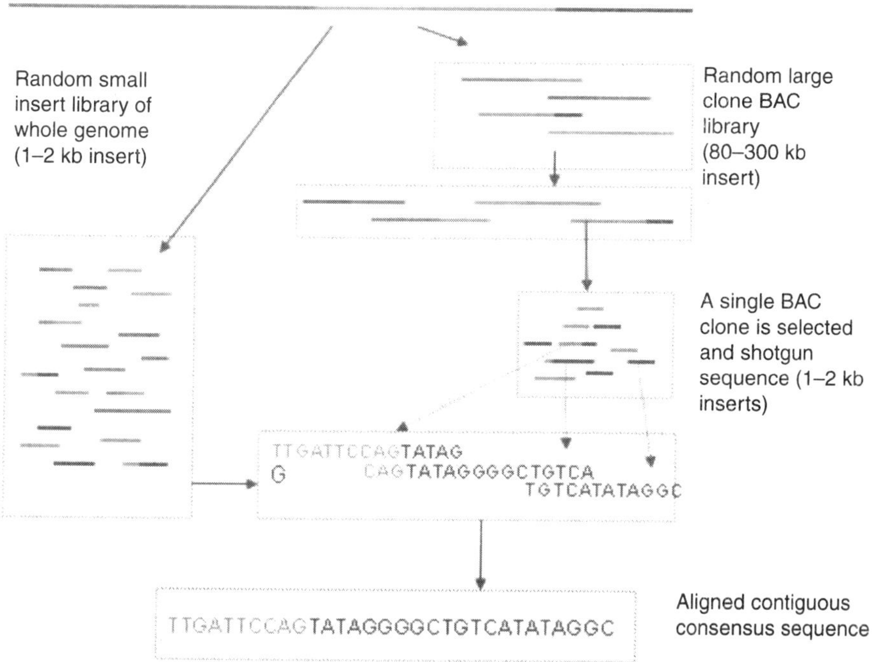

FIGURE 7.2 Schematic diagram of genome sequencing strategies.

Finishing is aided by the use of a graphical contig editor such as Consed.[30] For each assembly, Consed displays a list of the contigs and sequence reads. After selection of a contig, Consed displays the consensus sequence together with each sequence read that was assigned to that region by the assembler (Figure 7.3) The consensus sequence is displayed at the top of the assembly window, and the sequence reads appear in an

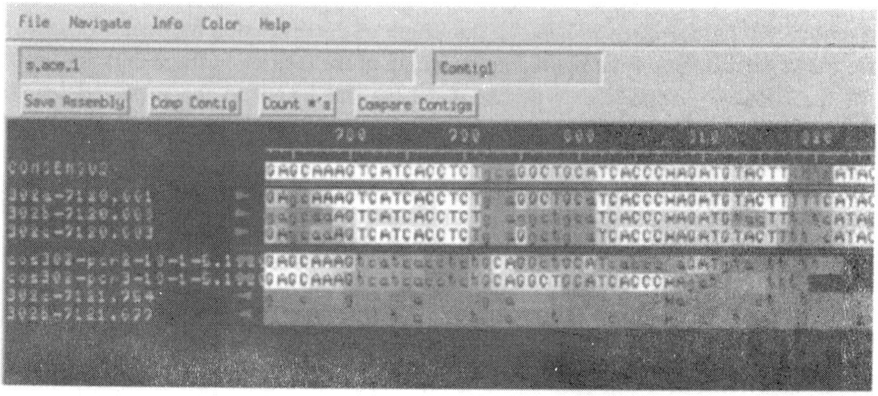

FIGURE 7.3 Consed — Sequence viewing and editing software.

alignment format just below the consensus. Important features such as base quality are color-coded for easy identification of low-quality regions. Other windows in Consed display the actual chromatograms of any desired sequence reads.

The additional work necessary to close a gap in the sequence assembly depends on the cause of the gap. Regions of low shotgun representation can often be resolved by sequencing the opposite end of the clone insert for reads adjacent to the low-quality region. This will generate sequence reads for both ends of the clone insert and will usually provide sufficient sequence coverage to allow the gap to be closed. If the gap is caused by vector bias, it will be necessary to clone the missing region into a different vector. For example, as we discussed above, M13 does not clone regions of repetitive DNA. Utilization of both M13 and plasmid clones can resolve problems caused by this cloning bias. In cases where the assembly algorithm has failed to derive the correct consensus sequence, the statistical parameters of the assembly algorithm can be manipulated to generate the desired assembly. For example, assembly algorithms assemble regions with tandem repeats poorly. These errors are easily iden-tified because of the abnormally large number of sequence reads assigned to a single region.

The most difficult class of problems encountered in finishing are regions that are intrinsically difficult to sequence, such as simple sequence repeats, homopoly-meric regions, or regions with secondary structure. These types of gaps are easily identified because of the uniformity with which the quality drops off. In other types of gaps, the low quality may be observed in only a few of the shotgun reads, but in areas that are difficult to sequence, all shotgun reads will exhibit the same low quality at the same point in the sequence. Resolution of these gaps requires a number of different strategies, each designed to deal with a particular sequencing problem. For example, specialized chemistry designed for simple sequence repeats can produce high-quality reads of repetitive regions. Larger repeats or secondary structures could require a targeted subcloning strategy designed to break up the difficult region prior to sequencing.

After finishing has produced a single contig, the consensus sequence is validated in a number of ways. First, the length of the sequence is compared to the expected length of the target sequence. The expected length is typically measured by a restriction digest of the target sequence. Second, the length of the restriction fragments observed for the target sequence is compared to the length of the restriction fragments predicted for the consensus sequence. Whereas matching the length to the expected length confirms that a sequence of the correct size has been generated, matching the predicted and observed restriction fragment lengths ensures that the general order of the sequence assembly is good. This procedure can be repeated with multiple restriction enzymes to verify the assembly even further.

For the small genomes of many bacteria and archaea, which contain contiguous sequences of 10 Mb or less with few repeat sequences, a straightforward whole genome shotgun strategy works well. The size and repetitive nature of eukaryotic genomes prevent the ready application of a whole-genome shotgun sequencing tech-nique, necessitating specialized strategies for sequencing these genomes. Depending on the level of completion required, eukaryotic genomes are sequenced by a mapping strategy (for high-quality sequences) or by a whole genome shotgun (WGS) strategy (for rough draft sequences), or by a hybrid of both. In this section, we discuss the shotgun strategy used to generate the first bacterial genome sequence, *Haemophilus influenzae*. The following section focuses on the two strategies used in eukaryotic genome sequencing.

The first genome completely sequenced by the shotgun method was that of *H. influenzae*, which was sequenced at the Institute for Genomic Research (TIGR) and published in 1995. As a bacterial genome, the repetitive DNA content was much lower than typical eukaryotes, and the size of the complete chromosome was only 1.8 million nucleotides. Despite that it was the first genome, the basic issues considered and strategies employed in sequencing are still representative of many genome current projects. Hence, a description of this elementary sequencing project is discussed in detail as a model.

Because the success of the shotgun method depends on a random selection of sequence templates, the TIGR team expended much care on template library construction. Two types of libraries were made for the *H. influenzae* project: a short-insert (1.6 to 2 kb) plasmid library and two long-insert (15 to 20 kb) phage libraries. The plasmid library was used as the primary sequencing template, whereas the libraries were reserved for finishing and validation (we discuss the utilization of the libraries below). For all libraries, the genomic DNA was mechanically sheared rather than digested by restriction enzymes. As mentioned previously, shearing the source DNA ensures a more random library sample than restriction digest.

Based on simple statistical calculations, we know that the probability that a nucleotide will be unsequenced in a shotgun sequencing project is $P_o = e^{-m}$, where m is the genome coverage. Sequencing random clones sufficient to cover the genome five times (fivefold coverage) results in a probability of 0.0067 that a nucleotide will be unsequenced. For the 1.8 million nucleotide *H. influenzae* genome, a fivefold shotgun coverage should result in approximately 12,000 unsequenced nucleotides, distributed randomly throughout the genome in sequence gaps between the contigs.

Given an average read length of 460 nucleotides (the limits of the sequencing technology at that time), approximately 19,000 sequence reads would be necessary to cover the *H. influenzae* five times. The TIGR team sequenced 19,687 short-insert templates using dye-primer sequencing chemistry ("forward reads"). To supplement these reads and to provide contig assembly information, 9297 templates were resequenced at the opposite end of the insert (known as "reverse reads"), also using dye-primer chemistry. The result is a pair of 460-nucleotide sequence reads that are known to be 700 to 1100 nucleotides apart, a significant advantage during finishing.

Automatic assembly of forward and reverse reads yielded 210 contigs. Because the parameters used in assembly are optimized for general use, local variations can occur due to differences in repeat or GC content. As a result, manual inspection of potential overlaps resulted in a reduction to only 140 contigs, which could not be further combined without additional sequencing reactions. The remaining gaps could be categorized into two types, depending on the orientation of paired sequence reads. If the forward reads at the end of one contig matched corresponding reverse reads from the same template at the end of a second contig, the TIGR team called the gap a "sequence gap." Because they are spanned by a single, small-insert plasmid, sequence gaps are known to be small (less than 1500 nucleotides), and known templates (the plasmid inserts spanning the gaps) are available for immediate, additional sequencing. For the *H. influenzae* genome, 98 gaps were sequence gaps.

The second type of gaps occurred when forward/reverse reads of the same template did not span the gap. The TIGR team labeled these gaps "physical gaps," because a clone insert that spanned the gap was not immediately available. Additional strategies were devised to identify and prepare suitable sequencing templates that spanned the physical gaps. For two gaps, protein sequences could be used to orient the adjacent contigs. This was possible because the gap occurred within a protein-coding gene, the 5′ end of which was sequenced on one contig and the 3′ end on another. By comparing the contig sequences to known protein sequences, these overlaps could be detected and PCR used to generate suitable sequencing templates.

The majority of the gaps were closed using one of two strategies. First, DNA fingerprinting and hybridization to oligonucleotides prepared from the ends of contigs revealed possible overlapping DNA segments, from which templates could be prepared. For example, if oligonucleotide probes prepared from the ends of two contigs hybridized to the same restriction fragment of the genomic DNA, it is likely that the hybridizing fragment contains the sequence spanning the gap between the two contigs. The other successful gap closure method was paired forward/reverse sequences from the libraries. Small gaps are more likely to be spanned by the large inserts (15 to 20 kb) of the libraries than the small inserts (~2 kb) of the plasmid library. Remaining physical gaps were closed by simple combinatorial PCR, systematically using oligonucleotide primers from each possible contig pair.

These strategies resulted in a single contig with a consensus sequence of 1,830,137 nucleotides. The assembly of the consensus sequence was validated using additional library forward/reverse sequence reads and restriction fragments and restriction site locations. Because the library has a known insert size of 15 to 20 kb, paired reads that deviate significantly from that distance would indicate a misassembly. Sizes of restriction fragments generated by three different enzymes could also be matched to the predicted sizes of the consensus sequence. Restriction site locations from restriction mapping could also be matched to the restriction sites found on the consensus sequence. For all of these validations, the *H. influenzae* consensus sequence matched the observed characteristics of the genome.

Although *H. influenzae* was the very first bacterial genome to be sequenced by the shotgun method, many of the techniques and strategies developed are the same as those used today for bacterial and archaeal genome projects. Library construction remains an important step to assure a random coverage of the genome. Sequencing both ends of the inserts and using different sequencing chemistries are both common strategies for gap closure and finishing. The major difference between a modern genome project and the *H. influenzae* project is rate. Facilitated by capillary sequencers and more powerful computers for assembly and finishing, a bacterial genome of the same size as *H. influenzae* can be completed in a fraction of the time and at a fraction of the cost.

Eukaryotic genomes differ from bacterial and archaeal genomes in several important ways that make genome sequencing projects more complex. Although several bacteria and archaea are known to have multiple chromosomes or "megaplasmids," a typical prokaryotic genome is a single, circular chromosome. In contrast, eukaryotic genomes are divided into several (sometimes many) linear chromosomes. Eukaryotic genomes are also much larger, on average: the largest bacterial genome sequenced to date is the 9 Mb chromosome from *Bradyrhizobium japonicum*.[31] At 3200 Mb, the human genome is approximately 355 times larger. The size and multiple chromosomes alone would challenge the assemble algorithms and finishers, but the sequence of eukaryotic genomes also contains a high repeat content, rendering cloning, sequencing, and assembly of particular regions difficult.

Because of these limitations, extra presequencing strategies must be developed to make the genome amenable to sequencing, and the definition of "finished genome" is often project specific. The most common strategy employed for eukaryotic genome sequencing is a "map-then-sequence" approach that seeks to obtain a physical map of the genome prior to the shotgun sequencing phase. The physical map is composed of ordered, overlapping large-insert clones anchored to markers in the genome, identified by genetic mapping. We discussed how this was done previously using fingerprinting approaches for large insert clones. Based on the order and orientation of the map, individual clones are selected and subjected to the standard shotgun sequencing and finishing, as described above. There are several advantages to this strategy, including an overall reduction in the number of repeats that need to be resolved during any particular finishing phase and the identification of problem regions before sequencing even begins.

Ideally, the physical map and finished genome sequence should cover the entire genome, but this level of accuracy is rarely achieved. Regions of concentrated repeats, such as telomeres and centromeres, do not clone well and are extremely difficult to sequence. Regions of high repeat content are called heterochromatin. Euchromatin has a much lower repeat concentration and is believed to contain the majority of the genes. As a result, most eukaryotic sequencing projects aim to completely sequence the euchromatic regions of the genome. For example, approximately one third of the fruit fly genome was heterochromatic and consequently not sequenced. Even when dealing with just euchromatin, small regions that are difficult to sequence can still occur, and most eukaryotic genome projects allow for a certain number of gaps in the "finished" product, provided the gap length is known. The public Human Genome Project allows an unspecified number of gaps, as long as their length is less than 150 kb.

After (or even during) construction of the physical map, sequencing can commence using the standard shotgun method. A sequence-ready clone is chosen, shotgun libraries are prepared, and the clone is sequenced. Additional clones are selected and sequenced with minimal overlap with the existing sequence. Finishing is limited to the regions that do not overlap with an existing, finished sequence.

With a physical map, another advantage becomes immediately apparent. Unlike a purely shotgun approach for small genomes, the sequencing of a mapped genome can be divided between more than one sequencing center. The genome can be divided into regions or even whole chromosomes and sequencing can take place in many different labs simultaneously. Additionally, even at a single sequencing center, sequencing can commence at different positions in the genome. As new sequence-ready clones are identified by physical mapping, shotgun sequencing and finishing can commence on other clones. Thus, the mapping and sequencing at a single lab can proceed simultaneously.

As mentioned previously, the physical framework strategy has been successfully applied to several prominent eukaryotic genome projects. The alternative strategy is the Whole-Genome Shotgun (WGS) or a mixture of mapping and WGS. WGS sequences are rarely finished to the same quality as a mapped-and-sequenced genome, but for certain genome projects, WGS provides a reasonable and affordable alternative to a completed sequence. Because of the unfinished quality of WGS sequences, they are often referred to as "draft" sequences.

Although the vast majority of genome sequence is obtained during the shotgun sequencing phase, the majority of effort (and money) is invested in the finishing phase. With the complete human genome sequence, finished genome sequences from other vertebrates may be an unnecessary expense for comparatively little information. Because genomes of mammals and vertebrates have some level of conserved gene content and order, the human genome sequence can serve as a reference template for other genome sequences. A WGS project could generate fourfold or fivefold coverage of the genome and still theoretically achieve 98% coverage of the euchromatic region of the genome. The sequence could then be compared to the finished draft of the human genome.

As with a standard shotgun sequencing project, a eukaryotic WGS begins with careful construction of genomic clone libraries. Celera's attempt to sequence the human genome by WGS showed that multiple clone libraries with a differing insert sizes are essential to the assembly of sequence contigs. For each library, both ends of the clone inserts must be sequenced to provide pairs of sequences a known distance apart on the chromosome. Paired end sequences are necessary to order and orient sequence contigs into "scaffolds." Finally, the actual assembly of eukaryotic WGS sequence reads usually proceeds in a multistep process, even with advanced computational hardware. Typically, the repeat sequences will be masked in the early assembly steps to prevent misassembly due to highly conserved repetitive sequences. Once contigs and scaffolds of unique sequences have been assembled, the repeat sequences can be added back to the assembly with the unique sequences acting as an assembly framework.

A third application of WGS is to use a mixed approach, with elements of both map-then-sequence and WGS. For example, a low-coverage WGS can be generated rapidly at a large genome sequencing center while finishing could be completed on a clone-by-clone basis at smaller sequencing centers. This hybrid approach was used for the *Drosophila* genome project. Celera Corp. provided a nearly 15-fold shotgun coverage of the *Drosophila* genome, and finishing was completed in a number of publicly funded laboratories.

Despite the advantages of WGS for some eukaryotes, the drawbacks are also significant. Experience with the rice genome has showed that contigs produced by WGS may be a poor substrate for gene identification.[32] In a detailed analysis of the complete sequence of rice chromosome 10, researchers predicted twice as many genes as had been predicted by previous WGS sequencing efforts of the same chromosome. Additionally, genes predicted from the WGS contigs were on average one third shorter than genes predicted from the finished sequence.

Several recent sequencing alternatives have arisen to the WGS rough draft approach for eukaryotic genomes. Eukaryotic genomes typically contain large amounts of repetitive DNA interspersed with genic regions. Approaches that focus on cloning and then sequencing only the gene rich regions of the genome can be more cost-effective and still provide highly usable genomic sequence information. One such approach takes advantage of differences in methylation patterns between repetitive and genic regions of the genome. In this technique, a WGS cloning approach is taken, but the shotgun clones are transformed into a methylation restrictive *Escherichia coli* host strain. Because repetitive regions of the genome are heavily methylated and the genic regions are not, the genic regions become overrepresented while the repetitive DNA is filtered out by the restrictive host strains of *E. coli*.[33] Unfortunately, the differential methylation pattern observed in plants is not observed in mammals, so this technique is only applicable in plant genomes.[34] Yet another approach is based on an old, but powerful technique (Cot analysis) for separating DNA based on sequence complexity. In this approach, heat-denatured and sheared genomic DNA is allowed to re-nature to a specific Cot value and then the single-stranded DNA is separated from the double-stranded DNA via hydroxyapatite chromatography.[35] This technique is based on the fact that sheared genomic DNA reassociates at a rate proportional to its representation in the genome. Hence, the highly abundant repetitive sequence and the low abundance genic sequence can be separated and the gene-rich fractions cloned separately to create libraries enriched for genic sequence. Unlike the methyl-filtration-based approach, the Cot-based cloning approach will work in both plant and animal systems. Gene-enriched libraries created with these approaches are then sequenced in a standard high-throughput manner. The resulting sequencing contigs can then be anchored to physical map and STC frameworks to spatially orient the gene rich sequencing contigs along the chromosomes.

Once the genome sequence has been completed to a previously defined set of quality standards, the biologically important features of the sequence can be identified in a process called annotation. Biologically important features identified in a typical genome project include genes and transposable elements. For bacterial or archaeal genomes, gene identification is relatively straightforward, while gene identification in eukaryotic genomes is much more difficult due to the presence of introns. Three basic methods are used for protein-coding gene identification: *ab initio* gene prediction, sequence similarity searches of public repositories/databases, and EST databases. All genome projects use both gene prediction and homology searching, while EST analysis is primarily a technique for eukaryotic gene identification.

The goal of *ab initio* gene prediction is to correctly identify the start and stop codons (and intron/exon boundaries, if relevant) of a gene given only the DNA sequence and statistical parameters derived from known genes. For the simple structure of a bacterial or archaeal genome, genes can be identified from open reading frames that match a known set of codon preferences. Glimmer is a popular program for prokaryotic gene prediction.[36] Statistical parameters considered for eukaryotic genes can include codon preferences, frequencies of longer combinations of codons, and attributes of intron/exon boundaries.[37] These statistical parameters are derived from analyses of a large number of experimentally identified and sequenced genes. Because eukaryotic gene prediction algorithms must correctly identify all possible exons, some of which can be very short, gene prediction in eukaryotic DNA is much less accurate than in prokaryotic DNA. Genscan and FGenesH are commonly used eukaryotic gene prediction software programs.

Genes can also be identified by their similarity to known sequences. Programs such as BLAST or FASTA compare a query sequence to a database of sequences.[38,39] For each sequence in the database, a similarity score is calculated, and the statistical significance of each similarity score can be estimated with a high degree of accuracy. If the query sequence matches a sequence in the database with a statistically significant similarity score, we can infer with confidence that the sequences are related either through a common biochemical or physiological function. Because the statistical properties of sequence similarity scores are well understood, sequence similarity searching is the most reliable computational method of identifying protein-coding genes, RNA genes, pseudogenes, and transposable elements. Because sequence similarity programs do not identify start and stop codons or intron/exon boundaries, additional manual editing is necessary to derive a gene prediction from the significant similarity to a known sequence.

A typical eukaryotic genome project will utilize every gene identification method during annotation. Several different *ab initio* gene prediction programs will be run on the genomic sequence. Additionally, sequence similarity searches will be performed on several different databases, including previously characterized protein sequences, known transposable elements, and ESTs from the target organism (and closely related species where available). The results of these computational analyses are then manually edited into a final annotation by a trained genome researcher. The annotation and the sequence can be deposited in one of the public DNA sequence databases, such as GenBank.

CONCLUSION

As consumers of genomic information, the majority of biologists are not directly involved in the generation of genome sequence data. Consequently, it is important for biologists to understand how genome sequence data are obtained to appreciate the advantages and limitations of using complete genomes. As we have explained, there is no single definition of "complete genome" to which all genome projects conform. Different levels of completion provide their own strengths and weaknesses. Sequences completed to the quality of most bacterial genomes and early eukaryotic genomes (yeast, *Caenorhabditis elegans, Arabidopsis*) give excellent insight into both the content and organization of the genome. For eukaryotes, such high-quality genome sequences require significant investment of time and funding. Draft sequences created by WGS are excellent for comparative genomics with close relatives (especially if the close relative has a high-quality completed sequence) and can provide insight into gene content. Draft sequences are probably unsuitable for comparative genomics with distantly related organisms and may not be adequate for *ab initio* gene prediction.

It is helpful to remember that the completed genome sequence is experimental data and subject to experimental error. Even high-quality, finished genome sequences will have error. Most genomic sequence that is subject to finishing is completed to an accuracy of 1 error in 10,000 nucleotides. With 3 billion nucleotides, we should expect 300,000 single-nucleotide errors in the human genome sequence. For most researchers, this level of error will hardly ever be noticeable, but occasionally researchers may stumble across one of these errors.

Finally, researchers should also keep in mind that efforts to annotate genomic sequence, and in particular to identify protein-coding genes, are subject to much higher and less quantifiable error rates than actually obtaining the sequence itself. For important research projects on particular genes or gene families, predicted genes should be subject to experimental verification, based on the evidence used to predict the gene. Genes predicted from a combination of EST similarity, *ab initio* predictions, and significant similarity to a closely related protein sequence will require less verification than those predicted from *ab initio* gene-prediction software alone.

At present, genome technology and sequencing continues to grow at an amazing rate. Many of the strategies and limitations discussed in this chapter could be resolved at some time in the future. As genomics continues to grow, our understanding of the chemical basis of life will also grow, providing new platforms and methods for understanding disease, inheritance, and evolution. The benefits of investing in genome sequencing will continue to be discovered for years to come.

REFERENCES

1. MD Adams, JM Kelley, JD Gocayne, M Bubnick, MH Polymeropoulos, H Xiao, CR Merril, A Wu, B Olde, RF Moreno. Complementary DNA sequencing: expressed sequence tags and human genome project. Science 252:1651–1666, 1991.
2. K Okubo, N Hori, R Matoba, T Niyama, A Fukushima, Y Kojima, K Matsubara. Large scale cDNA sequencing for analysis of quantitative and qualitative aspects of gene expression. Nat Genet 2:173–179, 1992.
3. DT Burke, G Carle, MV Olsen. Cloning of large segments of exogenous DNA into yeast by means of artificial chromosome vectors. Science 236: 806–812, 1987.
4. C Anderson. Genome shortcut leads to problems. Science 259:1684–1687, 1993.
5. DL Neil, A Villasante, RB Fisher, D Vetrie, B Cox, C Tyler-Smith. Structural instability of human tandemly repeated DNA sequences cloned in yeast artificial chromosome vectors. Nucleic Acids Res 18:1421–1428, 1990.
6. ED Green, HC Riethman JE Dutchik MV Olson. Detection and characterization of chimeric yeast artificial-chromosome clones. Genomics 11:658–669, 1991.
7. JC Venter, HO Smith, L Hood. A new strategy for genome sequencing. Nature 381:364–366, 1996.
8. M O'Connor, M Peifer, W Bender. Construction of large DNA segments in *Escherichia coli*. Science 244:1307–1312, 1989.
9. S-S Woo, J Jiang, BS Gill, AH Paterson, RA Wing. Construction and characterization of a bacterial artificial chromosome library of *Sorghum bicolor*. Nucleic Acids Res 22:4922–4931, 1994.

10. H Shizuya, B Birren, U-J Kim, V Mancino, T Slepak, Y Tachiiri, M Simoń. Cloning and stable maintenance of 300-kilobase-pair fragments of human DNA in *Escherichia coli* using an F-factor-based vector. Proc Natl Acad Sci USA 89:8794–8797, 1992.

11. JP Tomkins, H Miller-Smith, M Sasinowski, S Choi, H Sasinowska, M Verce, DL Freedman, RA Dean, RA Wing. Physical map and gene survey of the *Ochrobactrum anthropi* genome using bacterial artificial chromosome contigs. Microb Comp Genomics 4:203–217, 1999.

12. JP Tomkins, DG Peterson, TJ Yang, D Main, TA Wilkins, AH Paterson, RA Wing. Development of genomic resources for cotton (*Gosypium hirsutum*): BAC library development, preliminary STC analysis, and identification of clones associated with fiber development. Mol Breeding 8:255–261, 2001.

13. L Cai, JF Taylor, RA Wing, DS Gallagher, S-S Woo, SK Davis. Construction and characterization of a bovine bacterial artificial chromosome library. Genomics 29: 413–425, 1995.

14. N Willetts, R Skurray. Structure and function of the F factor and mechanism of conjugation. In *Escherichia coli* and *Salmonella typhimurium,* FC Neihardt, Ed., Cell Mol Biol 2:1110–1133, 1987.

15. JC Venter, MD Adams, EW Myers, PW Li, RJ Mural, GG Sutton, HO Smith, M Yandell, CA Evans, RA Holt, JD Gocayne, P Amanatides, RM Ballew, DH Huson, JR Wortman, Q Zhang, CD Kodira, XH Zheng, L Chen, M Skupski, G Subramanian, PD Thomas, J Zhang, GL Gabor Miklos, C Nelson, S Broder, AG Clark, J Nadeau, VA McKusick, N Zinder, AJ Levine, RJ Roberts, M Simon, C Slayman, M Hunkapiller, R Bolanos, A Delcher, I Dew, D Fasulo, M Flanigan, L Florea, A Halpern, S Hannenhalli, S Kravitz, S Levy, C Mobarry, K Reinert, K Remington, J Abu-Threideh, E Beasley, K Biddick, V Bonazzi, R Brandon, M Cargill, I Chandramouliswaran, R Charlab, K Chaturvedi, Z Deng, V Di Francesco, P Dunn, K Eilbeck, C Evangelista, AE Gabrielian, W Gan, W Ge, F Gong, Z Gu, P Guan, TJ Heiman, ME Higgins, RR Ji, Z Ke, KA Ketchum, Z Lai, Y Lei, Z Li, J Li, Y Liang, X Lin, F Lu, GV Merkulov, N Milshina, HM Moore, AK Naik, VA Narayan, B Neelam, D Nusskern, DB Rusch, S Salzberg, W Shao, B Shue, J Sun, Z Wang, A Wang, X Wang, J Wang, M Wei, R Wides, C Xiao, C Yan, A Yao, J Ye, M Zhan, W Zhang, H Zhang, Q Zhao, L Zheng, F Zhong, W Zhong, S Zhu, S Zhao, D Gilbert, S Baumhueter, G Spier, C Carter, A Cravchik, T Woodage, F Ali, H An, A Awe, D Baldwin, H Baden, M Barnstead, I Barrow, K Beeson, D Busam, A Carver, A Center, ML Cheng, L Curry, S Danaher, L Davenport, R Desilets, S Dietz, K Dodson, L Doup, S Ferriera, N Garg, A Gluecksmann, B Hart, J Haynes, C Haynes, C Heiner, S Hladun, D Hostin, J Houck, T Howland, C Ibegwam, J Johnson, F Kalush, L Kline, S Koduru, A Love, F Mann, D May, S McCawley, T McIntosh, I McMullen, M Moy, L Moy, B Murphy, K Nelson, C Pfannkoch, E Pratts, V Puri, H Qureshi, M Reardon, R Rodriguez, YH Rogers, D Romblad, B Ruhfel, R Scott, C Sitter, M Smallwood, E Stewart, R Strong, E Suh, R Thomas, NN Tint, S Tse, C Vech, G Wang, J Wetter, S Williams, M Williams, S Windsor, E Winn-Deen, K Wolfe, J Zaveri, K Zaveri, JF Abril, R Guigo, MJ Campbell, KV Sjolander, B Karlak, A Kejariwal, H Mi, B Lazareva, T Hatton, A Narechania, K Diemer, A Muruganujan, N Guo, S Sato, V Bafna, S Istrail, R Lippert, R Schwartz, B Walenz, S Yooseph, D Allen, A Basu, J Baxendale, L Blick, M Caminha, J Carnes-Stine, P Caulk, YH Chiang, M Coyne, C Dahlke, A Mays, M Dombroski, M Donnelly, D Ely, S Esparham, C Fosler, H Gire, S Glanowski, K Glasser, A Glodek, M Gorokhov, K Graham, B Gropman, M Harris, J Heil, S Henderson, J Hoover, D Jennings, C Jordan, J Jordan, J Kasha, L Kagan, C Kraft, A Levitsky, M Lewis, X Liu, J Lopez,

D Ma, W Majoros, J McDaniel, S Murphy, M Newman, T Nguyen, N Nguyen, M Nodell, S Pan, J Peck, M Peterson, W Rowe, R Sanders, J Scott, M Simpson, T Smith, A Sprague, T Stockwell, R Turner, E Venter, M Wang, M Wen, D Wu, M Wu, A Xia, A Zandieh, X Zhu. The sequence of the human genome. Science 291:1304–1351, 2001.

16. EW Myers, GG Sutton, AL Delcher, IM Dew, DP Fasulo, MJ Flanigan, SA Kravitz, CM Mobarry, KH Reinert, KA Remington, EL Anson, RA Bolanos, HH Chou, CM Jordan, AL Halpern, S Lonardi, EM Beasley, RC Brandon, L Chen, PJ Dunn, Z Lai, Y Liang, DR Nusskern, M Zhan, Q Zhang, X Zheng, GM Rubin, MD Adams, JC Venter. The genome sequence of *Drosophila melanogaster*. Science 287:2185–2195, 2000.

17. RA Holt, GM Subramanian, A Halpern, GG Sutton, R Charlab, DR Nusskern, P Wincker, AG Clark, JM Ribeiro, R Wides, SL Salzberg, B Loftus, M Yandell, WH Majoros, DB. Rusch, Z Lai, CL Kraft, JF Abril, V Anthouard, P Arensburger, PW Atkinson, H Baden, V de Berardinis, D Baldwin, V Benes, J Biedler, C Blass, R Bolanos, D Boscus, M Barnstead, S Cai, A Center, K Chaturverdi, GK Christophides, MA Chrystal, M Clamp, A Cravchik, V Curwen, A Dana, A Delcher, I Dew, CA Evans, M Flanigan, A Grundschober-Freimoser, L Friedli, Z Gu, P Guan, R Guigo, ME Hillenmeyer, SL Hladun, JR Hogan, YS Hong, J Hoover, O Jaillon, Z Ke, C Kodira, E Kokoza, A Koutsos, I Letunic, A Levitsky, Y Liang, JJ Lin, NF Lobo, JR Lopez, JA Malek, TC McIntosh, S Meister, J Miller, C Mobarry, E Mongin, SD Murphy, DA O'Brochta, C Pfannkoch, R Qi, MA Regier, K Remington, H Shao, MV Sharakhova, CD Sitter, J Shetty, TJ Smith, R Strong, J Sun, D Thomasova, LQ Ton, P Topalis, Z Tu, MF Unger, B Walenz, A Wang, J Wang, M Wang, X Wang, KJ Woodford, JR Wortman, M Wu, A Yao, EM Zdobnov, H Zhang, Q Zhao, S Zhao, SC Zhu, I Zhimulev, M Coluzzi, A della Torre, CW Roth, C Louis, F Kalush, RJ Mural, EW Myers, MD Adams, HO Smith, S Broder, MJ Gardner, CM Fraser, E Birney, P Bork, PT Brey, JC Venter, J Weissenbach, FC Kafatos, FH Collins, SL Hoffman. The genome sequence of the Malaria mosquito *Anopheles gambiae*. Science 298:129–149, 2002.

18. T Mozo, K Dewar, P Dunn, JR Ecker, S Fischer, S Kloska, H Lehrach, M Marra, R Martienssen, S Meier-Ewert, T Altmann. A complete BAC-based physical map of the *Arabidopsis thaliana* genome. Nat Genet 22:271–275, 1999.

19. M Chen, G Presting, WB Barbazuk, JL Goicoechea, B Blackmon, G Fang, H Kim, D Frisch, Y Yu, S Sun, S Higingbottom, J Phimphilai, D Phimphilai, S Thurmond, B Gaudette, P Li, J Liu, J Hatfield, D Main, K Farrar, C Henderson, L Barnett, R Costa, B Williams, S Walser, M Atkins, C Hall, MA Budiman, JP Tomkins, M Luo, I Bancroft, J Salse, F Regad, T Mohapatra, NK Singh, AK Tyagi, C Soderlund, RA Dean, RA Wing. An integrated physical and genetic map of the rice genome. Plant Cell 14:537–545, 2002.

20. MA Marra, TA Kucaba, NL Dietrich, ED Green, B Brownstein, RK Wilson, KM McDonald, LW Hillier, JD McPherson, RH Waterston. High throughput fingerprint analysis of large-insert clones. Genome Res 7:1072–1084, 1997.

21. CA Soderlund, S Humphray, A Dunham, L French. Contigs built with fingerprints, markers and FPC V4.7. Genome Res 10, 2000.

22. Y Ding, MD Johnson, WQ Chen, D Wong, Y-J Chen, SC Benson, JY Lam, Y-M Kim, H Shizuya. Five-color-based high-information-content fingerprinting of bacterial artificial chromosome clones using type IIS restriction endonucleases. Genomics 74:142–154, 2001.

23. JP Tomkins, DG Peterson, TJ Yang, D Main, EF Ablett, RJ Henry, LS Lee, TA Holton, D Waters, RA Wing. Grape (*Vitis vinifera* L.) BAC library construction, preliminary

STC analysis, and identification of clones associated with flavonoid and stilbene biosynthesis. Am J Enol Viticult 52:287–291, 2002.

24. B Ewing, P Green. Base-calling of automated sequencer traces using Phred. II. Error probabilities. Genome Res 8:186–194, 1998.

25. B Ewing, L Hillier, MC Wendl, P Green. Base-calling of automated sequencer traces using Phred. I. Accuracy assessment. Genome Res 8:175–185, 1998.

26. A. Bairoch., R Apweiler. The SWISS-PROT protein sequence database and its supplement TrEMBL. Nucleic Acids Res 28:45–48, 2000.

27. JS Gardiner, M Schroeder, H Polacco, Anchez-Villeda, Z Fang, M Morgante, T Landewe, K. Fengler, F Useche, M Hanafey, S Tingey, H Chou, R Wing, C Soderlund, EH Coe, Jr. Anchoring 9,371 maize expressed sequence tagged unigenes to the bacterial artificial chromosome contig map by two-dimensional overgo hybridization. Plant Physiol 134:1317–1326, 2004.

28. MT Ross, S LaBrie, J McPherson, V Stanton, Jr. Screening large-insert libraries by hybridization. In Current Protocols in Human Genetics, Boyl, Ed., New York: Wiley, 1999, 5.6.1–5.6.52.

29. X Huang, A Madan. CAP3: A DNA sequence assembly program. Genome Res 9:868–877, 1999.

30. D Gordon, C Abajian, P Green. *Consed:* a graphical tool for sequence finishing. Genome Res 8:195–202, 1998.

31. T Kaneko, Y Nakamura, S Sato, K Minamisawa, T Uchiumi, S Sasamoto, A Watanabe, K Idesawa, M Iriguchi, K Kawashima, M Kohara, M Matsumoto, S Shimpo, H Tsuruoka, T Wada, M Yamada, S Tabata. Complete genomic sequence of nitrogen-fixing symbiotic bacterium *Bradyrhizobium japonicum* USDA110. DNA Res 9:225–256, 2002.

32. SA Goff, D Ricke, T Lan, G Presting, R Wang, M Dunn, J Glazebrook, A Sessions, P Oeller, H Varma, D Hadley, D Hutchison, C Martin, F Katagiri, BM Lange, T Moughamer, Y Xia, P Budworth, J Zhong, T Miguel, U Paszkowski, S Zhang, M Colbert, W Sun, L Chen, B Cooper, S Park, TC Wood, L Mao, P Quail, R Wing, R Dean, Y Yu, A Zharkikh, R Shen, S Sahasrabudhe, A Thomas, R Cannings, A Gutin, D Pruss, J Reid, S Tavtigian, J Mitchell, G Eldredge, T Scholl, RM Miller, S Bhatnagar, N Adey, T Rubano, N Tusneem, R Robinson, J Feldhaus, T Macalma, A Oliphant, S Briggs. A draft sequence of the rice genome (*Oryza sativa* L. ssp. *japonica*). Science 296:92–100, 2002.

33. PD Rabinowicz, K Schutz, N Dedhia, C Yordan, LD Parnell, L Stein, WR McCombie, RA Martienssen. Differential methylation of genes and retrotransposons facilitates shotgun sequencing of the maize genome. Nat Genet 23:305–308, 1999.

34. PD Rabinowicz, WR McCombie, RA Martienssen. Gene enrichment in plant genomic shotgun libraries. Curr Opin Plant Biol 6:150–156, 2003.

35. DG Peterson, SR Wessler AH Paterson. Efficient capture of unique sequences from eukaryotic genomes. Trends Genet 18(11):547–550, 2002.

36. AL Delcher, D Harmon, S Kasif, O White, SL Salzberg. Improved microbial gene identification with GLIMMER. Nucleic Acids Res 27:4636–4641, 1999.

37. C Burge, S Karlin. Prediction of complete gene structures in human genomic DNA. J Mol Biol 268:78–94, 1997.

38. SF Altschul, TL Madden, AA Schäffer, J Zhang, Z Zhang, W Miller, DJ Lipman. Gapped BLAST and PSI-BLAST: a new generation of protein database search programs. Nucleic Acids Res 25:3389–3402, 1997.

39. WR Pearson. Flexible sequence similarity searching with the FASTA3 program package. Methods Mol Biol 132:185–219, 2000.

8 Sequence Reconstruction from Nucleic Acid Microarray Data*

Franco P. Preparata, Eli Upfal, and Samuel A. Heath

CONTENTS

INTRODUCTION

More and more complete genome sequences are being reported in the technical literature, of which the human genome is the most attention-capturing example.[1] Far from waning, the interest in the acquisition of nucleic acid sequences is bound to grow exponentially in time, both for comparisons with known genomes and for the augmentation of the genome library. Therefore, methods aimed at expediting the sequencing process have enormous implications for the health and life sciences and have naturally attracted considerable research interest.

The objective is obviously some revolutionary acceleration of the biochemical acquisition of nucleic acid data (hereafter, DNA) in the laboratory, to be passed on

* This work was partially supported by the National Science Foundation under Grant DBI-9983081. A preliminary version of this paper was presented as an invited lecture at the ENAR-IMS 2001 meeting in Chatlotte, NC, on March 27, 2001.

to conventional information processing, thereby parallelizing the inherently serial process of molecular migration that takes place in electrophoresis-based methods.

As is well known, *sequencing* is the acquisition, i.e., the read-out, of an unknown DNA sequence, herafter referred to as the *target*. This process is at present carried out by means of wet-lab techniques, which, despite their labor-intensive nature and the shortness of the specimens they can handle (<1000 bases), have enabled the extraordinary progress of molecular biology in the past two decades.

As an alternative to this approach, around 1990 several research groups[2–5] proposed to resort to the property that DNA sequences have to hybridize to their Watson/Crick complements, thereby opening up the possibility of the simultaneous acquisition of all relevant data in a single laboratory experiment.

The basic idea of this novel methodology is to deploy a set — frequently called a *library*— of short strings of nucleotides (A,C,G,T) (oligonucleotides) on some solid support, typically, but not necessarily, a glass substrate, called a "microarray" or "chip." The active area of the chip is structured as a matrix, in each region of which (called a *feature*) a very large number of copies of a specific oligonucleotide are implanted. The chip is immersed under controlled conditions within a solution of a suitably labeled target DNA sequence. A copy of the target DNA will bind (hybridize) to an oligonucleotide if the oligonucleotide is complementary, in the Watson/Crick sense, to one of its subsequences. The labeling of the target allows visualization of the chip features containing binding oligonucleotides, thereby yielding a method for automatically probing the target sequence for specific subsequences.

Although DNA microarrays are amenable to a wide variety of specialized applications, such as gene identification, detection of single-nucleotide polymorphisms, etc., the focus of this chapter is *de novo* sequencing, which is the most ambitious process, subsuming most of the other applications. In *de novo* sequencing the microarray library is complete; i.e., it contains oligonucleotides for all possible choices of the bases.

In summary, sequencing by hybridization (SBH) consists of two fundamental steps. The first, biochemical in nature, is the acquisition, by complementary hybridization with a complete library of probes, of all subsequences (of a selected pattern) of a given unknown target sequence; the set of such subsequences is called the sequence *spectrum*. The spectrum is basically the characteristic function of the set of subsequences of the target, because in current technology it only appears feasible to detect the absence or presence of a specific probe and not its multiplicity. The second step, combinatorial in nature, is the algorithmic reconstruction of the sequence from its spectrum.

Both steps offer a number of significant challenges to be overcome. In reality, the biochemistry of hybridization is anything but a yes/no phenomenon, and we address this important point at the end of the chapter. In what follows, however, we assume that hybridization is a "noiseless" process, which identifies all prescribed subsequences of the target.

This chapter, which is a digest of a variety of recent specialized results, focuses on the combinatorial aspect of the problem, namely, the identification of the most effective techniques to extract from the target sequence the information necessary for its reconstruction. This consists of two closely interacting items. The first is the

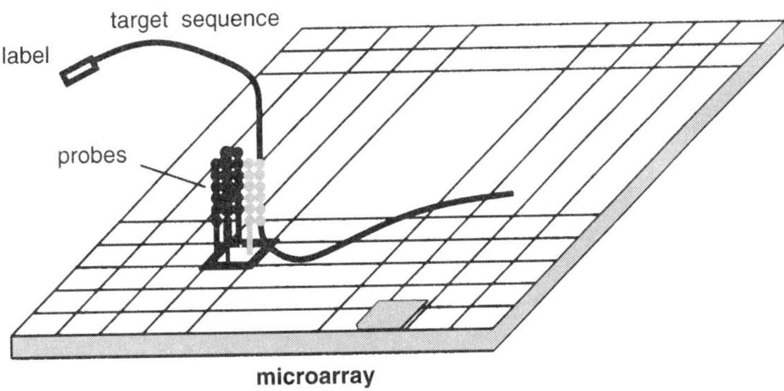

FIGURE 8.1 Illustration of microarray technology.

selection of the probing scheme, i.e., the types of subsequences of the target that are obtained in the acquisition step. The second is the design of the algorithms that effect the sequence reconstruction, that is, that create a *putative* sequence consistent with the obtained spectrum.

EARLY APPROACHES

Pioneering work on this topic, by Bains and Smith,[5] Lysov et al.,[2] Drmanac et al.,[4] and Pevzner,[3] focused on probing schemes (referred to here as "standard"), which use all 4^k k-mers (i.e., solid strings of k nucleotides), where k is a technology-dependent integer parameter.* Using probes in the form of k-nucleotide strings, standard approaches constructed the putative sequence as a path in a suitable directed graph G, where each path edge uniquely identifies a nucleotide. (Refer to Figure 8.1, where for a given short target sequence we have illustrated the relevant graph-theoretic notions.) In the earliest versions, the nodes of G were identified with the spectrum probes, and the successors of a node v were all the nodes whose probes overlapped in their *first* $(k-1)$ symbols (prefix) with the *last* $(k-1)$ symbols (suffix) of the v probe. Because the reconstruction had to account for all spectrum probes, in this modeling the path had to pass through all nodes of G (k-mer graph, Figure 8.2a) exactly once, a task well known to computer scientists as the construction of a Hamiltonian path (an "intractable" problem in general). Fortunately, the difficulty was only artificial, and we owe to Pevzner's[3] intuition the realization that the mentioned Hamiltonian path was an Eulerian path in disguise, a construct much easier to obtain. More strongly, Pevzner characterized the sequences consistent with a given spectrum, as those corresponding to Eulerian paths (i.e., paths using all edges exactly once) of a graph G' whose nodes are identified with the strings of $(k-1)$ symbols ($(k-1)$-mer graph, Figure 8.2b), and such that there is an edge from node u to node v if and only if the spectrum contains a probe whose prefix and suffix coincide, respectively, with u and v. This important characterization

---

* k is currently rather small, <10, but is expected to grow moderately.

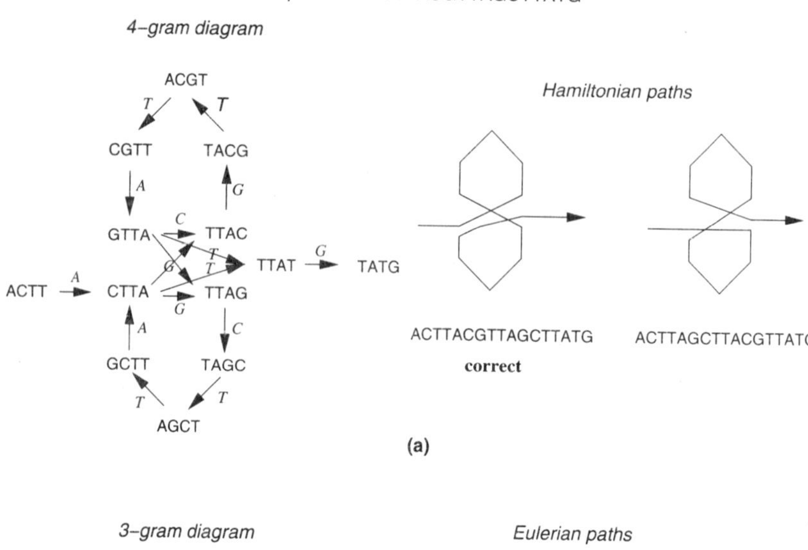

FIGURE 8.2 Illustration of Hamiltonian (a) and Eulerian (b) paths in the graph associated with a given target sequence; note that in both methods the reconstruction is ambiguous.

equated unambiguous sequence reconstruction with Eulerian path uniqueness in G'. Note that, differently from the k-mer graph, the corresponding $(k - 1)$-mer graph is Eulerian, i.e., all but the start and end nodes have identical numbers of incoming and outgoing arcs.

However, the combinatorial performance of the method was very discouraging, especially if contrasted with the so-called information-theoretic bound,[6] based on the following simple application of the pigeonhole principle. Each spectrum can be represented as a length-4^k binary vector, and there are 2^{4^k} such vectors. There are 4^m sequences of length m. Because there cannot be more sequences than there are vectors, the inequality $2^{4^k} \geq 4^m$ yields the bound:

$$m \leq 4^{k-\frac{1}{2}}$$

(8.1)

This inequality suggests that the length of a reliably reconstructible sequence should be $O(4^k)$. We critically revisit Equation 8.1 at the conclusion of this chapter.

However, probabilistic analysis based on random sequences of independent identically distributed symbols,[7,8] confirmed by empirical evidence from computer simulations, showed that k-nucleotide string probes could only achieve $m = O(\sqrt{4^k})$. In fact, in an influential survey paper,[9] noted biologist E. M. Southern observes, "The length of target that could be analysed is approximately equal to the square root of the number of oligonucleotides in the array."

This finding, coupled with well-known biochemical difficulties, cast serious doubts on whether such technology could ever become competitive with electrophoresis techniques. It was observed, on the other hand, that larger sequencing lengths could be achieved by introducing in the probing pattern a gap, to be realized by deploying all possible bases in the gap positions (so-called "degenerate bases"), with the adverse side effect of weakening the hybridization signal. Against this background, the next section introduces a novel method that achieves performances of the same order as the information theory bound, brings new vitality to the approach, and may revolutionize the sequencing practice.

THE "GAPPED" PROBING SCHEME

There are some revealing intuitive reasons behind the observed inadequacy of the standard method:

1. Sequence reconstruction is modeled as a path traversal in the $(k - 1)$-mer directed graph. An obvious feature of this process, and the source of ambiguity, is that any branching node (i.e., a node with two or more outgoing arcs, and an identical number of incoming arcs) contains no data pairing incoming to outgoing arcs.
2. The adoption of longer probing patterns (a larger value of k) may make an ambiguous branching disappear; however, a larger k implies an exponentially growing array cost. Therefore, there is intuition that a probing scheme that uses larger probe lengths (linking incoming to outgoing arcs of the above graph) without increasing the cost (i.e., the value of k), may afford an effective solution of the problem.
3. In sequence reconstruction by "extension" two consecutive probes of length k share $k - 1$ symbols. As a result of this dependence, very little information is provided by each additional probe.

These observations motivate a novel probing scheme, described below. Notationally, given two strings u and v over some finite alphabet, uv is the concatenation of the two strings.

Definition 1 *A probing pattern is a binary string (beginning and ending with a 1), i.e., a string of the form $1(0 + 1)^*1$ in the notation of regular languages. The length of the pattern is the number of its symbols, and its weight is the number of its 1's.*

Definition 2   *For $s + r = k$, an (s, r) probing scheme has direct and reverse patterns* $1^s(0^{s-1}1)^r$ *and* $(10^{s-1})^r 1^s$, *respectively (of weight k and length $(r + 1)s = v$).*

Definition 3   *An individual probe (an "oligonucleotide") is a string over the extended alphabet $\mathcal{A} = \{A, C, G, T, *\}$, where $*$ denotes the "wild card," i.e., a symbol that matches any symbol of the alphabet.* The convention is made that a probe occurs at position j of a sequence if j is the position of its rightmost symbol. Two strings over $\mathcal{A}$ of the same length, agree if they coincide in the positions where both have specified symbols (i.e., symbols different from $*$).*

Definition 4   *Given a sequence a over $\mathcal{A}$ the spectrum of a is the set of all of its probes, i.e., its subsequences conforming to the chosen probing pattern.*

Example 8.1.   For sequence $\mathbf{a}$ = CGGATACACTTGCAT and (direct) pattern 111001001, the spectrum is ACA**T**A(14), ATA**C**G(12), CAC**G**T(15), CGG**A**C(9), GAT**A**T(11), GGA**C**T(10), TAC**T**G(13) (probes listed lexicographically, their position within parentheses).

Definition 5   *The probe library associated with a given probing pattern is the set of the probes obtained by substituting each 1 of the pattern with natural bases in all possible ways and each 0 with a universal base.*

Sequence reconstruction, as in all other approaches to SBH, is accomplished through symbol-by-symbol extension from one end of the sequence to the other. Given an arbitrary sequence b (the current putative sequence), b_i denotes its ith symbol and $b_{(i,j)} = b_i\, b_{i+1}...b_j$.

To initialize the process we assume the presence of a segment of length $v - 1$ of the target sequence called a "seed" or a "primer." Such a segment can be derived from the spectrum itself, or, more expediently, we may assume that a standard primer is attached at the beginning of the target sequence. Similarly, to simplify termination, a standard primer of the same length may be attached to the other end of the target: its recognition signals termination of the reconstruction.

The advancing mechanism is as follows: The algorithm interrogates the spectrum with the query $q*$, where q is the $(v - 1)$-suffix of the current putative sequence. Such a query returns all feasible-extension probes contained in the spectrum. The following construct is the source of reconstruction ambiguities:

Definition 6   *A fooling probe is a feasible-extension probe for position i that occurs as a subsequence at position $j \neq i$ in the target sequence.*

For convenience of presentation, if we observe the reconstruction algorithm between two consecutive extensions of the putative sequence, we distinguish two modes of operation:

---

* Traditionally a wild card is called a "don't care." Physically, its realization is proposed by means of artificial universal bases,[10] i.e., bases that stack without binding. Such universal bases were heretofore a chemical curiosity, but may play a crucial role in microarray technology.

- **Extension mode.** The query $q*$ returns a single match and the sequence is extended by a single symbol.
- **Branching mode.** The query $q*$ returns more than one match (ambiguous branching). The algorithm attempts the extension of all paths issuing from the branching (and of all other paths spawned in turn by them) on the basis of spectrum probes. The breadth-first construction of such tree is pursued up to a maximum depth H (a design parameter), unless at some stage of this construction it is found that all surviving paths have a common prefix, which is then concatenated to the putative sequence.

Failure occurs when, operating in the branching mode, at depth H the common prefix mentioned above is empty.

The rationale of this advancing mechanism is that, whereas the correct path is deterministically extended, the extension of the spurious paths rests on the (probabilistic) presence of fooling probes in the spectrum. The parameter H should be chosen large enough to make the probability of spurious paths vanishingly small. The behavior of the described algorithm has been analyzed in some detail in Reference 11.

Since, except for trivial lengths of the target sequence, there is always a nonzero probability of ambiguous reconstruction, performance is naturally measured as the length m of sequences that are reconstructible with a given confidence level, under the standard hypothesis that the target sequence is generated by a maximum-entropy memoryless source (i.i.d. symbols). Although natural sequences do not quite comply with memoryless-source model (see Section "Concluding Remarks"), the latter has become the benchmark for comparative evaluations.

ANALYSIS OF RECONSTRUCTION FAILURES*

As discussed above, upon failure there are at least two "surviving" paths with H symbols beyond the branching, which reconstruct actual portions of the target sequence (i.e., their extension is fully supported by the spectrum). With this observation, we have two distinct failure modes, characterized as follows:

1. Failure Mode 1. *There are two paths identical except for their initial symbol* (corresponding to the branching).
2. Failure Mode 2. *There are two paths not satisfying the condition for Failure Mode 1.*

FAILURE MODE 1

This failure is caused by k fooling probes for the branching position, which are along the target sequence (with possible overlaps). In such case the spectrum does not permit disambiguation. We observe that the probability that a specific k-symbol probe does not occur at a specific sequence position is $(1 - 1/4^k)$ and, thus, the probability that it

---

* This section contains detailed technical material and may be skipped without loss of continuity.

does not occur at any position of a length-m sequence is $(1 - 1/4^k)^m$. We conclude that the probability that it occurs at least once in the sequence is

$$\left(1-\left(1-\frac{1}{4^k}\right)^m\right)$$

We find that for $m \geq 1$

$$1-e^{-\frac{m}{4^k}} \leq \left(1-\left(1-\frac{1}{4^k}\right)^m\right) \leq \frac{m}{4^k}$$

and note that for large values of m the (left) underestimate is much more accurate than the (right) overestimate. By similar reasoning, the probability that one of three specific sequences (causing the ambiguous extension) occurs in the sequence is approximately

$$\left(1-\left(1-\frac{1}{4^k}\right)^{3m}\right)$$

Therefore, ignoring the correction accounting for probe overlaps (see Reference 11), the probability that the k fooling probes occur at a specific reconstruction step can be expressed as

$$P = \left(1-e^{-\frac{3m}{4^k}}\right)\left(1-e^{-\frac{m}{4^k}}\right)^{k-1}$$

so that the probability that this event occurs at least once in the reconstruction process is

$$1-(1-P)^m \approx 1-e^{-mP} = 1-e^{-m\left(1-e^{-\frac{3m}{4^k}}\right)\left(1-e^{-\frac{m}{4^k}}\right)^{k-1}} \tag{8.2}$$

When appropriate, this rather accurate estimate is coarsely approximated as

$$P_1 = 3m\left(\frac{m}{4^k}\right)^k \tag{8.3}$$

Failure Mode 2

In this case, the tree of paths issuing from the branching contains the correct path and (at least) one competing (spurious) path. The latter begins with a $(v - 1)$-symbol segment, including or following the branching position, which is identical to an actual segment occurring in the sequence. Obviously, extension of this path is deterministically assured by probes guaranteed to belong to the spectrum and the reconstruction fails. (Such segments are referred to as *self-sustaining*.) The self-sustaining segment agrees, entirely or partially, with an equally positioned segment of the correct path, with the disagreements compensated for by fooling strings also occurring in the sequence.

Example 8.2 For reverse probing pattern 100100111, suppose the algorithm detects the following situation:

```
... A  C  G  A  G  T  C  (C  T  [G]  A  G  T  G  A  T  A  T  A  T ...
                             [T]  A  G  T  A  A)  T  C  T  G  G ...
```

where the pair [G][T] is the ambiguous branching, the top path represents the correct extension, and in the spurious bottom path, enclosed within parentheses, is the length 8 self-sustaining segment CTTAGTAA. This segment occurs *elsewhere* in the sequence. Clearly, indefinite extension of the spurious path is guaranteed by the spectrum. Segment CTTAGTAA is brought about by an appropriate collection of fooling probes, which compensate for disagreements between the two paths. Below the disagreements are evidenced within brackets.

```
                             0   1   2   3   4   5   6   7   8   9 ...
... A  C  G  A  G  T  C  (C  T  [G]  A  G  T  [G]  A  T  A  T  A  T ...
                            [T]  A  G  T  [A]  A)  T  C  T  G  G ...

1      C  *  *  G  *  *  C  T  T
2         G  *  *  T  *  *  T  T  A
3            A  *  *  C  *  *  T  A  G
4               -  C  *  *  T  *  *  T  A  A
5               T  *  *  T  *  *  G  T  A
```

The branching disagreement [G-T] is compensated for by probes 1 to 4, and disagreement [G-A] is compensated by probes 4 and 5. No other fooling probe is needed, because the required extending probes are guaranteed in the spectrum.

We conventionally denote the branching position as 0. The position-index immediately to the right of the self-sustaining segment is called the segment's *offset* and denoted J. Thus, $J \geq 0$.

The failure corresponding to $J = 0$ arises from the situation where there are two identical length-$(v - 1)$ segments occurring at different places in the target. Such an event is constructed by selecting two positions in the sequence (in

$$\binom{m}{2} \approx m^2/2$$

ways), of which the leftmost (encountered earlier in the reconstruction) identifies the correct path and the rightmost one the spurious path. The corresponding probability is therefore

$$\frac{m^2}{2}\frac{1}{4^{v-1}}\frac{3}{4} = m^2 \frac{3}{4^{v-2}\cdot 32} \tag{8.4}$$

since the $v - 1$ symbols of the self-sustaining segment are fully constrained and the branching symbol is selectable in three ways.

For $J > 0$, the positions of the two homologous segments are not interchangeable, so that there are about m^2 ways of selecting the event (rather than $m^2/2$), and its probability can be expressed as $m^2\pi_J$, for some coefficient π_J.

A detailed analysis of the terms π_J for $J > 0$ is rather cumbersome, so that we shall resort to appropriate approximations. The following informal observation will intuitively support the chosen approximations.

We begin by noting that the spectrum must contain a set of fooling probes necessary to compensate for the disagreements between the two competing segments (aligned on the two paths). Precisely, no probe is required at a position $0 \le j < J$ if and only if no disagreement (between the two alternative paths) occurs at positions $\{j, j - 1, \ldots, j - s + 1, j - 2s + 1, \ldots, j - v + 1\} \cap \{0, 1, \ldots, j\}$. Thus, a single disagreement may require the presence of several compensating fooling probes, and since a disagreement is three times as likely as an agreement, we may expect that there will be a fooling probe (with its rightmost symbol) at nearly every position in $[0, J - 1]$. If each position required a fooling probe, then we would have

$$\pi_J = \frac{3\alpha^J}{4^{v-1-J}}$$

However, not all positions demand a fooling probe (although, most of them do). Here we make the convenient simplification that "things are as if θJ positions do require a fooling probe," where parameter $\theta \le 1$ is a function of m and J, so that

$$\pi_J = \frac{3\alpha^{\theta J}}{4^{v-1-J}}$$

To obtain additional insight into parameter θ, a little reflection justifies the fact that, given a J-bit string describing the positions of agreement/disagreement, the positions of the required fooling probes are obtained by convolving the J-bit string with the reverse of the probing pattern: the nonzero terms of the convolution define the fooling probe positions. Almost all of the obtained convolutions have weight J, which would suggest $\theta \approx 1$. However, the few convolutions with weight $<J$ have much higher probability (since they invoke fewer fooling probes); this fact, for smaller values of α, decreases the value of θ. Thus, we expect θ to increase with α and to decrease with J. Detailed computer analysis suggests that, as a coarse approximation, we may take $\theta = 0.9$, independent of α and J for the most interesting probing patterns (for example, reverse (4,4) probes).

We conclude that

$$\sum_{J=1}^{v} \pi_J = \frac{3a}{4^{v-2}} \cdot \frac{1-(4a^q)^v}{1-4a^q} \tag{8.5}$$

For $J > v$, terms π_J are essentially negligible. However, within the stated approximation, $\theta(v-1)$ fooling probes are required for the self-sustaining segment; in addition, by the same argument, we may say that a fooling probe is required for each of positions $0, 1, \ldots, J - v$, so that

$$\pi_J = 3\alpha^{J-v+1}\alpha^{\theta(v-1)} = \frac{3}{4^{v-1}}\alpha^{J-v+1}(4\alpha^\theta)^{v-1}$$

and

$$\sum_{J>v} \pi_J = \frac{3\alpha^2}{4^{v-2}} \cdot \frac{(4\alpha^\theta)^{v-1}}{4(1-\alpha)} \tag{8.6}$$

Putting Equations 8.4 through 8.6 together, we obtain the following estimate of the probability of Mode 2 failure:

$$P_2 = \frac{3m^2}{4^{v-2}}\left(\frac{1}{32} + \alpha\frac{1-(4\alpha^\theta)^v}{1-4\alpha^\theta} + \alpha^2\frac{(4\alpha^\theta)^{v-1}}{4(1-\alpha)} \right) \tag{8.7}$$

Results of the described analyses have been found to be in excellent agreement with extensive simulations using random sequences.

PERFORMANCE OPTIMIZATION

We repeat below, for convenience, the estimates of the probabilities of Mode 1 and Mode 2 failures as obtained in the preceding section (with the choice $\theta = 0.9$):

$$P_1 = 3m\left(\frac{m}{4^k} \right)^k \tag{8.8}$$

$$P_2 = \frac{3m^2}{4^{v-2}}\left(\frac{1}{32} + \alpha\frac{1-(4\alpha^\theta)^v}{1-4\alpha^\theta} + \alpha^2\frac{(4\alpha^\theta)^{v-1}}{4(1-\alpha)} \right) \tag{8.9}$$

The functions $1 - P_1$ and $1 - P_2$ are, respectively, plotted in Figure 8.3 and Figure 8.4 in the range [0.9, 1] for $k = 0$, with independent variables r and m. We note that, whereas P_1 depends exclusively on the parameter k, P_2 strongly depends on v, and, consequently, on s and r. Indeed, Failure Mode 2 is the basis for performance

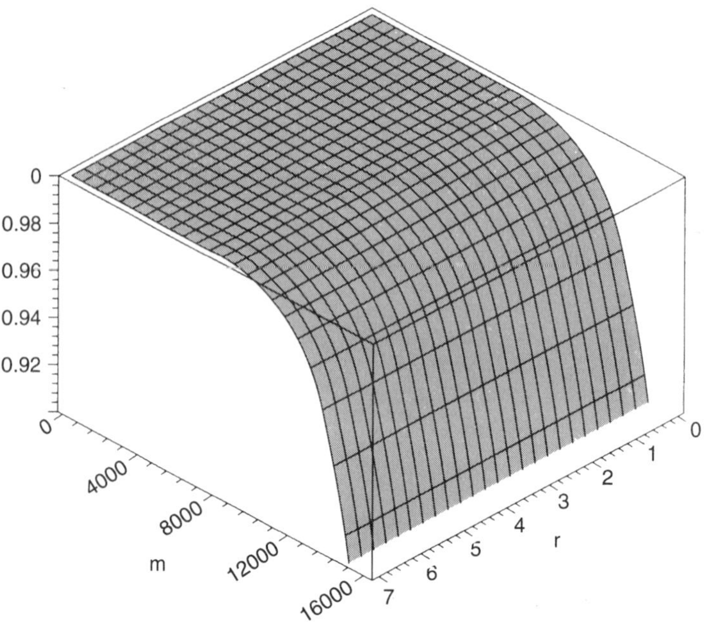

FIGURE 8.3 Diagram of the function $1 - P_1(m, r)$ for a (4,4) probing scheme.

optimization; this is not surprising, as the choices $s = 1$ and $s = k$ both yield as a special case the standard method, whose performance is known to be very poor.

The analogous plot of $1 - P_1 - P_2$ (the probability of successful reconstruction) in the range [0.9, 1] is displayed in Figure 8.5; this diagram illustrates that small or large values of r yield designs not substantially different from the standard one, and that the best performance is achieved for $r \approx s$: indeed, designs (4, 4) and (5, 3) have identical best performance. We now attempt a quantitative assessment of this performance.

Simple analysis of Equations 8.8 and 8.9 shows that Failure Mode 1 is dominant in the range of interest (where $P_1 + P_2$ is significantly > 0). Thus, for a small value ε, we wish to determine the solution m^* of the equation

$$P_1(m) = \varepsilon$$

Using approximation (8.3), we have

$$m = \left(\frac{\varepsilon}{3} 4^{k^2} \right)^{\frac{1}{k+1}}$$

$$= 4^{k-1-\frac{1}{k+1}} \sqrt{\log \frac{4\varepsilon}{3}}$$

which shows that we have a gap of a factor a little larger than 2 between achievable performance and bound (Equation 8.1).* Although from the viewpoint of asymptotic

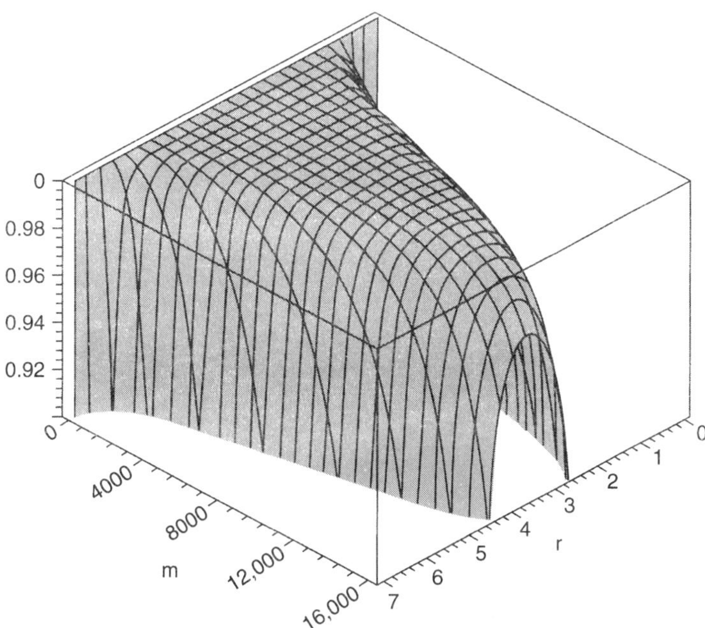

FIGURE 8.4 Diagram of the function $1 - P_2(m, r)$ for a (4,4) reverse probing scheme.

analysis the issue of optimality is settled, further algorithmic improvements are practically very important.

Exploring possible improvements, we note that the outlined algorithm, upon detection of failure, makes no use of information provided by the currently constructed prefix of the putative sequence. Indeed, upon failure the algorithm cannot discriminate between two competing sets of probes, either of which may be fooling. Intuition suggests, however, that only probes pertaining to the spurious alternative are guaranteed to be scattered along the sequence, whereas those pertaining to the correct extension are mostly concentrated around the current position. Therefore, further discrimination is provided by the relative counts in the prefix of the probes for the two alternatives. We call "polling"[12] a decision based on this count, and we could show by probabilistic analysis (confirmed by empirical evidence) that a further gain of about 20% in performance can be obtained over the algorithm not using the polling provision, thereby achieving performance $m \approx 0.2 \cdot 4^k$.

CONCLUDING REMARKS

We have analyzed in some detail the combinatorics of sequence reconstruction in SBH. There are, however, several additional aspects of the problem that we should address in order to place the topic in the appropriate perspective. These aspects concern the

---

* This can be readily verified from the approximate equation $3m\alpha^k = \varepsilon$, with the initial estimate $m = 0.25 \cdot 4^k$ and $\varepsilon = 0.1$.

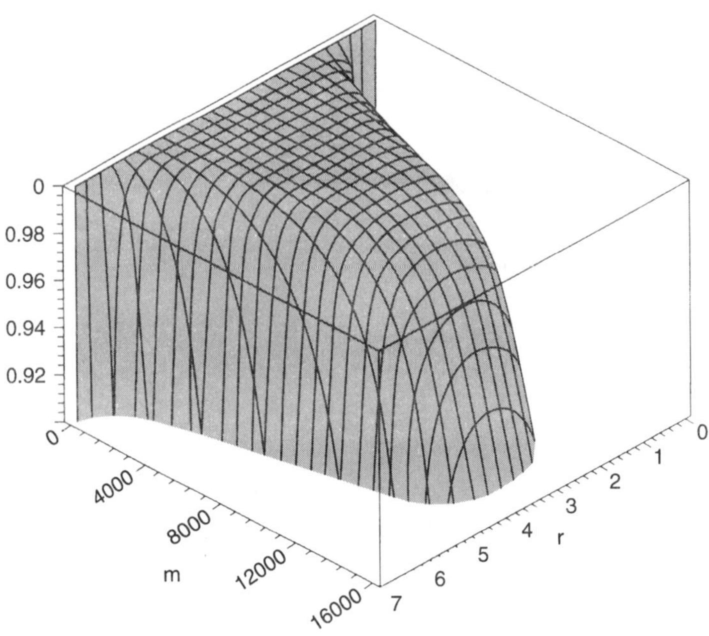

FIGURE 8.5 Probability of correct sequence reconstruction for a (4,4) reverse scheme as a function of m and r.

biochemistry of DNA hybridization, the gauge of optimality (the information-theory bound), and the nature of actual genomic sequences.

THE BIOCHEMISTRY OF DNA HYBRIDIZATION

All combinatorial analyses of SBH make some drastic simplifications of hybridization, which is usually portrayed as a noiseless process. In this modeling, string match/mismatch correspond, respectively, to perfect hybridization or no hybridization at all. The reality is much more complicated. First of all, Watson/Crick-complementary binding (match) is the result of the individual binding of the base pairs involved; despite the obviously additive character of binding energy, there is some interaction between adjacent base pairs. Second, a mismatch (a pairing of two noncomplementary bases) produces only a limited destabilization, which does not qualitatively alter the nature of the process. DNA match/mismatches have been carefully modeled and experimentally observed,[13] and parameters are available to evaluate rather accurately the binding energy of a specific DNA duplex (a pair of equal-length strings).

A less drastic simplification of hybridization assumes independence between adjacent positions and distinct binding energies for strong (C-G) and weak (A-T) base pairs (normally, in a simplistic 2/1 ratio). But even these simplifications lead to a substantial spread in the distribution of the binding energies of complementary duplexes of a given length. Because binding energy is a monotonic function of the temperature of annealing (related to the *melting* temperature), the resulting spread of melting temperatures for the different microarray probes considerably complicates the biochemical experiment. Presumably, the melting temperatures of each microarray feature should be individually measured.

This brief discussion points to experimental difficulties that either should be overcome in the laboratory or may themselves modify the combinatorial model of SBH. The issue of noisy SBH has been addressed in the literature,[5,7,8,14,15] with a convenient model that assumes that false positives and false negatives are Poisson processes. Clearly, false negatives are much more detrimental than false positives, as they fatally interrupt the reconstruction process, while false positives basically add to the fooling probe pool. The mentioned biochemical difficulties have so far prevented SBH from becoming a truly competitive technology for *de novo* DNA sequencing.

A GAUGE OF OPTIMALITY

The standard practice is to compare SBH methods on the basis of their conventional performance, expressed as the length of random sequences that are reconstructed with a prescribed confidence level. However, it is desirable to evaluate a method against some absolute gauge of optimality. Such a gauge is, currently, the information-theory bound (Equation 8.1) introduced earlier.

In all fairness, this bound is unsatisfactory for a number of reasons. First, it simply states that for $m \geq 4^{k-1/2}$ there are pairs of sequences with identical spectra, a fact that may be of no practical consequence if such sequences were a rarity. Second, it takes into consideration "illegal" spectra, that either contain more distinct probes than there are symbols in the target sequence, or such that their probes do not correspond to all subsequences of a target sequence.

More significant is perhaps a gauge related to the sequential nature of the reconstruction of the putative sequence, as carried out by all algorithms dealing with this problem. Such a process may be viewed as the construction of a rooted tree, each of whose paths (from the root) is fully supported by the spectrum. Basically, one such path, reproducing the target sequence, is deterministically extended, and spawns spurious paths in a random fashion. Each spurious path may in turn spawn additional spurious paths, and the phenomenon may appear as a branching process, except for the fact that successive extensions are not truly independent.

Non-overlapping probes are independent in our memoryless model. Suppose that m has a value sufficiently large for the tree extension to become significant, say, $m \geq 4^{k-1}$. In such case $\alpha > 0.2212\ldots$ It is easily shown that the maximum overlap of two distinct (s, r)-gapped probes is $s - 1$, so that the overlap of two probes at the same site involves the constraint of at least $k - s + 1 = r + 1$ additional symbols. We conclude that for $r \geq 3$, $\alpha >> 1/4^{r+1}$, so that we may reasonably ignore probe overlap so that probes may be considered independent.

Let us denote conventionally as 0 the position where a branching occurs, i.e., where the correct path (subtree) deviates from the spurious paths (subtrees). Let random variable n_j denote the number of leaves of one of the spurious subtrees at position j, and let p_j be the probability that a tree node has j descendants. Obviously, $n_0 = 1$ and the generating function $P(\zeta) = \Sigma_{i=0}^{4} p_i \zeta^i$ has the expression $(1-\alpha + \alpha\zeta)^4$

in our model (i.i.d. symbols). Denoting $P_j(\zeta)$ the generating function of the distribution of n_j, it is well known that $P_{j+1}(\zeta) = P(P_j(\zeta))$ (see Feller,[16] p. 272). Moreover, unbounded growth of tree depth j has nonzero probability for values of $\{p_0, p_1,\ldots, p_4\}$ for which the equation

$$\zeta = P(\zeta)$$

has a real solution <1 in addition to the trivial solution $\zeta = 1$. Since $P(\zeta)$ is (downward) convex, such condition occurs when $P'(1) > 1$, i.e.,

$$P'(1) = 4\alpha(1 - \alpha + \alpha\zeta)^3 \big|_{\zeta=1} = 4\alpha > 1$$

which is equivalent to $\alpha > 1/4$, or

$$m > 4^k \ln \frac{4}{3} \approx 0.288 \cdot 4^k$$

Thus, $m \approx 0.288 \cdot 4^k$ is the critical value for potential explosion of the number of paths. With respect to this bound, using the results presented earlier, the performance gap is narrowed to the interval $[0.2 - 0.288] \cdot 4^k$.

PROCESSING NATURAL DNA

As mentioned earlier, the common standard for performance evaluation is the ensemble of sequences generated by maximum-entropy memoryless quaternary source. This standard enables comparative evaluations of different methods, but does not necessarily model the reality of natural sequences, which indeed deviate, sometimes substantially, from this random abstraction. Deviation from memoryless randomness follows some general, but sometimes elusive, pattern: short genomes (typically, prokaryotic) appear sufficiently random, whereas more complex genomes (typically, eukaryotic) show marked deviations, especially in the noncoding or intronic regions of the genome. Loss of randomness may consist of unequal symbol frequencies, presence of source memory, nonstationary source behavior, and the occurrence of "repeats," sometimes in the form of long, nearly identical segments sufficiently far apart, sometimes in the form of long concatenations of short periods. The latter feature is frequent in higher organisms, and subtle algorithmic provisions are needed to detect and control such periodicities.

REFERENCES

1. Issue of *Science* devoted to "The Human Genome." Science, 291(5507), 1145–1434, 2001.
2. Yu.P. Lysov, V.L. Florentiev, A.A. Khorlin, K.R. Khrapko, V.V. Shih, and A.D. Mirzabekov, Sequencing by hybridization via oligonucleotides. A novel method. Dokl Acad Sci USSR, 303, 1508–1511, 1988.

3. P.A. Pevzner, 1-Tuple DNA sequencing: computer analysis. J Biomol Struct & Dyn, 7(1), 63–73, 1989.
4. R. Drmanac, I. Labat, I. Bruckner, and R. Crkvenjakov, Sequencing of megabase plus DNA by hybridization. Genomics, 4, 114–128, 1989.
5. W. Bains and G.C. Smith, A novel method for DNA sequence determination. J Theor Biol, 135, 303–307, 1988.
6. M.E. Dyer, A.M. Frieze, and S. Suen, The probability of unique solutions of sequencing by hybridization. J Comp Biol, 1, 105–110, 1994.
7. P.A. Pevzner, Yu.P. Lysov, K.R. Khrapko, A.V. Belyavsky, V.L. Florentiev, and A.D. Mirzabekov, Improved chips for sequencing by hybridization. J Biomol Struct Dyn, 9(2), 399–410, 1991.
8. P.A. Pevzner and R.J. Lipshutz, Towards DNA-sequencing by hybridization. Proceedings 19th Symp on Mathem Found of Comp Sci, LNCS-841, 1984, 243–258.
9. E.M. Southern, DNA chips: analysing sequence by hybridization to oligonucleotide on a large scale. Trends Genet, 12(3), 110–115, 1996.
10. D. Loakes and D.M. Brown, 5-Nitroindole as a universal base analogue. Nucleic Acids Res, 20, 4039–4043, 1994.
11. F.P. Preparata and E. Upfal, Sequencing-by-hybridization at the information-theory bound: an optimal algorithm. J Comp Biol, 7(3/4), 621–630, 2000.
12. S.A. Heath and F.P. Preparata, Enhanced sequence reconstruction with DNA microarray application. Proceedings of COCOON 2001, Guilin, China, 2001, 64–74.
13. J.J. SantaLucia, A unified view of polymer, dumbells, and oligonucleotide DNA nearest-neighbor thermodynamics. Proc Natl Acad Sci USA, 95, 1460–1465, 1998.
14. R.J. Lipshutz, Likelihood DNA sequencing by hybridization. J Biomol Struct Dyn, 11, 637–653, 1993.
15. K. Doi and H. Imai, Sequencing by hybridization in the presence of hybridization errors. Genome Informatics, 11, 53–62, 2000.
16. W. Feller, An Introduction to Probability Theory and Its Applications. New York: John Wiley & Sons, 1960.

9 Ancient DNA

William Goodwin

CONTENTS

INTRODUCTION

The first publications reporting the analysis of ancient DNA were greeted with a great deal of interest from several academic disciplines and also the general public, excited by the potential of addressing previously unanswerable questions. From this initial interest and excitement, the discipline of ancient DNA analysis was born.

The technological catalyst was the development of the polymerase chain reaction (PCR) technology,[1] which for the first time allowed extremely small amounts of degraded DNA to be analyzed. This opened up for the first time the possibility of studying DNA extracted from a plethora of samples, which could address questions of evolutionary, archaeological, and historical importance.

The reality of what could be achieved by analyzing ancient DNA did not live up to the initial expectations. Many of the early reports that claimed that DNA could be extracted from plant and animal remains millions of years old, which generated headlines in both the popular and scientific press, could not be substantiated and have since been explained by contamination from modern sources. These early erroneous claims placed the whole discipline of ancient DNA analysis in a poor light.

The contamination was a negative aspect of the PCR; this method, which opened the possibility of analyzing small amounts of degraded material, does not distinguish between modern contaminating DNA and the endogenous DNA that is of interest.

Despite these setbacks, an increased understanding of the processes involved in DNA degradation led to improved technical processes for extracting and amplifying ancient DNA. In addition to the technical advances, the establishment of rigid criteria that have to be satisfied before ancient DNA can be accepted as being from the sample of interest and not the result of contamination provide a more structured framework in which to work. Because of these advances a large body of reliable work is now accumulating, which is being used to address a number of questions.

This chapter discusses the analysis of ancient DNA; what constitutes ancient DNA is debatable and the chapter could equally be entitled "Degraded DNA." The methodology and theory behind the analysis of DNA that is 50,000 years old is largely the same as that used for DNA only a few years old if the DNA is equally degraded, even if the questions asked in the study may be different. In the chapter the processes that are important in the preservation of DNA are discussed and an overview of the controls that have to be satisfied before the results of any study can be accepted are addressed. In the second part of the chapter some specific example covering areas from human evolution, taxonomy, archaeology are discussed.

THE PROCESS OF DNA DECAY AND DNA PRESERVATION

DNA DEGRADATION

When an organism dies, its DNA will start to degrade. Nucleases that are compartmentalized within the living cell are released upon cell breakdown and start to degrade the endogenous DNA. Colonizing bacteria and fungi will continue the enzymatic breakdown. This process is especially rapid in soft tissues that rapidly putrefy unless the process is arrested by low temperatures, desiccation, or chemical environments that inhibit the action of the nucleases. Even when the endogenous DNA is in a relatively stable environment the DNA will continue to break down over time due largely to the effects of hydrolysis and oxidation (Figure 9.1).[2] Hydrolytic damage will result in the removal of bases; purines are particularly susceptible to this process and depurination is one of the main routes of ancient

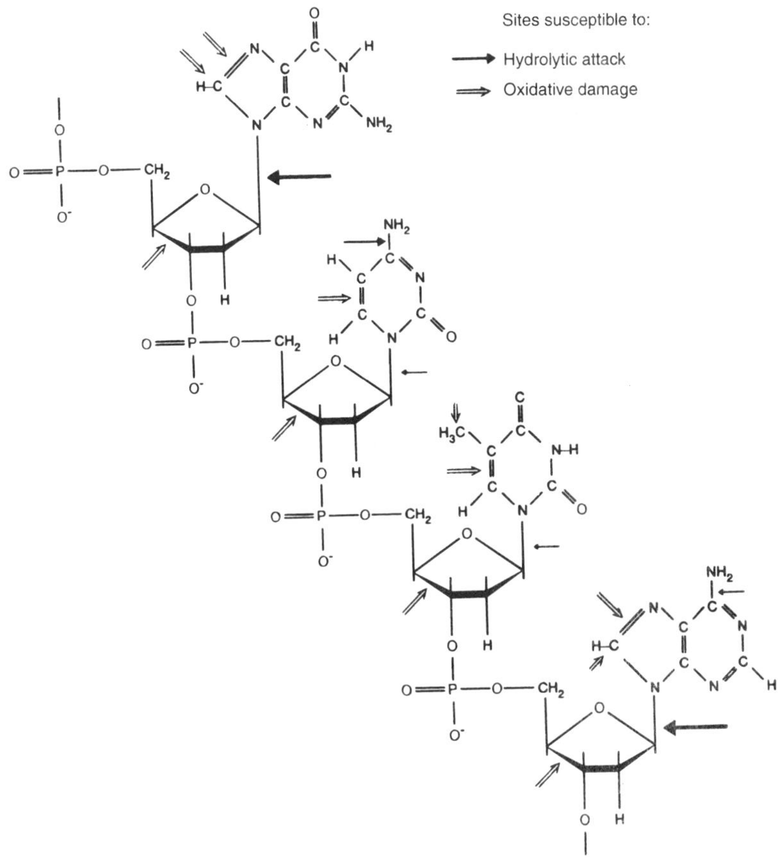

FIGURE 9.1 The degradation of DNA. The major sites at which DNA degradation proceeds either through oxidative or hydrolytic reactions are indicated. (Permission to reproduce modified figure from Lindahl, 1993.)

DNA degradation. Oxidative damage leads to lesions in the sugar-phosphate backbone of the DNA molecule and chemical alterations of the bases; the oxidative damage is mediated through the effects of both direct and indirect ionizing radiation.[2]

JUMPING PCR AND DNA DAMAGE

DNA degradation and damage, in addition to limiting the length of any fragment of DNA that might reasonably be found in a fossil sample, also complicates the PCR amplification process. The DNA damage can lead to two processes that can produce erroneous and, at times, misleading results. These errors can occur through two processes: jumping PCR and the incorrect incorporation of residues caused by DNA damage.

The phenomena of jumping PCR can knit partial PCR products together resulting in hybrid molecules.[3] If extension occurs from a primer but is terminated prematurely due to DNA damage, the resulting truncated PCR product can then act as primer in the next round of PCR. The priming now starts farther downstream of the original

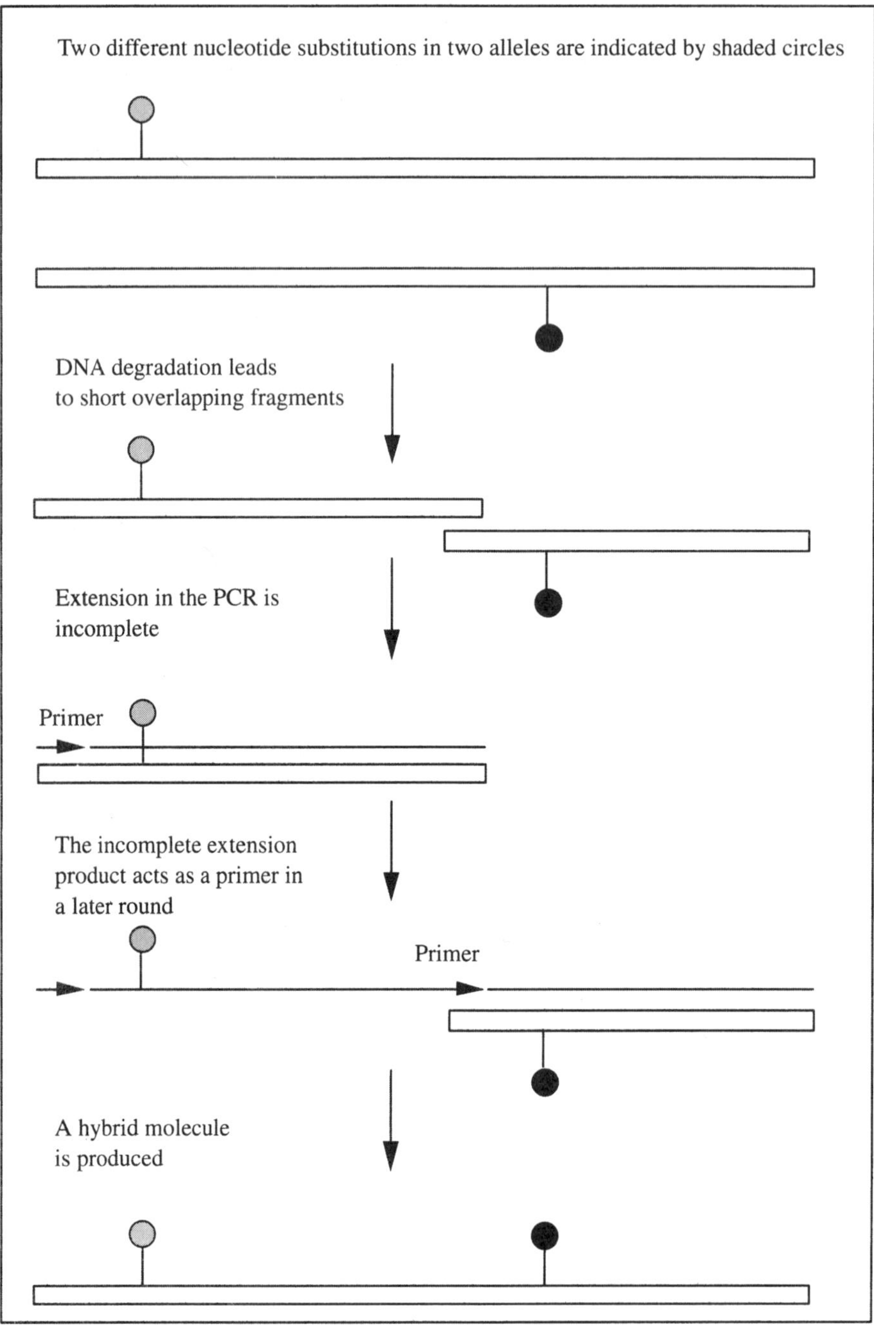

FIGURE 9.2 When two variants of an allele are present in a sample of degraded DNA, it is possible to form a hybrid molecule through the process of jumping PCR.

primer site and can lead to the formation of a hybrid molecule (Figure 9.2). This can lead to problems in interpreting data, particularly from loci that may well have two different alleles within any given individual.

Damaged DNA is also more problematic for the *Taq* (*Thermo aquaticus*) DNA polymerase to copy; the purine bases, guanine (G) and adenine (A), are particularly prone to hydrolytic attack leaving the DNA template with gaps in the sequence of bases. If the damage is severe enough, PCR-mediated amplification may prove impossible; in one study oxidation-mediated chemical changes in the pyramiding bases were shown to be positively correlated with the inability to amplify endogenous DNA.[4] Deamination, in particular of the cytosine residue, has been shown to be common in both the DNA of living organisms and fossils.[5] The deamination of the cytosine base will lead to detection of transitions in the PCR products that are artifacts of the DNA damage rather than reflections of the endogenous sequence. After deamination of the cytosine base, *Taq* DNA polymerase will incorporate deoxyadenosine residues at a position where a deoxyguanosine would have been incorporated prior to the deamination, leading to an incorrect cytosine (C) to thymine (T) and G to A transitions.

ASSESSMENT OF MOLECULAR PRESERVATION

Many samples that could be considered interesting to analyze will not contain any endogenous DNA. In such cases there is little merit in undertaking destructive analysis that is expensive and time-consuming and has no possibility of yielding results. Whenever multiple samples are available for analysis, it is advantageous to determine which of the samples provides the greatest probability of yielding ancient DNA. Through empirical observations and an improved understanding of the process of DNA degradation, the likelihood of a sample yielding DNA can be estimated. Although consideration of the environment in which the sample has been found, along with a chemical analysis, can virtually exclude the possibility of finding ancient DNA, it is important to realize that a positive assessment is no guarantee that endogenous DNA will be recovered from a given sample; rather it is an indication that the sample could potentially harbor ancient DNA. In addition to estimating whether obtaining ancient DNA is a possibility, positive indicators, environmental and chemical, also provide confidence in any results when ancient DNA is extracted and analyzed.

ENVIRONMENTAL

The environment plays a large role in the rate of DNA degradation. Low temperature is generally considered to be the most important single factor in the preservation of ancient DNA. Temperature directly controls the rate of chemical reactions; generally a chemical reaction rate doubles with an increase of 10°C. It is therefore not surprising to find that most of the successful analyses involving ancient DNA have been with samples that are from cooler climates. Beyond the general observations, a more detailed survey of a number of fossils recovered from Pleistocene (an epoch dating 10,000 to 1.64 million years ago) sites demonstrated a positive correlation between the thermal age of the fossils and the recovery of endogenous DNA. As was expected, lower that the average temperatures were positively correlated with a higher probability of retrieving ancient DNA.[6] (Thermal age is the number of years required at a constant temperature of 10°C to produce the equivalent degradation that would have occurred based on its thermal history.) The effect of temperature on the chemical preservation of DNA has been demonstrated directly in studies where higher levels of chemical damage in ancient DNA have been correlated directly with higher environmental temperatures.[4]

Although temperature is the most important single factor, other environmental factors have to be taken into consideration when estimating if DNA could be present in a sample, including air and soil humidity, soil pH, average temperatures in different earth layers, and microbial-mediated decay.[6,7] The interplay of these and other factors makes predicting the preservation of material from a site based on environmental information complex, and the information can act only as a guide rather than being definitive. This point is illustrated by the variation in gross and molecular examination of human bodies recovered from the same archaeological site and therefore sharing the same thermal history and also most other factors in involved in the degradation of DNA.[8] The variations in microenvironments make precise predictions of ancient DNA preservation very difficult.

CHEMICAL

The direct chemical assessment of a sample can provide information on the degree of diagenetic changes that has occurred in a sample. This has the advantage over the assessment based only on the environment history of measuring the actual sample that will be used in the ancient DNA analysis and therefore removing variables such as different microenvironments. It does have the disadvantage, however, that it is also a destructive technique; careful consideration therefore must be given before valuable samples are analyzed.

Proteins that are present in all biological matter are more stable and easier to analyze than the DNA and therefore provide a good proxy for assessing DNA degradation. Assessment of changes in the proteins allows a measure of diagenetic change, which in turn provides an estimation of the amount of ancient DNA degradation and modification that is likely to have occurred. The most widely used method has been the measurement of different forms of amino acid. With the exception of glycine, amino acids can exist in the form of two optical isomers, D and L. In living organisms the L-enantiomer is exclusively used in protein biosynthesis. However, when the amino acids are no longer part of a living organism (i.e., upon death), they undergo racemization to the D-enantiomer; eventually the two forms will reach equilibrium at which point they will be present at equal levels. Measurement of the racemization of aspartic acid found that when the D/L ratios were below 0.08 DNA could be extracted and that generally with lower D/L ratios longer PCR products could be generated. Samples with D/L ratios above 0.08 yielded no endogenous ancient DNA.[9] Collagen content and composition in bone[7,10] and flash pyrolysis with gas chromatography and mass spectrometry[11] are other methods that have shown the potential to be good indicators of the preservation of endogenous DNA.

LIMITATIONS OF ANCIENT DNA ANALYSIS

The unfounded optimism of the early days of ancient DNA analysis, which led to hopes of extracting DNA from a plethora of samples stretching back in time millions of years, has all but died. Early reports, the most spectacular of which included the recovery of DNA from Miocene plant material 17 million to 20 million years old,[12] from insects that had been embedded in amber up to 120 million to 135 million years,[13,14] and finally from dinosaurs dating to the Cretaceous period (a geological period 65 million to 146 million years ago),[15,16] have all proved impossible to repeat.

With an increased understanding of the process of DNA degradation and therefore the limitations of its application, it is now generally accepted that it is very unlikely that endogenous DNA will be recovered from any samples older than 50,000 to 100,000 years, even with extremely favorable environmental conditions. The only way that information can be gained on the DNA sequences older than around 100,000 years is through inference, using extant sequences as a guide.

SUBSTRATES FOR ANCIENT DNA ANALYSIS

The most abundant source of material for ancient DNA analysis is bone. This is normally all that is left of a vertebrate after decomposition. In exceptional circumstances soft tissue is also available. This occurs when a body or part of a body has become mummified, either naturally through desiccation or chemical environment or else through human intervention such as the numerous mummies from Egyptian antiquity and an enormous amount of animal skins preserved in museum displays. Organisms, in particularly mammoths and other Pleistocene megafauna, that are periodically recovered from melting ice or from areas of permafrost where they have been entombed since death periodically provide another source of soft tissue.

However, even when preserved soft tissue is available for analysis, bone is normally the material of choice. The bone material acts as a valuable harbor for endogenous DNA; in addition to its resistance to putrefaction, the hydroxyapatite mineral in the bone stabilizes the DNA molecules and therefore aids in their preservation (see Lindahl[2]). Bone also has other advantages over soft tissue. Because relatively large pieces of bone are recovered, it is possible to remove the outer layers of the bone and therefore remove any contaminating DNA. In many cases it has also been easier to isolate DNA from bone material without co-extracting PCR inhibitors that prevent any analysis. Teeth are also highly desirable as a source of ancient DNA. In addition to the advantages that teeth share with bone samples, the tooth enamel acts as a natural barrier to bacterial and fungal invasion, which further helps to preserve the endogenous DNA. They are also relatively easy to clean; vigorous chemical treatment can be used to clean the surfaces without damaging the internal DNA.

In the early days of ancient DNA analysis amber was thought to be a good potential source of ancient DNA. Amber provides a good potential harbor for DNA because of the desiccation of the material and the barrier to atmospheric oxygen, and the prospects of well-preserved material led to several attempts to analyze insects that had been entombed in amber, but these all failed to yield ancient DNA. Amber, while protecting specimens to some degree, has since been shown to be insufficient to stop diagenetic changes to the encased material, particularly over the long time periods that the material has often been in the amber.[17]

Coprolites (fossilized feces) have also shown themselves to be an unlikely, if limited, source of ancient DNA and a valuable resource for studying the diets of other climatic periods.[18,19]

When dealing with botanical samples, seeds have proved to be the most promising source. Desiccation is a normal and controlled part of seed formation and therefore seeds have an immediate advantage over other sources of botanical material in that they are designed to harbor DNA for a period of months and even years. Ancient DNA has been reported to have been extracted from material from Greece dating back as far as the early Bronze age (see Brown[20]).

TECHNICAL PROCEDURES IN ANCIENT DNA ANALYSIS

DNA EXTRACTION

Once a sample has been assessed in terms of its age, molecular preservation, and environmental history, the endogenous DNA must be extracted. At all times during the extraction process rigorous precautions must be followed to minimize the possibility of contamination with exogenous DNA and controls are included to maximize the possibility of detecting any potential sources of contamination. Ideally, a laboratory that is dedicated to ancient DNA analysis should be used.

The exact processes depend on the nature of the material that is being analyzed. Bone material has the advantage over many other potential samples in that the external surface, which may well contain contaminating DNA, can be removed. After the outer layer has been removed, the bone can be further treated with agents that will destroy DNA on the surface; common treatments include washing in strong detergents and sodium hydroxide solutions and treatment with intense ultraviolet (UV) light. When these steps are not possible, then care must be taken to use samples that have a low chance of being contaminated.

The extraction method varies depending on the sample but is usually a variant of techniques commonly used when analyzing bone samples. The sample is ground to a fine powder and then dissolved in a 0.5 M EDTA solution; the addition of proteinase K aids the process. Nonbone samples are often powered by grinding in the presence of liquid nitrogen and then incubated in solutions containing detergents and also proteinase K to break down the cellular material and place the ancient DNA into solution. Separating the endogenous DNA from all types of ancient samples can be problematic due of the DNA becoming chemically linked to protein components. The addition of the chemical PTB (N-phenacylthiazolium bromide), which is a reagent that cleaves glucose-derived protein cross-links, to the DNA extraction has proved helpful in recovering ancient DNA from both coprolite and bone material.[18,21] Once the ancient DNA is in solution, it is then most commonly further extracted using phenol and chloroform before the DNA in the aqueous phase is concentrated using filter centrifugation. Further cleanup procedures may have to be undertaken on the extract to remove inhibitors of the PCR reaction.

A vital part of any extraction involving ancient DNA samples is that negative controls have to be set up for all stages of the extraction procedure to monitor for external contamination. Even in the most stringently controlled environmental contamination can still occur.

The extracts from fossil samples cannot be easily quantified as the levels of DNA are normally very low; large amounts of DNA will sometimes be present but the source of this DNA is virtually always bacterial or fungal rather than from the sample. Competitive PCR is one method than can be used to estimate the number of molecules that are present in an extract,[22] and the development of real-time quantitative PCR is now a viable alternative.[23]

PCR AMPLIFICATION AND SEQUENCING

In fossil samples only a small number of chemically modified molecules can normally be recovered.[4] While early studies attempted to analyze ancient DNA directly without an amplification phase,[24,25] the low number of starting molecules made such analysis extremely difficult and the technique was of limited scope. PCR circumvented the problem of a low number of starting molecules; the technique is extremely powerful: in theory, one single molecule can be amplified several billion times and generate enough product to analyze directly. Over the last 13 years the PCR-mediated amplification of selected target loci has proved to be an extremely powerful technique in ancient DNA analysis, as in virtually all areas of molecular biology.

However, the power of the technique can also be a problem as any contaminating DNA that enters the DNA extract or PCR reaction will also be amplified and it will not be possible to distinguish endogenous DNA from contaminating molecules. Because of chemical damage to the endogenous DNA, any contaminating DNA that is present in the PCR reaction may well be amplified preferentially. In the resulting DNA extract, if any endogenous DNA is present, in most circumstances only a few molecules will be present.

Multiple PCR amplifications from ancient DNA extracts are an important step to undertake when the number of target molecules in an ancient DNA extract is very low. When the number of target molecules is limited, any errors that are introduced into the PCR product at an early point in the amplification process will appear in most or all of the cloned PCR products and will therefore appear to be the actual sequence of the endogenous DNA. By undertaking multiple amplifications, ideally from both the same and duplicate extractions, the risk of the same errors being incorporated into the PCR products is low.

A possible source of errors during PCR amplification of ancient DNA is through the deamination of cytosine, which leads to C to T and G to A substitution because the deaminated deoxycytidine residues in the template are read by the *Taq* polymerase as deoxythymidine residues. Treatment of the template with *N*-glycosylase removes the deaminated cytosine from the template; a strand break then occurs through a hydrolysis reaction. The errors are therefore not incorporated into the template.

Once the PCR amplifications have been carried out, the PCR products can be sequenced. If there are enough molecules, then the PCR products may be directly sequenced, however, when analyzing ancient DNA it is standard practice to subclone the PCR products and sequence several different molecules.[21,26–28] This allows detection of errors introduced in the PCR products during the amplification that have been introduced either because of template damage or infidelity of the *Taq* polymerase enzyme. Another critical aspect of the subcloning of the PCR products is that it allows mixtures to be detected. Even with all the precautions and controls in place, it is still not uncommon to detect products from more than one source, endogenous DNA (one hopes!) along with contaminating sequences. The problems of contamination are particularly acute when working with hominid remains as contamination from modern humans is very difficult to eliminate.[21,26,27]

AUTHENTICATION

The measures that are required to satisfy researchers that they have in fact analyzed ancient DNA rather than some contamination are numerous. In addition to the care and the negative controls that are included in the extraction and amplification procedures, there are other factors that must be considered. These include the appropriate behavior of the DNA; for example, if large amplicons several hundred base pairs (bp) long can be generated, then the results should be treated with suspicion. Also as progressively shorter PCR products are generated, then the amount of PCR product should increase, because the number of small molecules should always be greater than the number of large molecules.[26] Table 9.1 lists the steps that are required before recovered DNA can be considered to be ancient DNA rather than contamination.

One of the most important controls is that the work should be repeated in a second laboratory. This is particularly true when dealing with hominid remains, which are much more prone to contamination.

TABLE 9.1
Steps That Should, When Possible, Be Included in the Analysis of Ancient Material

Preliminary assessment
    Thermal age
    Morphological preservation
    Amino acid racemization
    Collagen composition
Extraction, amplification, and analysis
    Cleaning of bone surface
    Multiple extractions in an environment free of contaminating DNA
        Appropriate behavior of the PCR reactions
    Multiple PCR amplifications
    Subcloning of PCR products and sequencing of several individual products
Independent analysis in a second laboratory
Establishing that the results are phylogenetically viable

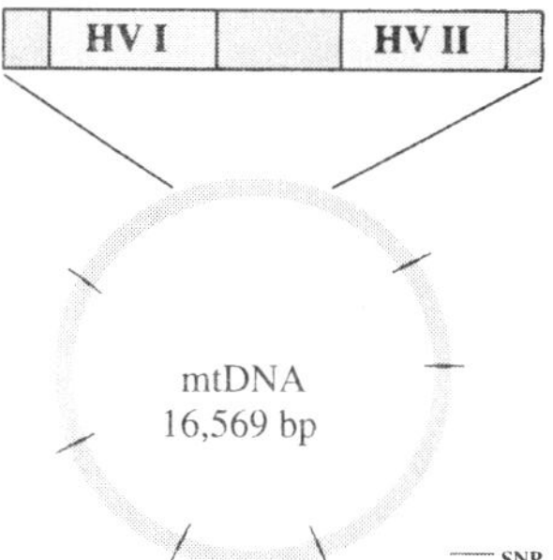

FIGURE 9.3 The human mitochondrial genome is composed of 16,569 nucleotides; there are two highlighted regions of the molecule called the hypervariable regions I and II. These have a higher rate of mutation than the rest of the molecule and provide a relatively compact area to analyze in order to assess differences between individuals. Around the rest of the molecule are sites, referred to as SNPs (single nucleotide polymorphisms), that are very useful for reconstructing the geneology of the molecule. When analyzing ancient DNA the SNPs are more difficult to analyze than the hypervariable regions as they are dispersed throughout the genome (there are many more SNPs than the ones shown). When looking at differences between rather than within species, more slowly evolving regions of DNA are commonly used, in particular the regions that encode for the cytochrome *b*, 12S and 16S genes.

A final verification should be that the results make sense phylogenetically. Although this criterion can be contentious in some cases, there are other cases where it is clearly applicable; if the analysis is of, for example, a mammoth, the sequence should look more similar to known relatives such as the elephants than to more distant relations. This is unfortunately not applicable when handling ancient human remains, as any contamination is likely to be from modern day humans.

TARGET LOCI

When analyzing ancient DNA, only one locus has so far proved to be very useful. This is the mitochondrial genome (mtDNA). There are two main reasons mtDNA has been the loci of choice when dealing with ancient material. First, it is present in at high copy number; there are approximately 500 to 1000 copies of the mito-chondrial genome in each cell compared to two copies of the nuclear genome (which may be different from each other in any given individual).[29] The large number of molecules increases the probability of enough DNA being present in an archaeological sample to allow a successful analysis. It also has the advantage when it is used to compare divergent populations that nucleotide substitutions occur frequently (relative to the nuclear genome), particularly in two regions termed the hypervariable regions I and II (HV I and HV II; Figure 9.3), and therefore differences between separated populations accumulate rapidly. It is also maternally inherited, which allows the genealogy of the mtDNA genome to be interpreted without having to account for recombination.

Analysis of the hypervariable regions has proved powerful when examining, evolutionarily speaking, recent events such as the relationship of modern humans and Neanderthals (see below). When examining more distant events, then more slowly evolving regions of the mitochondrial genome can be utilized. Three loci that are commonly used are the 12S and 16S rRNA genes and the cytochrome *b* coding region.

There are limitations with only using one locus when examining the genetic relationships between different populations as all loci individually are prone to chance events, such as genetic drift, which may change their frequencies in a way that is not reflective of the changes occurring at a population level. One way to increase the powers of ancient DNA analysis would be to analyze more loci. However, only the mitochondrial loci occur in multiple copies and in most cases there is very little chance of analyzing nuclear genetic material. So it remains a limitation of the studies. Some of the fauna recovered from permafrost does offer the potential for examining nuclear loci. Single copy nuclear genes from both Alaskan and Siberian mammoths have been successfully analyzed,[30] demonstrating that the potential exists for analysis of nuclear as well as mitochondrial genomes.

APPLICATIONS OF ANCIENT DNA ANALYSIS

With an increased awareness of the pitfalls of ancient DNA analysis along with a better understanding of the process of DNA degradation, a number of studies have now been published that satisfy the criteria that are needed before the results of any ancient DNA analysis can be widely accepted.

The second half of this chapter examines some of the applications of ancient DNA analysis. The first of these concerning the relationship of the Neanderthals to the modern European population is discussed in some detail to give an overview of the processes that are important in ancient DNA analysis. Brief overviews of work undertaken in different areas are then discussed.

HUMAN EVOLUTION AND THE PLACE OF THE NEANDERTHALS

Background

The relationship of the Neanderthals to modern humans has been the source of heated debate ever since the first specimen was identified as a Neanderthal after it was recovered from the Feldhofer Cave in 1856 (specimens had been discovered in Belgium and Gibraltar before this but only identified as Neanderthals after the Feldhofer specimen). Neanderthals had occupied Europe for hundreds of thousands of years; 40,000 years ago the first anatomically modern humans entered Europe and approximately 10,000 years after the first modern humans entered Europe the last Neanderthals disappeared. The mechanism of the Neanderthal extinction has been the source of much argument, with several competing hypotheses. The multiregional hypothesis advocates that an ancestral population of *Homo erectus* (that spread around the world from Africa around 1 million years ago) evolved into modern humans regionally, with gene flow between geographically distinct populations. The out-of-Africa model proposes that modern humans evolved in Africa around 150,000 years ago and subsequently spread throughout the world, replacing all earlier hominid species, including the Neanderthals. The intermediate hybridization and assimilation hypotheses predict that the modern European population is a product of mixing between the Neanderthals and the modern humans from Africa.

Studying the Neanderthal mtDNA pool directly has allowed questions regarding the genetic composition of the Neanderthals to be addressed.

Samples

Mitochondrial DNA now been successfully extracted and analyzed from three Neanderthal specimens recovered from the Feldhofer Cave in Germany, the Mezmaiskaya Cave in the Caucasus Mountains, and the Vindija Cave in Croatia.[21,27,28] All these samples had the benefit of dating from the end of the period of Neanderthal occupation, approximately 40,0000 years for the Feldhofer and Vindija samples and 30,000 years for the Mezmaiskaya sample. All three samples were also from areas of relatively low average temperature.[6] Other Neanderthal samples that had been examined before from warmer areas in southern Europe had displayed too much diagenetic change to justify DNA analysis.[10] Despite having only three sequences that are separated geographically and temporally it has been possible to examine some aspects of the Neanderthal mtDNA pool.

DNA Extraction and Sequence Analysis

In all three successful extractions very similar DNA extraction protocols were used. Bone material was used in all three cases; the procedure used is shown in Figure 9.4. The DNA extracts were amplified using primers that amplified regions of the mitochondrial hypervariable region that were then subcloned into a plasmid vector. Multiple clones from separate PCRs were sequenced. The results of the sequencing of the Mezmaiskaya Neanderthal are shown in Figure 9.5. The consensus sequence is derived by selecting only the substitutions and insertions that occur in the majority of the clones from at least two independent reactions. There are a number of substitutions that occur in only one cloned PCR product, which can be explained as either due to errors introduced through the infidelity of the *Taq* DNA polymerase or due to damage in the DNA template. The analysis of the Mezmaiskaya Neanderthal was unusual in that enough PCR product could be generated to allow direct sequencing, but subcloning and sequencing the PCR products was still necessary to detect errors introduced during the amplification, as well as the presence of more than one type of mtDNA (endogenous and also modern contaminating mtDNA).

The Neanderthal sequences have been analyzed in a number of different ways. The Cambridge Reference Sequence (CRS) acts as a reference sequence for all mtDNA analysis, and comparison to this indicated the degree of variation in comparison to modern sequences. The information displayed in Figure 9.6 is in comparison to the CRS.

The Feldhofer, Mezmaiskaya, and Vindija Neanderthals contain 27, 22, and 22 substitutions relative to the CRS, respectively; all of them contained one insertion at position 16,263 (see Figure 9.1). The three Neanderthal sequences share 18 substitutions and one insertion with respect to the CRS.

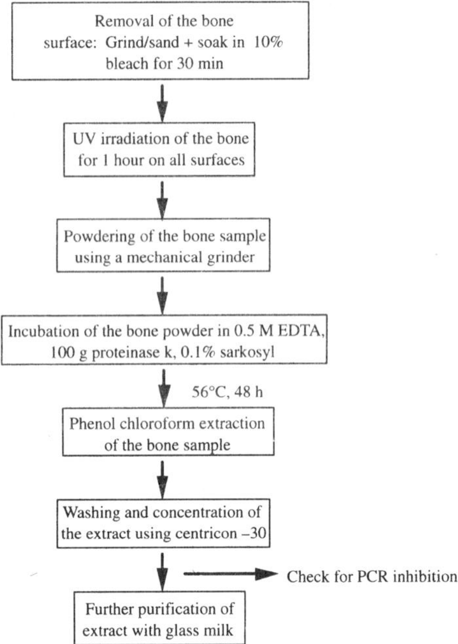

FIGURE 9.4 Flow diagram showing the methodology that is commonly used for extracting DNA from bone samples, including the extraction from the Mezmaiskaya Neanderthal. With the Mezmaiskaya Neanderthal bone no PCR inhibition was detected with the extracted sample after the centricon-30 purification stage. Therefore the final clean using glass milk (or other related method, e.g., Höss and Pääbo[50]) was not carried out.

	16086	16118	16129	16139	16148	16156	16169	16182	16183	16189	16209	16223	16230	16234	16243	16244	16250	16256	16261	16262	16263.1	16278	16299	16311	16320	16344	16362	16365	16393
CRS	T	G	G	A	C	G	C	A	A	T	T	C	A	C	T	G	C	C	C	C	-	C	A	T	C	C	T	C	C
Direct 1	C	.	A	T	T	A	T	C	C	C	C	T	G	T	.	A	.	A											
P1	C	.	A	T	T	A	T	C	C	C	C	T	G	T	.	A	.	A	.	T	A	T							
P2	C	.	A	T	T	A	T	C	C	C	C	T	G	T	.	A	.	A	.	T	A	T							
P3	C	A	A	T	T	A	T	C	C	C	C	T	G	T	.	A	.	A	.	T	A	T							
Direct 2												T	G	T	.	A	.	A	.	T	A	T	G	C	T	T	C	.	.
577.1												T	G	T	.	A	.	A	T	T	A	T	G	C	T	T	C	.	T
557.2												T	G	T	.	A	T	A	.	T	A	T	G	C	T	T	C	.	.
581.2												T	G	T	.	A	.	A	.	T	A	T	G	C	T	T	C	.	.
581.3												T	G	T	C	A	.	A	.	T	A	T	G	C	T	T	C	T	.
Mezmaiskaya	C	.	A	T	T	A	T	C	C	C	C	**T**	**G**	**T**	**.**	**A**	**.**	**A**	**.**	**T**	**A**	**T**	**G**	**C**	**T**	T	C	.	.

FIGURE 9.5 DNA sequences of the PCR fragments obtained by direct sequencing (Direct 1 and 2) are shown along with several cloned PCR products generated during the analysis of the Neanderthal from Mezmaiskaya Cave. The sequence that could be duplicated in a second laboratory is shown in bold within the compiled Mezmaiskaya sequence.

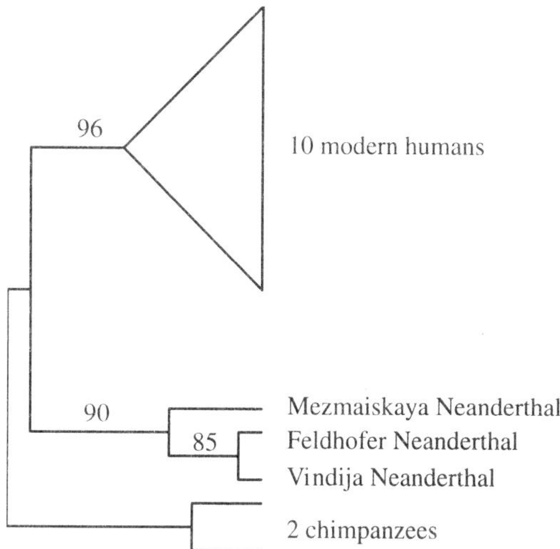

FIGURE 9.6 Phylogenetic analysis of Neanderthal and modern human DNA. A phylogenetic tree produced using parsimony analysis, showing the relationship between the three Neanderthals and a selection of 10 modern humans. The result is typical of several data sets and is also similar to that found when using other phylogenetic tree-building algorithms. The numbers on the branches indicate the strength of the different branches (100 is the highest possible value).

The data have been analyzed using phylogenetic analysis. The results of this have all come to the same conclusion, using methods that incorporated maximum-likelihood, parsimony, and distance analyses. All the different types of analysis identify the Neanderthal and modern human sequences as different lineages (Figure 9.5).

That the three sequences share a large number of substitutions and an insertion and that they all group together after phylogenetic analysis act as a powerful means of verifying that the three sequences are from the Neanderthals and are not a product of some peculiar contamination or a result of amplifying highly damaged DNA.

The Age of Divergence

The number of substitutions that have accumulated in modern humans and the Neanderthal mtDNA lineages since they split has be used to estimate the time of the divergence using the substitution rate as a molecular clock. The genetic distance between modern and Neanderthal mtDNA was used to date the split between the modern humans and the Neanderthals mtDNA and also to estimate the age of the Neanderthal mtDNA lineage.[31] The split between the modern human and Neanderthal lineages has been dated to approximately 600,000 years (365,000 to 853,000) while the age of the most recent common ancestor of the eastern and western Neanderthals is 151,000 to 352,000 years ago. Using the same data modern humans are estimated to have had a common ancestor 106,000 to 246,000 years ago[28] (Figure 9.7).

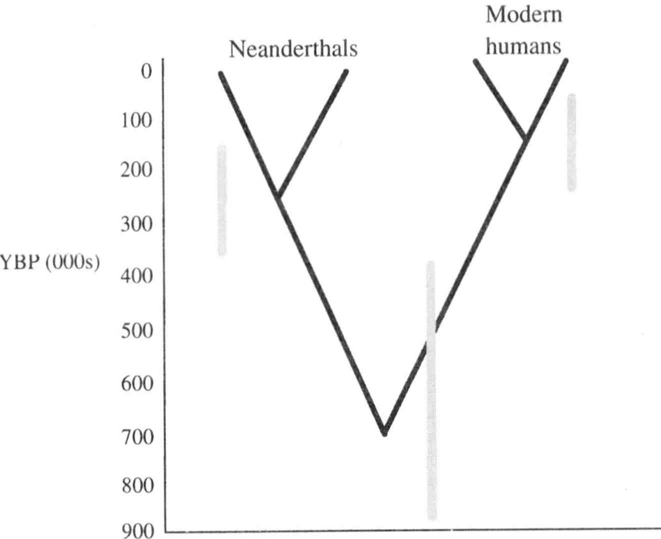

FIGURE 9.7 Using the mitochondrial DNA as a molecular clock, it has been possible to date the times (years before present) of the most recent ancestors of modern humans, Neanderthals, and also the common ancestor of modern humans and Neanderthals (the 95% confidence intervals are shown by the vertical lines).

Neanderthal Diversity

The three Neanderthals have provided an indication of the diversity that is present in the Neanderthal lineage. Using only three specimens there is a 50% probability that the deepest split in the Neanderthal lineage has been detected (probability of sampling the deepest split = $n - 1/n + 1$, where n is the number of sampled specimens). It is therefore unlikely that a Neanderthal specimen will be found that is highly diverse from the three specimens analyzed to date.

The three Neanderthals differ at 8 ± 3.46 positions, which is similar to the levels of diversity that are found in modern humans. This contrasts to the high levels of diversity found in chimpanzees and gorillas, which have much greater levels of sequence diversity.[21] The low levels of diversity found in modern humans have been interpreted as reflecting a rapid growth of a small population,[32] indicating that the demography of the Neanderthals could have been similar to that of modern humans.

Admixture

The analyses undertaken on the samples to date have shown that the Neanderthal and modern mtDNA form distinct lineages and that the modern mtDNA pool is derived entirely from the ancestral modern humans. The absence of mtDNA from the Neanderthal lineage in the modern gene pool does not exclude the possibility that Neanderthals and modern humans may have interbred. Most of the lineages that were present in the modern human population 30,000 years ago (approximately 1500 generations) will have been lost through the process of genetic drift.[33] This point is further illustrated by the fact that most of the lineages that are present in the present European population can be attributed to lineages that date to less than 20,000 years ago.[34]

PHYLOGENY AND POPULATION GENETICS USING EXTINCT ANIMAL POPULATIONS

Since the first report of ancient DNA analysis from an extinct animal, the quagga,[24] studies have been undertaken on several species. These include among others the marsupial wolf,[35,36] the saber-tooth cat,[37] the moa[38,39] and the moa-nalo,[40] the cave bear,[41–43] and the giant ground sloth.[4] The main aim of these studies has been to examine the evolutionary relationship between extinct and extant taxa. The cave bear studies have an advantage over most other types of study that there are a lot of remains available and a number of these are from areas of low temperatures and therefore the potential for DNA survival is greater. Given the potential for extracting ancient DNA, the cave bears offer the possibility of studying populations rather than isolated individuals, and therefore features of the population such as the diversity prior to periods of extinction and the differentiation of populations in response to climatic change can be addressed. It has also been suggested that the information from some of the studies should influence conservation strategies,[42] although there is no universal consensus on how the data should be incorporated into policy.[44]

One of the most striking studies carried out on ancient DNA involves the complete mitochondrial sequencing of two New Zealand moa genera.[39] The complete sequencing improved the precision of the phylogenetic relationships among ratite birds (including ostrich, kiwi, moa, rhea, cassowary, and elephant bird). The data from the extinct and extant samples were used to date the age of the divergence and speciation events of the ratite taxa to the Late Cretaceous. This date supported the hypothesis that the breakup of the Gondwana continent (in the Cretaceous) was the event that led to the physical barriers between ratite populations and therefore acted as the catalyst for the proliferation of ratite species. The data by providing a new perspective on Cretaceous biogeography indicate the potential value of using ancient DNA in conjunction with extant taxa to resolve important biogeographical issues.

THE ORIGINS AND SPREAD OF AGRICULTURE

The shift of human lifestyles from hunter-gatherers to agriculturists over the last 10,000 years has been dramatic. In keeping with the importance of this transition, the origin and spread of agriculture have been important areas of study for archaeology in the last century. Molecular genetics has acted as another source of information in this multidisciplinary research area and has been used to address questions regarding both the domestication of animals and the development of crops.

Molecular genetics can address two aspects of animal domestication and crop development. Did the domestication/development of particular species occur as a single discrete event or did they occur several times, and what were the wild progenitors of the modern breeds and varieties?

One of the most comprehensive studies examining domestication undertaken to date[45] examined the origin of European cattle. By analyzing 392 extant animals from Europe, Africa, and the Near East along with 4 extinct wild British oxen, the extinct British oxen could be excluded as a likely source of the domesticated European cattle. The genetic data from the extant populations suggest that the Near East is a much more likely source of the European cattle stock. The data again illustrate the power of combining the analysis of extant and extinct populations. Earlier ancient DNA studies on European cattle[46] had suggested that a recent expansion of European cattle from a diverse wild progenitor species had occurred. In the course of similar research examining the domestication of the domestic horse, several Pleistocene horses were examined and the variation found within overlapped with the variation found in modern horses, indicating that much of the mtDNA diversity entered the domestic horse gene pool early on in domestication.[47]

The development of crop plants, corn and barley in Europe and the Near East, rice in the Far East, and corn in the Americas, was an extremely important phase in the transition from hunter-gatherer to agricultural lifestyles. As with the examination of livestock, the histories of the development of these crops have received much attention. The seeds offer a potential source of DNA and are relatively commonly found at archaeological sites; desiccated and charred remains have yielded ancient DNA (see Brown[20]). Analysis of remains from different sites in Europe has provided some information on the timing of some of the key stages in the development of wheat, in particular the first appearances of hexaploid wheat.[48]

FUTURE DIRECTIONS FOR ANCIENT DNA

The realization that ancient DNA is very unlikely to survive longer than 100,000 years means that efforts should now be focused on samples that fall within this time interval, in particular, samples recovered from environments with relatively low average temperatures where there is a higher likelihood that the ancient DNA will have survived.

Further advances in understanding the chemistry of the substrates will also lead to the possibility of improving the efficiency of ancient DNA extractions and opening new types of samples to analysis; the successful analysis of coprolites is one example where this has already occurred.[18,19]

There is great potential for continued studies of extinct populations that should provide insights into population dynamics and diversity over time. The Neanderthals of Northern Europe and the Caucasus also offer the potential to study the population genetics of this extinct hominid. The Pleistocene megafauna remains that have been recovered from the arctic permafrost along with remains from high-altitude caves offer important sources of material that are amenable to population-level studies. The material recovered from permafrost, in addition to the relatively large number of samples available, also offers the potential for examining more informative nuclear loci.[49]

Finally, an important aspect of ancient DNA analysis is that the improved technical procedures that have been made possible through a better understanding of the process of degradation of ancient DNA will continue to feed into and enhance other fields including forensics and conservation biology, where small amounts of DNA from problematic substrates are also commonly encountered.

REFERENCES

1. Saiki RK, Scharf S, Faloona F, Mullis KB, Horn GT, Erlich HA, Arnheim N. Enzymatic amplification of beta-globin genomic sequences and restriction site analysis for diagnosis of sickle-cell anaemia. Science 230, 1350–1354, 1985.
2. Lindahl T. Instability and decay of the primary structure of DNA. Nature 362, 709–715, 2003.
3. Pääbo S, Irwin DM, Wilson AC. DNA damage promotes jumping between templates during enzymatic amplification. J Biol Chem 265, 4718–4721, 1990.
4. Höss M, Jaruga P, Zastawny TH, Dizdaroglu M, Pääbo S. DNA damage and DNA sequence retrieval from ancient tissues. Nucleic Acids Res 24, 1304–1307, 1996.
5. Hofreiter M, Jaenicke V, Serre D, von Haeseler A, Pääbo S. DNA sequences from multiple amplifications reveal artifacts induced by cytosine deamination in ancient DNA. Nucleic Acids Res 29, 4793–4799, 2001.
6. Smith CI, Chamberlain AT, Riley MS, Cooper A, Stringer CB, Collins MJ. Not just old but old and cold? Nature 410, 771–777, 2001.
7. Ovchinnikov IV, Gotherstrom A, Romanova GP, Kharitonov VM, Linde K, Goodwin W. Not just old but old and cold? M Reply. Nature 410, 772–772, 2001.
8. Hagelberg E, Bell LS, Allen T, Boyde A, Jones SJ, Clegg JB. Analysis of ancient bone DNA—techniques and applications. Philos Trans R Soc B 333, 399–407, 1991.
9. Poinar HN, Höss M, Bada JL, Pääbo S. Amino acid racemization and the preservation of ancient DNA. Science 272, 864–866, 1996.
10. Cooper A, Poinar HN, Pääbo S, Radovcic J, Debenath A, Caparros M, Barroso-Ruiz C, Bertranpetit J, Nielsen-Marsh C, Hedges REM, Sykes B. Neandertal genetics. Science 277, 1021–1024, 1997.
11. Poinar HN, Stankiewicz BA. Protein preservation and DNA retrieval from ancient tissues. Proc Natl Acad Sci USA 96, 8426–8431, 1999.
12. Golenberg EM, Giannasi DE, Clegg MT, Smiley CJ, Durbin M, Henderson D, Zurawski G. Chloroplast DNA-sequence from a Miocene magnolia species. Nature 344, 656–658, 1990.
13. Desalle R, Gatesy J, Wheeler W, Grimaldi D. DNA-sequences from a fossil termite in Oligomiocene amber and their phylogenetic implications. Science 257, 1933–1936, 1992.
14. Cano RJ, Poinar HN, Pieniazek NJ, Acra A, Poinar GO. Amplification and sequencing of DNA from a 120–135-million-year-old weevil. Nature 363, 536–538, 1993.
15. Woodward SR, Weyand NJ, Bunnell M. DNA-sequence from Cretaceous period bone fragments. Science 266, 1229–1232, 1994.
16. Wang HL, Yan ZY, Jin DY. Reanalysis of published DNA sequence amplified from cretaceous dinosaur egg fossil. Mol. Biol. Evol. 14, 589–591, 1997.
17. Stankiewicz BA, Poinar HN, Briggs DEG, Evershed RP, Poinar GO. Chemical preservation of plants and insects in natural resins. Proc R Soc Lond B Biol 265, 641–647, 1998.
18. Poinar HN, Hofreiter M, Spaulding WG, Martin PS, Stankiewicz BA, Bland H, Evershed RP, Possnert G, Pääbo S. Molecular coproscopy: dung and diet of the extinct ground sloth *Nothotheriops shastensis*. Science 281, 402–406, 1998.
19. Poinar HN, Kuch M, Sobolik KD, Barnes I, Stankiewicz AB, Kuder T, Spaulding WG, Bryant VM, Cooper A, Pääbo S. A molecular analysis of dietary diversity for three archaic Native Americans. Proc Natl Acad Sci USA 98, 4317–4322, 2001.
20. Brown TA. (1999) How ancient DNA may help in understanding the origin and spread of agriculture. Philos Trans R Soc B 354, 89–97, 1999.

21. Krings M, Capelli C, Tschentscher F, Geisert H, Meyer S, von Haeseler A, Grossschmidt K, Possnert G, Paunovic M, Pääbo S. A view of Neandertal genetic diversity. Nat Genet 26, 144–146, 2000.

22. Handt O, Krings M, Ward RH, Pääbo S. The retrieval of ancient human DNA sequences. Am J Hum Genet 59, 368–376, 1996.

23. von Wurmb-Schwark N, Higuchi R, Fenech AP, Elfstroem C, Meissner C, Oehmichen M, Cortopassi GA. Quantification of human mitochondrial DNA in a real time PCR. Forensic Sci Int 126, 34–39, 2002.

24. Higuchi R, Bowman B, Freiberger M, Ryder OA, Wilson AC. DNA-sequences from the Quagga, an extinct member of the horse family. Nature 312, 282–284, 1994.

25. Pääbo S. Molecular-cloning of ancient Egyptian mummy DNA. Nature 314, 644–645, 1985.

26. Handt O, Richards M, Trommsdorff M, Kilger C, Simanainen J, Georgiev O, Bauer K, Stone A, Hedges R, Schaffner W, Utermann G, Sykes B, Pääbo S. Moleculargenetic analyses of the Tyrolean Ice Man. Science 264, 1775–1778, 1994.

27. Krings M, Stone A, Schmitz RW, Krainitzki H, Stoneking M, Pääbo S. Neandertal DNA sequences and the origin of modern humans. Cell 90, 19–30, 1997.

28. Ovchinnikov IV, Gotherstrom A, Romanova GP, Kharitonov VM, Liden K, Goodwin W. Molecular analysis of Neanderthal DNA from the northern Caucasus. Nature 404, 490–493, 2000.

29. Robin ED, Wong R. Mitochondrial-DNA molecules and virtual number of mitochondria per cell in mammalian-cells. J Cell Physiol 136, 507–513, 1988.

30. Greenwood AD, Capelli C, Possnert G, Pääbo S. Nuclear DNA sequences from late Pleistocene megafauna. Mol Biol Evol 16, 1466–1473, 1999.

31. Tamura K, Nei M. Estimation of the number of nucleotide substitutions in the control region of mitochondrial DNA in humans and chimpanzees. J Mol Evol 10, 512–526, 1993.

32. Harpending HC, Batzer MA, Gurven M, Jorde LB, Rogers AR, Sherry ST. Genetic traces of ancient demography. Proc Natl Acad Sci USA 95, 1961–1967, 1998.

33. Nordborg M. On the probability of Neanderthal ancestry. Am J Hum Genet 63, 1237–1240, 1998.

34. Richards M, Macaulay V, Hickey E, Vega E, Sykes B, et al. Tracing European founder lineages in the near eastern mtDNA pool. Am J Hum Genet 67, 1251–1276, 2000.

35. Thomas RH, Schaffner W, Wilson AC, Pääbo S. DNA phylogeny of the extinct marsupial wolf. Nature 340, 465–467, 1989.

36. Krajewski C, Buckley L, Westerman M. DNA phylogeny of the marsupial wolf resolved. Proc R Soc Lond B Biol 264, 911–917, 1997.

37. Janczewski DN, Yuhki N, Gilbert DA, Jefferson GT, O'Brien SJ. Molecular phylogenetic inference from saber-toothed cat fossils of Rancho-la-brea. Proc Natl Acad Sci USA 89, 9769–9773, 1992.

38. Cooper A, Mourerchauvire C, Chambers GK, von Haeseler A, Wilson AC, Pääbo S. Independent origins of New Zealand moas and kiwis. Proc Natl Acad Sci USA 89, 8741–8744, 1992.

39. Cooper A, Lalueza-Fox C, Anderson S, Rambaut A, Austin J, Ward R. Complete mitochondrial genome sequences of two extinct moas clarify ratite evolution. Nature 409, 704–707, 2001.

40. Sorenson MD, Cooper A, Paxinos EE, Quinn TW, James HF, Olson SL, Fleischer RC. Relationships of the extinct moa-nalos, flightless Hawaiian waterfowl, based on ancient DNA. Proc R Soc Lond B Biol 266, 2187–2193, 1999.

41. Hanni C, Laudet V, Stehelin D, Taberlet P. Tracking the origins of the cave bear (*Ursus spelaeus*) by mitochondrial-DNA sequencing. Proc Natl Acad Sci USA 91, 12336–12340, 1994.
42. Leonard JA, Wayne RK, Cooper A. Population genetics of Ice Age brown bears. Proc Natl Acad Sci USA 97, 1651–1654, 2000.
43. Loreille O, Orlando L, Patou-Mathis M, Philippe M, Taberlet P, Hanni C. Ancient DNA analysis reveals divergence of the cave bear, *Ursus spelaeus,* and brown bear, *Ursus arctos*, lineages. Curr Biol 11, 200–203, 2001.
44. Pääbo S. Of bears, conservation genetics, and the value of time travel. Proc Natl Acad Sci USA 97, 1320–1321, 2000.
45. Troy CS, MacHugh DE, Bailey JF, Magee DA, Loftus RT, Cunningham P, Chamberlain AT, Sykes BC, Bradley DG. Genetic evidence for near-eastern origins of European cattle. Nature 410, 1088–1091, 2001.
46. Bailey JF, Richards MB, Macaulay VA, Colson IB, James IT, Bradley DG, Hedges REM, Sykes BC. Ancient DNA suggests a recent expansion of European cattle from a diverse wild progenitor species. Proc R Soc Lond B Biol 263, 1467–1473, 1996.
47. Vila C, Leonard JA, Gotherstrom A, Marklund S, Sandberg K, Liden K, Wayne RK, Ellegren H. Widespread origins of domestic horse lineages. Science 291, 474–477, 2001.
48. Schlumbaum A, Jacomet S, Neuhaus JM. Coexistence of tetraploid and hexaploid naked wheat in a neolithic lake dwelling of central Europe: Evidence from morphology and ancient DNA. J Archaeol Sci 25, 1111–1118, 1998.
49. Greenwood AD, Castresana J, Feldmaier-Fuchs G, Pääbo S. A molecular phylogeny of two extinct sloths. Mol Phylogenet Evol 18, 94–103, 2001.
50. Höss M, Pääbo S. DNA extraction from Pleistocene bones by a silica based purification method. Nucleic Acids Res 21, 3913–3914, 1993.
51. Lindahl T. The Croonian Lecture, 1996: Endogenous damage to DNA. Philos Trans R Soc Lond B Biol Sci 351, 1529–1538, 1996.

10 Forensic DNA Sequencing

Terry Melton and Victor W. Weedn

CONTENTS

INTRODUCTION

DNA sequencing underlies all forms of forensic DNA testing, historically permitting description of the genomic variation of all typing systems. Routine DNA typing was originally performed using fragment sizing for restriction fragment length polymorphisms (RFLPs) and now is performed by fragment sizing for short tandem repeats (STRs). These methods are powerful, rapid, and relatively inexpensive. Sequencing protocols in forensic casework are applied exclusively in mitochondrial DNA (mtDNA) analysis[1–5] in those special cases where routine DNA typing is not helpful—mostly commonly for hairs and skeletal remains.

Although the STR loci could be sequenced, fragment length analysis for each locus is considered a more practical and cost-effective DNA typing technology than sequence analysis. The basic instrumentation is the same, reagent costs are greater for sequencing, and sequence analysis is not as intuitive, facile, or fast. Moreover, STR analysis permits multiplexing many genetic loci simultaneously, where sequencing does not. Presumably, if sequencing technology became sufficiently fast and inexpensive, it might compete with other forensic tests.

MtDNA is the most polymorphic region within the human genome and unlike other regions (with the arguable exception of the human leukocyte antigen [HLA] region) can by itself (in isolation) be used for forensic identity testing. However, as the polymorphisms are within a single region that can be sequenced, they are not genetically independent. Thus, the frequency rates of the mtDNA polymorphic sites cannot be multiplied together. MtDNA haplotype profiles are not as powerful as current STR multiplex profiles. The high discriminatory rates of RFLP and STR

217

analysis are achieved precisely because the frequency rates of the disparate loci are independent and the frequencies can be multiplied together.

Unlike nuclear DNA, mtDNA does not contain significant repetitive DNA sequences; instead, variation is present as single nucleotide polymorphisms and interrogation cannot be based on simple fragment sizing like other forensic DNA markers. Hybridization assays can capture this sequence information, but sequencing has generally been preferred for mtDNA analysis to efficiently capture the large range of possible polymorphic sites. Nonetheless, dot/blots were originally used by Mark Stoneking to perform in mtDNA analysis. In Europe a multiplex solid-phase fluorescent minisequencing system was developed to rapidly survey ten substitution polymorphisms and two length polymorphisms in HV1 and HV2.[6] More recently, a rapid hybridization assay using 33 immobilized probes has been commercialized by Roche Molecular Systems as the "Linear Array Mitochondrial DNA HVI/HVII Region—Sequence Typing Kit."

These new hybridization assays are intended to be used as screening tools, rather than replace current sequencing methods. Such systems capture most, but not all polymorphic sites and thus sacrifice an already limited discriminatory potential. Also, dot/blots are problematic with respect to mixtures and heteroplasmy, whereas sequencing generates information that helps to interpret the quality of the data that is difficult to assess in a dot/blot assay; e.g., it is possible to use neighboring peaks to interpret a given sequencing peak. Furthermore, reverse sequencing can verify the sequencing results. Last, shoulder regions are themselves polymorphic, complicating probe assays. Even pyrosequencing assays involving short stretches of DNA sequence of up to 100 bp in length[7] have been used. Nonetheless, when these screening methods fail to make an exclusion, full-length DNA sequencing is performed to confirm the match and derive statistics from a full database search of the complete profile. Hybridization assays may also be used to augment the mtDNA sequence information by capturing polymorphic sites away from the area of sequencing.

HISTORICAL PERSPECTIVE

Commercial entities dedicated to forensic DNA testing (Forensic Science Associates, Lifecodes, Cellmark Diagnostics) opened their doors to casework in 1986 and 1987 and government laboratory testing began thereafter (the FBI began casework in December of 1988 and Virginia began the first state crime lab testing in March of 1989). Although sequencing was performed for research and validation purposes from the inception of forensic DNA typing, it was only later that sequencing was used as the routine analytic casework technique, in the specific instance of mitochondrial analysis.

Mitochondrial DNA analysis began in academic settings and a few government laboratories within a few years after the inception of nuclear forensic DNA typing. Much of the early groundwork and many of the investigators, such as Mark Stoneking and Svante Pääbo, come from the Molecular Evolution Laboratory of the late Allan Wilson's laboratory at the University of California at Berkeley during the 1970s and 1980s. MtDNA sequencing was first used in a forensic application by Mary Claire-King and Chuck Ginther in the late 1980s to reassociate family kindred of the "disappeared" in Guatemala. The first case of mtDNA victim identification was performed by Mark Stoneking to identify a body in the southwestern U.S. in 1990. Erika Hagelberg at Cambridge used the technique in 1991 to identify a Nazi war criminal, Joseph Mengele. Victor Weedn, Deborah Fisher, Rhonda Roby, and Mitchell Holland at the Armed Forces DNA Identification Laboratory (AFDIL) then began to apply the technique routinely and systematically to skeletal remains of the war dead; AFDIL first used mtDNA sequencing in a case for the first Persian Gulf War in 1991. The Defense Science Board (DSB), led by Joshua Lederberg, reviewed the mtDNA identification of remains performed by the AFDIL. The DSB concluded that the methodology was a sound method of performing such identifications and that adequate quality assurance methods were employed to assure public confidence in the results. By the mid-1990s only AFDIL, the FBI, and a commercial lab, LabCorps, were performing forensic mtDNA sequencing casework in the U.S. In 1998, Mitotyping Technologies', LLC, opened as the only commercial laboratory devoted to this activity. Celera used high-throughput sequencing techniques to assist dentifications of the World Trade Center disaster. Mitochondrial DNA is being performed by more public and private laboratories, but most crime laboratories find that it is too expensive, time-consuming, and specialized to perform. In 2003, the FBI sponsored the development of four regional laboratories.

SPECIAL FORENSIC CONSIDERATIONS

In no other DNA sequencing effort is the outcome as fraught with potential pitfalls or rewards as in forensic testing. Forensic DNA analysis may result in a criminal charge, exonerate the accused, identify a missing individual, open the door to solving cold cases, or begin postconviction relief for the falsely incarcerated. When such analyses can so profoundly affect the lives of defendants and victims, both reliability of method and quality control assume orders of magnitude more importance than in a nonforensic approach to DNA sequencing. Most research laboratories are not prepared to handle forensic casework.

Forensic testing differs from clinical testing in a number of significant ways. The evidentiary specimens unlike clinical specimens are not pristine, but instead have been exposed to the environment for various lengths of time. Many forensic samples have a history of severe environmental insult, with exposure to cycles of heat and cold, moisture, ultraviolet radiation, and acidic soil. The evidentiary specimen may be minute and is nonreplenishable; thus retesting may not be possible. Last, the questions to be addressed are usually different as the clinical test will be for the presence or absence or quantity of a given analyte whereas the forensic test is to demonstrate the uniqueness of the specimen and its linkage to the crime.

Of course, forensic testing carries the potential that it might be used in a legal proceeding as evidence and receive judicial scrutiny. One implication is that documentation is paramount. The documentation begins with chain-of-custody documentation that accounts for the custody of the evidentiary specimen from receipt through testing. Identification of the source of the reference specimens as being of the person represented must specifically be documented.

TABLE 10.1
Examples of Validation Studies Performed in Forensic Laboratories

Reproducibility of known samples (cell lines and NIST standards)
Assessment of mixtures (detection of heteroplasmy and mixed templates)
Lower-level sensitivity of instrumentation and chemistry
Accidental cross-species detection (primer specificity)
Behavior of compromised samples (dirt, heat, light, acid soil)
Performance using different tissue types: hair, bone, blood, saliva, organ, fingernails, etc.
Cleaning of samples (satisfactory removal of deliberate contamination)
Interinstrument variation (comparison of multiple sequencers and thermocyclers)
Interlaboratory sample exchanges (retest samples analyzed by another lab)
External review of standard operating procedures by outside experts
Testing of all new reagents, equipment, and methods prior to introduction in casework

---

The procedures performed must also be well documented. The procedures must also have been validated prior to casework in the given laboratory, even though others have previously scientifically validated the procedure. Internal validation studies that assess the sensitivity, accuracy, and reproducibility of a laboratory's specific analytic system with its particular instrumentation and quality guidelines must be documented. Table 10.1 shows the kinds of validation studies a forensic laboratory must perform prior to instituting forensic DNA sequencing.

Forensic DNA testing laboratories must conform to quality assurance standards otherwise unparalleled in the molecular biology field. Accrediting bodies such as the American Society of Crime Laboratory Directors/Laboratory Accreditation Board (ASCLD/LAB) or the National Forensic Science Training Center (NFSTC) determine if laboratories perform testing according to their specifications. These accrediting bodies appear to be moving to ISO-based formats. Furthermore, procedural guidelines specific to mtDNA analysis have been promulgated by the FBI DNA Advisory Board (DAB) and recently modified by its Scientific Working Group on DNA Analysis Methods (SWGDAM). Although the guidelines are purportedly voluntary, they operate as *de facto* legal forensic evidentiary standards.

These forensic standards mandate certain college-level coursework as well as rigorous on-the-job training of technicians and analysts to guarantee that personnel handling evidence are familiar with the nuances of forensic sample handling, for example, chain-of-custody and courtroom testimony. Semiannual proficiency testing is required of each analyst; the College of American Pathologists, Collaborative Testing Service and Quality Forensics sells external proficiency surveys for this purpose. Regular audits, both internal and external, investigate a laboratory's adherence to all appropriate guidelines.

Where a typical DNA sequencing laboratory would focus largely on gathering and storing DNA sequence data for its projects, only a small proportion of the forensic DNA sequencing laboratory's effort is concerned with this. Instead, security of evidence, client confidentiality, laboratory cleanliness, prevention and tracking of contamination. data storage and preservation, file maintenance, and documentation of quality assurance and control consume at least one half of the forensic lab's resources on a daily basis. Only samples from one case are handled at a time.

A particular concern for forensic mtDNA analysis is recognition and avoidance of contamination. Typically, the analysis is performed specifically in cases where exquisite sensitivity is needed—beyond that of routine polymerase chain reaction (PCR) testing. The sensitivity of a mitochondrial analysis is one to two orders of magnitude greater than that of a nuclear DNA analysis, such that even a few skin cells from fabric may be co-extracted and co-amplified with the stain of interest. The sample preparation should be performed in space dedicated to "low copy number" specimens. Many controls and much redundancy are built into forensic protocols that would permit recognition of any contamination. Furthermore, forensic DNA sequencing laboratories frequently maintain their own DNA sequence databases, which include the sequences of their own laboratory staff personnel as well as those of all samples ever handled by the lab. These databases serve as a starting point for identifying laboratory sources of contamination that may be observed during casework.

FORENSIC mtDNA APPLICATIONS

Mitochondrial DNA analysis is primarily applied in certain types of forensic cases. Specifically, mtDNA analysis is performed where nuclear DNA is unavailable (e.g., hair analysis), highly degraded (e.g., skeletal remains), present in trace quantities (e.g., fingerprint residues), and where family reference specimens are available only from distant kindred (e.g., Czar Nicholas II).

Naturally shed (telogen) hairs will not usually yield sufficient nuclear DNA for typing. On the other hand, plucked hairs with hair roots are suitable for routine nuclear STR analysis. Mitochondrial DNA is generally well preserved in the medulla of the shaft, protected by the keratinous cuticle and cortex, but the nuclear DNA was destroyed during the process of epithelial cell keratinization. Shed hairs are not uncommonly found at crime scenes; the average adult has approximately 100,000 scalp hairs and loses approximately 60 to 100 hairs per day. Furthermore, pubic hairs are frequently found in cases of sexual assault.

DNA will be degraded during decomposition through putrefaction by endogenous enzymes, including DNases, and through bacterial action. Nonetheless recently skeletonized remains provide plentiful nuclear DNA for routine typing. Aged bone, on the other hand, generally requires mtDNA analysis. Many mtDNA particles may be found in the hundreds to thousands of mitochondria in the cytoplasm of each cell. This high copy number of mtDNA relative to nuclear DNA is the main reason for successful DNA typing using mtDNA when nuclear DNA is unsuccessful. The enamel of teeth and the calcified matrix of dense cortical bone provide some level of protection from the elements. Some believe that the circular nature of mtDNA also provides some minor added protection from enzymatic action.

As mtDNA is maternally inherited without recombinant crossover events, distant maternal relatives, in the absence of a mutation event, will have exactly matching mtDNA sequences. By contrast, identification of human remains using nuclear DNA testing, if technically possible at all, is more complex in that multiple family members need to be compared to the missing individual and each other via assembly of a "kinship" tree.

Blood, saliva, and semen are often collected off complex, dirty substrates such as clothing or bedding, or contain mixtures such as those encountered in sexual assault, and will often show these mixtures in the mtDNA sequences. Mixtures are very problematic for mtDNA analysis. Therefore, samples that can be cleaned of external contaminants prior to testing, such as hairs and bones, are the best candidates for mtDNA analysis, and rarely give mixed profiles.

Nonetheless, because of the characteristic high copy number of mtDNA, mtDNA may prove to be of value in the newly burgeoning area of so-called "trace" DNA or "low copy number" (LCN) DNA testing. It is now known that admixed with the traditional fingerprint residues are traces of DNA. Pushing the sensitivity of traditional PCR techniques can permit LCN typing of fingerprint residues. Thus, the trigger of a gun or the handle of a knife could yield biologic evidence of the perpetrator. As background contamination is an issue, this testing is controversial and used by some laboratories for investigatory purposes rather than for use in court as probative evidence.

Forensic DNA sequencing has also been applied to the identification of nonhuman samples. Some early studies of mtDNA profiles in domesticated cats and dogs suggested that significant mtDNA control region variation occurred in these species, and that dog and cat hairs found at crime scenes could, therefore, be very probative evidence.[8] Subsequent studies have demonstrated that while there is high mtDNA diversity in dogs and cats in the homologous hypervariable control region, it is much more limited than that of humans. Because of this, in most cases mtDNA can be used only as an exclusionary tool in animal casework. However, Savolainen et al.[9] have recently shown that certain short tandem repeat regions of the mtDNA molecule in dogs and wolves also possess internal nucleotide variation, virtually individualizing different animals. Although this method cannot be used on hair evidence due to high levels of heteroplasmy, it may be useful for blood and tissue. Cytochrome *b*, a gene coded for by the mtDNA genome, has been used to identify different species of animals and birds in forensic testing, particularly wildlife investigations.[10] Sequencing of a 981 bp amplicon fragment from conserved primer pairs with high homology to many species permits species identification.

FORENSIC mtDNA SEQUENCING

The mtDNA particle, a 16.5-kb organellar molecule that codes for 37 different proteins, tRNAs, and rRNAs, is typically present in hundreds to thousands of copies per cell. Interindividual human mtDNA sequence variation is predominantly found in the noncoding region that is made up of two hypervariable regions that flank the origin and is known as the displacement loop (D-loop) or the "control region." Approximately one third of this 1.1 kb region has been observed to contain nucleotide substitutions relative to a published standard reference sequence, known as the Cambridge Reference Sequence (CRS) or Anderson Sequence.[11] This reference sequence has recently been modified (RCRS).[12,13]

Forensic mtDNA sequencing analysis for identification purposes captures the interindividual sequence variation at approximately 700 bp in the noncoding control region. The bulk of human nucleotide variation exists between nucleotide positions (nps) 15998–16400 (hypervariable region 1, HV1) and nucleotide positions 30–407 (hypervariable region 2, HV2).

The most common strategy is to use the PCR to amplify these two regions from the sample's extracted DNA template. Evidentiary materials, which are automatically presumed to have minimal and/or degraded DNA, are amplified in four fragments, two for each hypervariable region. Figure 10.1 shows one typical amplification strategy carried out on hair and bone samples. Between 32 and 40 cycles of PCR are used, depending on the template, which typically cannot be quantified pre-PCR as there is no sufficiently sensitive method for detecting the few copies of mtDNA template present in such specimens. Reference blood or buccal swab samples, which have abundant mtDNA, may simply require amplification of HV1 and HV2 alone with as few as 30 or 32 cycles. In fact, the entire control region (nps 15998–407) is often amplified as a single long fragment for databasing high-content reference mtDNA specimens. More than 40 cycles of PCR and any form of nested PCR are not recommended due to the increased likelihood of amplifying non-authentic products (contaminants) from equipment or reagents.

Samples with significant DNA degradation have a template that is broken or cut into short fragments. In these cases, closely spaced PCR primers will successfully amplify the abundant but fragmented mtDNA, which otherwise cannot be captured using the standard approach that generates 250 to 300 base pair amplicons. Overlapping amplicons of 80 to 160 bp in size can be sequenced to provide complete coverage of HV1 and HV2 in what has been called either an "ancient DNA approach"[14] or "mini primer set sequencing."[15] This method was pioneered by molecular anthropologists to capture DNA sequence data from fossilized remains like Neanderthal skeletons,[16] but can be applied in an identical fashion to any biological material, especially crime scene hairs.

Different electrophoretic sequencing instruments are used by forensic laboratories with equal success. In North America, Applied Biosystems 373 and 377 acrylamide gel plate systems are being replaced by its capillary gel electrophoresis 310

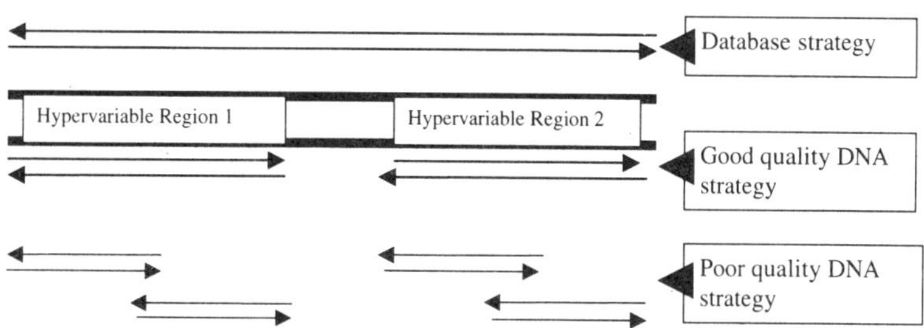

FIGURE 10.1 Typical amplification and sequencing strategy for mtDNA analysis. Two hypervariable regions are targeted in large or small amplicons, depending on the quality of DNA template in the sample and the purpose of sequencing (testing or databasing).

(single channel) or 3100 (16 channel) instruments. Other manufacturers of sequencing instrumentation are Beckman and Pharmacia, the latter more commonly used in Europe. Goals of forensic sequence data collection are (1) quality, (2) quantity, and (3) redundancy. Quality of data is first and foremost the most important consideration of a forensic laboratory, and good-quality data permits efficient and confident identification of the nucleotide sequence that characterizes a particular sample. A full profile of 600 to 800 double-stranded base pairs is most desirable; maximum data from the two mtDNA hypervariable regions strengthen the match. Redundancy, or the double-stranded data/overlapping fragment approach, confirms by multiple "looks" the nucleotide substitutions that characterize a particular sample in several independent PCR reactions. Sequencing protocols are optimized during the laboratory's validation period to reduce the level of noisy background, increase signal strength, and remove chemistry artifacts.

Different laboratories have chosen different sequencing chemistries to suit their instrumentation and protocols. Most laboratories are using *taq*-based cycle-sequencing protocols in kits from various manufacturers. The FBI mtDNA unit uses the Applied Biosystems D-rhodamine chemistry, because while overall it is less sensitive than Applied Biosystems Big Dyes chemistry, it satisfies FBI sensitivity requirements derived from capillary electrophoresis (CE) quantification measurements of first-round PCR products.[17] Other laboratories use the Big Dyes chemistry with success. Overall, Sanger's dideoxyterminator sequencing, rather than dye-labeled primer sequencing, is the method of choice. This is probably true because many different primer sets may be called upon in sample-specific situations to address different control region sequence phenomena.

Most laboratories use 25 amplification cycles with predefined input template amounts based either on first-round post-PCR agarose gel quantification (qualitative) or CE quantification (quantitative). Nested cycle sequencing using internal primers is not generally used, as most laboratories use first-round primers for sequencing as well, and rely on internal sequencing primers only for special situations such as length heteroplasmy or degraded template. Most mtDNA control region primer sequences published in the refereed scientific literature may be used for forensic analysis; as a result there are no proprietary issues surrounding primer design and sequence. These primer sequences have been derived largely from early human evolutionary and population genetics studies and are shared throughout the forensic and academic community.

Included among published primers are those closely spaced forward and reverse oligonucleotides to use for "ancient DNA" approaches, which cover all of hypervariable region 1 in four to five amplicons and overlap by approximately 10 bp. The AFDIL has developed additional "mini-primer sets" to cover hypervariable region 2.[15] While this approach has been very successful in recovering degraded mtDNA template in forensic testing, a commonly encountered limitation is that the DNA extraction material is often exhausted before a complete sequence profile can be developed. This has led to strategies whereby forensic labs choose to consume their finite amount of template in a site-directed approach, confirming observed unique or rare nucleotide substitutions from resequencing multiple PCR reactions, especially in samples with low quantities of DNA, since these sites are most informative and statistically robust in characterizing a sample. Typically, confirmation of unique or rare sites in multiple PCR reactions gives a high degree of confidence in the authenticity of a profile.

One important factor in forensic sequencing not often appreciated elsewhere in the sequencing community is that cycle sequencing product cleanup methods must attempt to preserve as much of the amplicon as possible for sequencing. When mini-primer sets (small amplicons) overlap by only 10 bases, a cleanup method that removes 20 to 40 bases from the 5′ end of the single-strand product or that diminishes the quality of sequence data in this region will be unacceptable. Otherwise there will be no overlap among amplicons. Various methods for product cleanup include generic, inexpensive methods such as ethanol precipitation or kits such as Edge Gel Filtration Cartridges (Edge Biosystems).

The regions typically sequenced from the mtDNA molecule have several commonly recognized characteristics that require additional compensatory steps to resolve. When the dual goals of obtaining maximum data and consuming minimal amounts of sample are paired in forensic testing, dealing with sequence artifacts such as site heteroplasmy and length heteroplasmy usually means that a one-size-fits-all protocol will not be useful.

Automation is most likely to be successful only where input template quantity can be easily titrated, whereas extraction, amplification, and data analysis may be the most rate-limiting steps, due to individual evidentiary sample needs. Traditional forensic standards require that each DNA sequence be evaluated for quality and edited by two experienced analysts, which takes approximately 30 min per sample per analyst. As a result of this specific challenge, expert sequence analysis software is being developed that will assist the forensic community in shortening DNA sequence analysis time overall by allowing some degree of automated base calling with respect to the RCRS.

FORENSIC mtDNA SEQUENCE INTERPRETATION

A basic tenet of forensic analysis is that the defendant should be given the benefit of the doubt to avoid the conviction of an innocent person. Thus, evidence is interpreted with extreme conservatism. Sequences acceptable for other purposes may not necessarily be acceptable for forensic purposes.

After mtDNA sequence data are collected, forensic examiners determine if the quality is acceptable for interpretation. There should be limited artifact background (noise). Sequence data should be of uniformly high quality and should lack ambiguity when comparing nucleotide positions on forward and reverse strands and overlapping fragments. Evidence of clear-cut nucleotide differences on different strands when aligning the data with alignment software should result in reanalysis at the bench (if necessary, re-extraction, re-amplification, and/or resequencing). Available software packages such as Sequencher, Sequence Navigator, and Lasergene DNA Star are all satisfactory for analyzing mtDNA data. The most important factor in successful forensic data analysis is the experience of the examiner, who will quickly learn both to recognize mixtures, hypervariable regions 1 and 2 length heteroplasmy, as well as identify hypermutable "hot spots" for site heteroplasmy. Community practice requires all interpretations to be reviewed technically by another experienced examiner, effectively looking at every nucleotide base on the electropherogram traces.

Concordance between the examiners is required prior to reporting results, and all base-calling conflicts should be resolved to the satisfaction of both. Documentation is also administratively reviewed.

The goal of forensic DNA sequencing in criminal casework is to develop clear inculpatory or exculpatory data from biological samples such as hair, blood, semen, and saliva found at crime scenes and their comparison samples from known individuals who may be connected to the crime. In missing persons cases, which are often linked to criminal investigations, the object of testing is to determine whether human remains are consistent with those of the missing individual by comparison to a family member or a personal effect, such as a toothbrush, containing biological material. The conclusion of the test is that "Person X can be excluded as the contributor of Sample Y" or "Person X cannot be excluded as the contributor of Sample Y," or in the case of a missing person, "Remains Y can/cannot be excluded as having come from a relative of Person X." An inconclusive outcome is possible, but fortunately is rare due to the high degree of diversity found in the mtDNA. The forensic scientist works with prosecutors, law enforcement, defense attorneys, and the court to guide their understanding and application of these outcomes.

The extent of mtDNA sequence variation is not known and may never be completely known, but in fact, the rare variant mtDNA haplotypes or sequences, estimated to be in the tens of thousands, give forensic mitochondrial DNA testing its primary statistical power. Currently, a novel type observed in casework would not be randomly selected from at least 99.94% of North American individuals. This

---

TABLE 10.2
Individuals of North American Forensic Significance in the FBI DNA Sequence Database

Population	Subpopulation	N
African origin	African American	1148
	Sierra Leone	109
	Egyptian	75
Caucasian	Caucasian	1655
	India	19
Hispanic		686
Asian	Japan	163
	Korea	182
	Thailand	52
	China/Taiwan	329
	Guam	87
	Pakistan	8
Native American	Navajo	146
	Apache	180

Note: Most sequences cover both hypervariable regions.

---

estimate is a 95% upper bound frequency based on the size of a forensic database maintained by the FBI ($N = 4839$; Table 10.2).[4] Therefore, for the foreseeable future, mtDNA sequencing is the method of choice to assay the rare, valuable variation present in this small genome.

Mitochondrial DNA haplotypes vary in frequency; all ethnic populations show a distribution with a few high-frequency haplotypes and a majority of rare haplotypes (Figure 10.2). Within the pool of Caucasian mtDNAs found in Europe and North America, there is a single type that occurs at "high" frequency, found in approximately 7% of these individuals. This type is characterized by a nucleotide substitution with respect to the CRS at position 263 (A to G in the light strand orientation) and one or more light strand C insertions at positions 309 and 315 in the hypervariable region 2 homopolymeric C-stretch.

When a high-frequency haplotype is observed in casework and results in a match between questioned and known samples, some degree of uncertainty regarding the true source of the sample will naturally exist. Two additional regions 3′ to the two hypervariable regions that contain somewhat limited control region nucleotide variation, called Variable Region 1 (nps 16471–16562) and Variable Region 2 (nps 424–548), can further test the presumed match by identifying nucleotide differences between the two samples, obviating the match. Research is under way to investigate nucleotide positions within other portions of the mtDNA genome, especially coding regions, that will aid in discriminating subtypes of this and other somewhat common types recognized in current databases. Ultimately some form of SNP assay may be used to add limited sequence data for additional match-testing.

It is well recognized that ethnic classification is correlated with mtDNA haplotype clusters, or "haplogroups" (but by no means is always predictive of phenotype).[18,19] This ethnicity correlation is helpful in identifying common profiles that need further investigational studies to discriminate them.

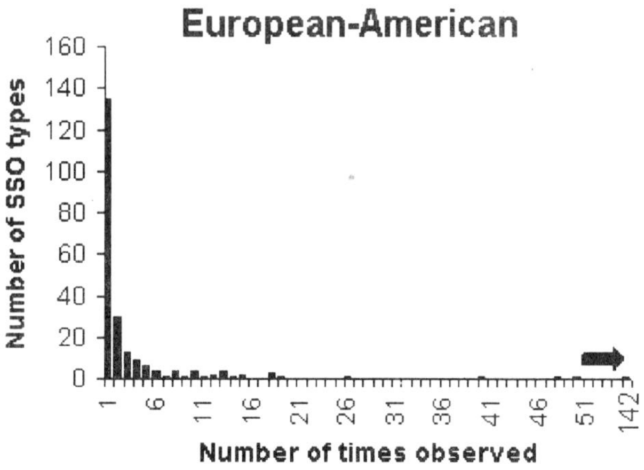

FIGURE 10.2 Distribution of haplotype frequencies in a European-American database.

MITOCHONDRIAL DNA HETEROPLASMY

The baseline state of an organism's tissue is mitochondrial homoplasmy, wherein each mitochondrial DNA molecule has an identical sequence over the entire length. In reality, however, probably each individual is heteroplasmic to some degree, meaning that two or more populations of different mtDNA molecules, one the "wild-type" dominant population and the other(s) minor, inhabit the mitochondria. The minority mtDNA populations start as single copies that have sustained mutational changes at single nucleotide positions. Some may rise to detectable frequency in certain tissues through mtDNA replication, while the vast majority will not. This heteroplasmic state is well recognized in mitochondrial genetic disease, where the dosage of mutated or nonwild-type molecules deleteriously affecting coding regions may be correlated with the severity of the syndrome.

In forensic testing, mtDNA site heteroplasmy is observed at frequencies ranging from approximately 1% of blood samples to 15% of hairs,[5,20] and usually appears as two nucleotide peaks (C plus T, or A plus G, and more rarely as purine-pyrimidine combinations) at a single position in the sequenced region (Figure 10.3). The existence of two or more heteroplasmic sites in a single forensic sample has been reported, but appears to be rare.[21] In casework, a heteroplasmic position should be confirmed by sequencing both the light and heavy strands of DNA to observe it in both orientations, and may also be confirmed by sequencing multiple or overlapping amplicons. Site heteroplasmy is most commonly observed at so-called "fast sites,"

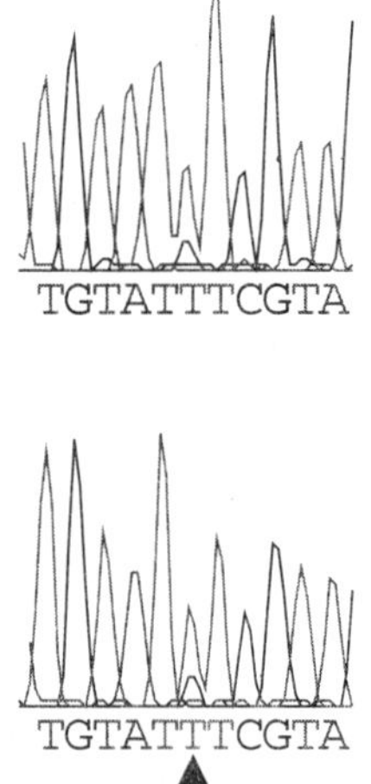

FIGURE 10.3 Example of mtDNA site heteroplasmy showing the C/T mixture on the light and heavy strands (both orientations; the heavy strand has been reverse-complemented). This is position 16093 in HV1, one of the most common "hot spots," where T is the RCRS nucleotide and this individual has T/C.

which have been observed in phylogenetic studies to have mutated multiple times in human history.

Heteroplasmy appears to vary among tissues. Higher levels of site heteroplasmy in hairs is explained by a small bottleneck during individual, monoclonal hair development that allows mutational changes to arise to detectable levels from a small number of starting copies of mtDNA. In contrast, blood originates from a polyclonal source (bone marrow) and is mixed throughout the body, making detectable site heteroplasmies less likely to be sampled. The overwhelming majority of mtDNA molecules in an individual are identical.

Length heteroplasmy is a very commonly observed phenomenon in mtDNA control region sequencing in two regions, the area surrounding a small homopolymeric cytosine (C) stretch in hypervariable region 1 where a T-to-C transition at position 16189 sometimes occurs, and in a longer C stretch around position 309 in hypervariable region 2.[22] In casework these areas exhibit some degree of length heteroplasmy, from mild and almost undetectable to severe and uninterpretable, in around 30% of samples. The presumed mechanism for this phenomenon is that the polymerase involved in mtDNA replication cannot faithfully reproduce the correct ancestral number of C molecules, which results in populations of molecules in the individual with 7 Cs, 8 Cs, 9 Cs, etc. (Figure 10.4). First-round PCR amplification determines which proportions of these templates will appear in the sequence data, and repeated amplifications of these templates show remarkable fidelity in replicating the approximate ratios of the populations of varying lengths. Therefore, it is believed that sequence data represent well the approximate proportions of different length templates. In sequencing these templates containing different populations with varying numbers of cytosines, the sequence often cannot be read 3′ to the C stretch due to the overlapping template molecules. To correct this problem, the PCR template should be sequenced with internal primers downstream of the C stretch, and also sequenced in both directions on both light and heavy strands. This will allow evaluation of the sequence 3′ to these complex regions.

MIXTURES

When a mixture is present for mtDNA, it is currently impossible to sort out the linkage phase of the mixed haplotypes without cloning and then sequencing the clones, a prohibitively expensive proposition. Mixed haplotypes are not necessarily additive—when a major discrepancy in proportion of the mixture components exists (at least 80:20), nucleotide base "dropout" of the minor component is often observed. Alternatively, if the mixture components are approximately equal, then nucleotide peaks at the variant positions may shift back and forth in height in different PCR products, making it impossible to assign linkage phase. Multiple haplotypes may be generated by the large numbers of permutations of the mixed sites, for example, a sequence displaying 5 mixed sites results in 32 possible mtDNA haplotypes ($5^2 = 32$). In addition, different primer pairs have variable performance under the same conditions, complicating the comparison of overlapping regions.

(a)

(b)

(c)

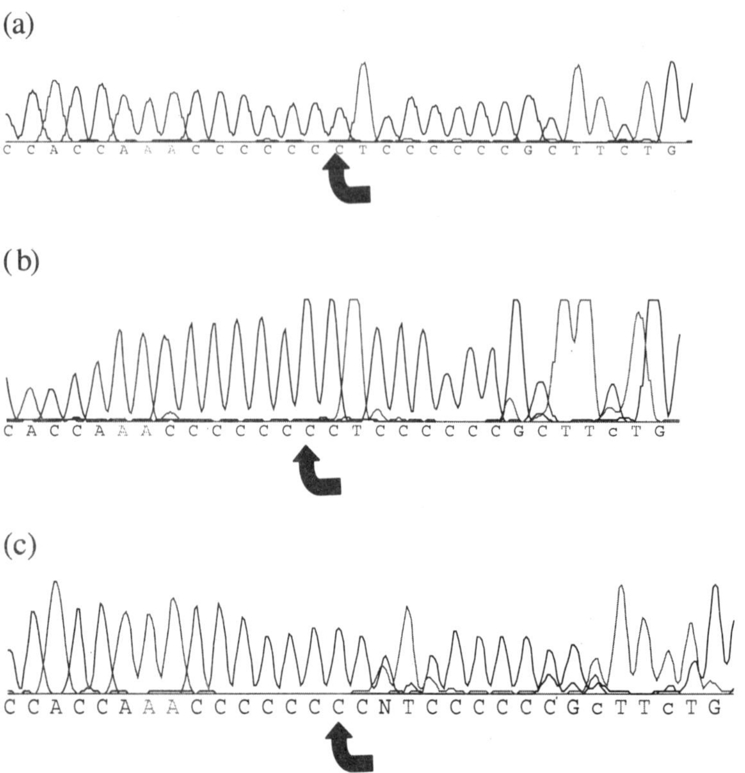

FIGURE 10.4 Mitochondrial DNA HV2 homopolymeric C stretch: (a) no length hetero-plasmy, (b) length heteroplasmy showing less severe effect on sites downstream (3′ end, light strand) of the length variants, with a single C insertion after position 309 relative to the RCRS being dominant (309.1); (c) length heteroplasmy showing more severe effect on sites down-stream (3′, light strand) of the length variants, with two C insertions after position 309 relative to the CRS dominant (309.1, 309.2). Position 309 is denoted by the arrow.

REFERENCES

1. Wilson M, DiZinno JA, Polanskey D, Replogle J, Budowle B. Validation of mitochon-drial DNA sequencing for forensic casework analysis. Int J Legal Med 108:68–74, 1995.
2. Carracedo A, D'Aloja E, Dupuy B, Jangblad A, Karjalainen M, Lambert C et al. Reproducibility of mtDNA analysis between laboratories: a report of the European DNA profiling group (EDNAP). Forensic Sci Int 97:165–170, 1998.
3. Carracedo A, Bär W, Lincoln P, Mayr W, Morling N, Olaisen B et al. DNA Com-mission of the International Society for Forensic Genetics: guidelines for mitochon-drial DNA typing. Forensic Sci Int 110:79–85, 2000.
4. Holland MM, Parsons TJ. (1999) Mitochondrial DNA sequence analysis: validation and use for forensic casework. Forensic Sci Rev 11(1):21–50, 1999.

5. Melton T, Nelson K. Forensic mitochondrial DNA analysis: two years of commercial casework experience in the United States. Croatian Med J 42(3):298–303, 2001.

6. Morley JM, Bark JE, Evans CE, Perry JG, Hewitt CA, Tully G. Validation of mitochondrial DNA minisequencing for forensic casework. Int J Legal Med 112:241–248, 1999.

7. Andreasson H, Asp A, Alderborn A, Gyllensten A, Allen M. Mitochondrial sequence analysis for forensic identification using pyrosequencing technology. BioTechniques 31(2):2–9, 2001.

8. Savolainen P, Lundeberg J. Forensic evidence based on mtDNA from dog and wolf hairs. J Forensic Sci 44(1):77–81, 1998.

9. Savolainen P, Arvestad L, Lundeberg J. A novel method for forensic investigation: repeat-type sequence analysis of tandemly repeated mtDNA in domestic dogs. J Forensic Sci 45(5):990–999, 2000.

10. Zehner R, Zimmerman S, Mebs D. RFLP and sequence analysis of the cytochrome *b* gene of selected animals and man: methodology and forensic application. Int J Legal Med 111:323–327, 1998.

11. Anderson S, Bankier AT, Barrell GB, de Bruijn MHL, Coulson AR, Drouin J, Eperon IC, Nierlich DP, Roe BA, Sanger F, Schreier PH, Smith AJ, Staden R, Young IG. Sequence and organization of the human mitochondrial genome. Nature 290:457–465, 1981.

12. Andrews RM, Kubacka I, Chinnery PF, Lightowlers RN, Turnbull DM, Howell N. Reanalysis and revision of the Cambridge reference for human mitochondrial DNA [letter]. Nat Genet 23(2):147, 1999.

13. Wallace DC, Lott MT. MITOMAP: A human mitochondrial genome database, http://www.mitomap.org, 2003. [Revised Cambridge Reference Sequence: http://www.mitomap.org/mitomap/mitoseq.html]

14. Pääbo S, Higuchi RG, Wilson AC. Ancient DNA and the polymerase chain reaction: the emerging field of molecular archaeology. J Biol Chem 264:9709–9712, 1989.

15. Gabriel MN, Huffine EF, Ryan JH, Holland MM, Parsons TJ. Improved mtDNA sequence analysis of forensic remains using a "mini-primer set" amplification strategy. J Forensic Sci 46(2):247–253, 2001.

16. Krings M, Stone M, Schmitz RW, Krainitzke H, Stoneking M, Pääbo S. Neanderthal DNA sequences and the origin of modern humans. Cell 90:19–30, 1997.

17. Isenberg AR, Moore JM. Mitochondrial DNA analysis at the FBI Laboratory. Forensic Science Communications 1(2), 1999, http://www.fbi.gov/fbilibrary/forensicscience-communications/backissues.

18. Connor A, Stoneking M. Assessing ethnicity from human mitochondrial DNA types determined by hybridization with sequence-specific oligonucleotides. J Forensic Sci 39:1360–1371, 1994.

19. Melton T, Clifford S, Kayser M, Nasidze I, Batzer M, Stoneking M. Diversity and heterogeneity in mitochondrial DNA of North American populations. J Forensic Sci 46:46–52, 2001.

20. Calloway CD, Reynolds RL, Herrin GL, Anderson WW. The frequency of heteroplasmy in the HVII region of mtDNA differs across tissue types and increases with age. Am J Hum Genet 66:1384–1397, 2000.

21. Budowle B, Allard MW, Wilson MR. Critique of interpretation of high levels of heteroplasmy in the human mitochondrial DNA hypervariable region I from hair. Forensic Sci Int 126:30–33, 2002.

22. Stewart JEB, Fisher CL, Aagaard TJ, Wilson MR, Isenberg AR, Polanskey D et al. Length variation in HV2 of the human mitochondrial DNA control region. J Forensic Sci 46(4):862–870, 2001.

Index